S. Shivaji

Human Ocular Microbiome

Bacteria, Fungi and Viruses in the Human Eye

Springer

S. Shivaji
Microbiology Department
L V Prasad Eye Institute
Hyderabad, India

ISBN 978-981-19-1753-0 ISBN 978-981-19-1754-7 (eBook)
https://doi.org/10.1007/978-981-19-1754-7

This Springer imprint is published by the registered company Springer Nature Singapore Pte Ltd.
The registered company address is: 152 Beach Road, #21-01/04 Gateway East, Singapore 189721, Singapore

To all scientists and clinicians contributing towards eye care and health

Foreword

While a number of advances were made in biological research, the area of 'Microbiome' is certainly one of the most exciting ones. The knowledge about the trillions of microorganisms that inhabit the human body, the concept of human microbiome and its role in normal and disease states has opened up new doors.

The plethora of diseases with which microbiome is associated, the protection it confers against a multitude of infections, and its role in immunity has become a subject of great importance and interest in recent years.

While a lot of literature already exists on the gut microbiome, the concept of niche microbiomes is relatively nascent. One such is the emerging scientific work on ocular microbiome. A few groups around the world were intrigued by this concept and began investigating about the existence of an ocular surface microbiome and how it may influence the various infective and inflammatory conditions of the eye.

While numerous books were written on the gut microbiome, there is no book that provides a comprehensive coverage on the subject of ocular microbiome. Dr. Shivaji and his group are one of the early groups engaged in this research that has yielded some valuable insights. From the historical discovery of microbes to the compounding factors that influence the microbiomes will be discussed. Various chapters also highlighted the concept of core genera, dysbiosis and discriminating genera. Most importantly, the functional relevance of the dysbiotic microbiomes, mycobiomes and viromes would be analysed with respect to an ocular disease.

The conclusion provided a positive note that the ocular microbiome, mycobiome and virome may indeed help to understand the basis of some idiopathic ocular diseases.

Overall, the structure of the book and the lucid way the entire subject is presented by a respected authority in the field like Dr. Shivaji will be very valuable to ophthalmologists, ocular and general microbiologists and students of medicine and biology. I compliment Dr. Shivaji for bringing out this publication which is so timely.

L V Prasad Eye Institute,
Hyderabad, Telangana, India

Gullapalli N. Rao

Preface

The Beginning of Microbiome Research

Every inch of the human body harbours an enormous number of microorganisms (bacteria, fungi, viruses, protozoans, etc.) which together constitute the human 'microbiome'. The most abundant are those colonising the gastrointestinal tract termed the 'gut microbiome' and a conservative estimate is that this number exceeds 10^{13} bacteria with an estimated 3.3 million genes. This enormous microbiome, over thousands of years, has co-evolved along with the host and plays a crucial role in digestion and energy generation, in conferring protection against pathogenic bacteria, in the development of the host immune system, in the production of vitamins and short-chain fatty acids, etc. But the flip side is that imbalance or alteration in the gut microbiome referred to as 'dysbiosis' has become a hallmark of several intestinal diseases (like obesity, Crohn's disease, type 1 diabetes mellitus, type 2 diabetes mellitus, colorectal cancer, gastric cancer, etc.) and extra-intestinal diseases (such as cancers, muscular dystrophy, vaginosis, neuro-developmental and neuro-degenerative diseases). Thus, the gut microbiome is a subject of intensive research and serves as the forerunner to explore niche specificity in the microbiome in various parts of the body like the skin, urogenital, gastrointestinal, oral and nasal regions. This niche-specific microbial footprint appears to have some predictive value in relation to various diseases. For instance, low diversity in the gut microbiome was associated with obesity and inflammatory bowel disease, and the presence of *Clostridium difficile* was associated with irritable bowel syndrome. These pioneering studies on gut microbiomes of different niches of the human body were also an impetus to the present proposal on the ocular microbiome.

The Human Ocular Microbiome

The origins of this proposal on the ocular microbiome could be traced to the Human Microbiome Project and the multitude of questions that emerged such as: does the ocular surface possess a microbiome? does the ocular microbiome play a role in disease physiology? does the ocular surface microbiome vary from the adjacent skin microbiome? does the ocular surface microbiome influence immune tolerance? can ocular surface microbes be used to overcome eye diseases? etc. A reasonable number of studies have addressed some of the above queries and thus there is a need to capture all this information in a comprehensive book. Fungi and viruses which constitute the mycobiome and virome of the ocular surface are also included. Further, the cellular and molecular basis of pathogenesis of eye diseases is discussed.

Contents of the Book *Human Ocular Microbiome: Bacteria, Fungi and Viruses in the Human Eye*

Over the years there has been an exponential growth in the number of research publications and books on the 'Microbiome' dominated by books on the gut microbiome. But, books on the microbiome of a given body niche are very limited and confined to books on the skin, urogenital tract, lung, oral microbiome, brain, etc. A book dedicated to ocular microbiome has not been published and such a book would add to the above list of niches. The proposed book would be structured as follows:

Chapter 1. Introduction: From the Gut Microbiome to the Ocular Surface Microbiome: Are They Associated?
Chapter 2. The Cultivable Bacteria of the Human Eye
Chapter 3. The Microbiome of the Human Eye
Chapter 4. Antimicrobial Resistance in Ocular Bacteria
Chapter 5. The Cultivable Fungi of the Human Eye
Chapter 6. The Mycobiome of the Human Eye
Chapter 7. Viruses of the Human Eye
Chapter 8. Virome of the Healthy Human Eye
Chapter 9. Virome of the Diseased Human Eye
Chapter 10. Cellular and Molecular Basis of Ocular Diseases
Chapter 11. What We Need to Know About the Human Ocular Microbiome

This book would document the historical discovery of microbes (bacteria, fungi and viruses) on the ocular surface both by the conventional cultivable method and NGS technologies in the healthy eye and the diseased eyes (with respect to several ocular diseases including keratitis, uveitis, endophthalmitis, blepharitis, conjunctivitis, etc.). Confounding factors that influence the microbiomes (bacteria), mycobiomes (fungi) and viromes (viruses) of the ocular surface would be discussed. The chapters would also highlight the concept of core genera, dysbiosis and discriminating genera. Most importantly, the functional relevance of the dysbiotic microbiomes, mycobiomes and viromes would be analysed with respect to an ocular disease and wherever possible cellular and molecular mechanisms with respect to ocular diseases would be discussed. The book concludes by highlighting the gaps in our understanding of the ocular microbiome, mycobiome and virome which may eventually unravel the basis of the little understood ocular idiopathic diseases.

Target Audience and Importance

- This book would cater to the needs of ophthalmologists, ocular microbiologists, general microbiologists and students of medicine and biology.

It would help to bridge the gap between conventional microbiologists and molecular microbiologists.

Hyderabad, India S. Shivaji

Acknowledgements

My heartfelt thanks to Drs D. Balasubramanian and Gullapalli Nageswara Rao who catalysed my transition from basic research related to psychrophilic microorganisms from extreme cold habitats to research related to the microbiology of the human eye. I am extremely grateful to the support extended unknowingly by the clinicians and scientists of L V Prasad Eye Institute, Hyderabad, India. I readily acknowledge my students, project fellows and postdoctoral fellows who shared my zeal and enthusiasm in my newly found interest. Special thanks to Drs Konduri Ranjit and Shalem Raj who did a meticulous job in compiling the references and in formatting the figures. The affection and moral support from my granddaughter, my sons, daughters-in-law and my wife made my task all the easier. Happy reading!

Contents

About the Author

S. Shivaji, Director Emeritus and Senior Scientist at the Brien Holden Eye Research Centre at LVPEI, Hyderabad, was the Director of BHERC till 2020. His research focuses on the connection between gut and ocular microbiomes with eye diseases. Earlier, at the CCMB, he explored the microbial diversity of Antarctica, Arctic, Himalayas, and Stratosphere. He is a fellow of National Academy of Sciences, Indian Academy of Sciences, Telangana Akademi of Sciences and Association of Microbiologists of India. He is a recipient of National Antarctic Award, National Polar Sciences and Cryosphere Award, E K Janaki Ammal Award and others. He is ranked among the top 2% of scientists in the world and has published >350 research papers and two books.

Abbreviations

Ab	Antibacterial
Aap	Accumulation-associated protein
ADEP	Acyldepsipeptide antibiotic
AHC	Acute haemorrhagic conjunctivitis
AHLs	Acyl-homoserine lactones
adRP	Autosomal dominant RP
adCSNB	Autosomal dominant congenital stationary night blindness
arRP	Autosomal recessive RP
Ai	Anti-inflammatory
AI	Auto-inducer
AIP	Auto-inducer peptide
AK	Acanthamoeba keratitis
AMD	Age-related macular degeneration
AMP	Antimicrobial peptides
AMR	Antimicrobial resistance
aPA	Acanthamoeba plasminogen activator
APC	Antigen-presenting dendritic cells
ARN	Acute retinal necrosis
BHP	Biofilm-associated homologue protein
BK	Bacterial keratitis
BK-CR	Corneal scraping from bacterial keratitis individuals
BK-SW	Conjunctival swabs from bacterial keratitis individuals
BRiSK	Biome representational in silico karyotyping
Can	Collagen-binding adhesin
CbpA	Choline-binding protein A
CCL	Inflammatory chemokine
CCR	Chemokine receptor
CD11a	Cluster of designation molecule 11a
CD14	Cluster of differentiation 14, is a protein of the innate immune system

CFA	Complete Freund's adjuvant
CFO	Calcofluor white
Cfu	Colony forming units
CL	Contact lens
CMV	Cytomegalovirus
Co	Commensal
CoNS	Coagulase negative *Staphylococcus*
crAssphage	Cross assembly phage
CsA	Cyclosporin A
CXCL1	T cell–attracting chemokines
CXCR	Chemokine receptor
DC	Dendritic cell
DED	Dry eye disease
DHFR	Dihydrofolate reductase
DHPS	Dihydropteroate synthase
DM	Diabetes mellitus
DR	Diabetic retinopathy
EAU	Experimental autoimmune uveitis
EBV	Epstein-Barr virus
EDE	Evaporative dry eye
EDTA	Ethylenediamine tetra-acetic acid
EFG1	Transcription factor
EKC	Epidemic keratoconjunctivitis
EPS	Extra-polymeric substance
ERK	Extracellular signal–related kinase
Fas	Apoptotic marker
FasL	Apoptotic marker
FES	Floppy eyelid syndrome
FK	Fungal keratitis
FK-CR	Corneal scraping from fungal keratitis individuals
FK-SW	Conjunctival swabs from fungal keratitis individuals
FLASH	Fast length adjustment of short reads
FnBPs	Fibronectin binding proteins
GF	Germ-free
GVD	Gut virome database
HAART	Highly active antiretroviral therapy
HC	Healthy control
HCEC	Human corneal epithelial cells
HERV	Human endogenous retroviruses
HIV	Human immune-deficiency virus
HLAB27	Human leukocyte antigen B27
hMGEC	Human meibomian gland epithelial cells
HPV	Human papillomavirus
HSL	Homoserine lactone

HSM	Heparin surface modified
HSV	Herpes simplex virus
ICAM	Intercellular adhesion molecule
ICK	Infectious crystalline keratopathy
IgAIp	IgAI protease
IFN	Interferon
IK	Infectious keratitis
IL	Interleukin
IOL	Intra-ocular lens
IRBP	Interphotoreceptor retinoid-binding protein
ISK	Immune stromal keratitis
ITS	Internal transcribed spacer region of nuclear ribosomal gene
IVCM	In vivo confocal microscopy
JNK	C-Jun N-terminal kinase
LASL	Linker-amplified shotgun library
LDA	Linear discriminant analysis
LFA-1	Lymphocyte function-associated antigen 1
LEC	Lax eyelid condition
LES	Lax eyelid syndrome
LPCB	Lactophenol cotton blue
LPS	Lipopolysaccharide
LR	Low-responder
LSCD	Limbal stem cell deficiency
LuxR	The response regulator
MAP-K	Mitogen-activated protein kinase
mAPC	Mature antigen-presenting cell
MBP	Mannose-binding protein
MD2	Myeloid differential protein-2
MDA	Multiple displacement amplification
MG	Meibomian gland
MGD	Meibomian gland dysfunction
MHCII	Major histocompatibility complex II
MIC	Minimum inhibitory concentration
MIP133	A collagenase
mPCR	Multiplex PCR
MCV	Merkel cell polyomavirus
MCPyV	Merkel cell polyomavirus
MMP	Matrix metalloproteinases
Mø	Resident macrophages
MPTP	1-methyl-4-phenyl-1,2,3,6-tetrahydropyridine
MSRV	Multiple sclerosis-associated retrovirus
MSCRAMMs	Bacterial microbial surface components recognising adhesive matrix molecules
NanA	Neuraminidase A

NETS	Neutrophil extracellular traps
NF-κB	Nuclear factor kappa B
NGAL, Lcn2	Neutrophil gelatinase-associated lipocalin
NGS	Next generation sequencing
NK	Natural killer cells
NMDS	Non-metric multidimensional scaling
NO	Nitric oxide (NO)
NR	Non-responder
nS	Non-susceptible recipient
NSK	Necrotising stromal keratitis
OC	Orbital cellulitis
oGVHD	Ocular graft-versus-host disease
OSM	Ocular surface microbiome
OTU	Operational taxonomic unit
Pa	Pathogenic
PavA	Pneumococcal adherence and virulence factor A
PCF	Pharyngoconjunctival fever
PCoA	Principal coordinate analysis
PCR	Polymerase chain reaction
PDT	Photodynamic therapy
PfbA	Fibronectin binding protein A
PFR	Post-fever retinitis
Pi	Pro-inflammatory
PIA	Polysaccharide intercellular adhesin
PIP	Prolactin inducible protein
Ply	Pneumolysin
PMMA	Poly(methyl methacrylate)
PMN	Neutrophil
PORN	Progressive outer retinal necrosis
PpmA	Putative protease maturation protein A
PQS	*Pseudomonas* quinolone signal
Pr	Probiotic
PsaA	Pneumococcal surface adhesin A
PspA	Pneumococcal surface protein A
PspC	Pneumococcal surface protein C
QIME	Quantitative insights into microbial ecology
qPCR	Quantitative polymerase chain reaction
QS	Quorum sensing
QSM	Quorum sensing molecules
R	Responder
RASL	Random amplified shotgun library
rDNA	The gene for 16S rRNA
RNS	Reactive nitrogen species
ROS	Reactive oxygen species

RP	Retinitis pigmentosa
SEM	Scanning electron microscopy
SAP6	Hyphal-associated gene 6
SEN	A small DNA virus related to TTV
SIsA	Streptococcal lipoprotein rotamase A
SJS	Stevens–Johnson syndrome
SNP	Single nucleotide polymorphism
SPF	Specific pathogen-free
SS	Sjögren's syndrome
ssDNA	Single-stranded DNA (ssDNA)
TCR	T cell receptor
T1DM	Type 1 diabetes mellitus
T2DM	Type 2 DM
TGF	Transforming growth factor
TEN	Toxic epidermal necrolysis
Th1	T helper 1 cells
Th17	T helper 17 cells
TLR	Toll-like receptor
TNF-α	Tumour necrosis factor alpha
Treg	Regulatory T cells
TTVs	Torque teno viruses
UVT	Uveitis
V3-V4	Variable regions of 16S rRNA gene
VC	Viral control group
VK	Viral keratitis
VKC	Vernal keratoconjunctivitis
VLP	Virus-like particles
VZV	Varicella zoster virus
ZmpC	A metalloprotease

Chapter 1
Introduction: From the Gut Microbiome to the Ocular Surface Microbiome—Are They Associated?

Abstract The first chapter is an introduction to the gut microbiome and its unique features: the enormous number, the diversity and canonical functions. It also highlights the concept of the core gut microbiome and dysbiosis in the gut microbiome and dwells into the evidence that dysbiosis (changes in diversity, abundance and functions of microbes) in the gut microbiome is associated with several systemic diseases including ocular diseases. This chapter concludes on a positive note that dysbiosis in the gut microbiome would unravel changes in the gut bacterial microbiome in diversity, abundance and functions of the microbes and help to evolve strategies to mitigate inflammatory diseases of the eye by the application of probiotic therapy and/or faecal microbiota transplantation.

Keywords Gut microbiome · Ocular surface microbiome · Dysbiosis · Ocular disease · Intestinal disease

1.1 Introduction

An array of microorganisms (fungi, bacteria, archaea, viruses, protozoans and other microorganisms) which together are referred to as the human microbiome (Sender et al. 2016; Qin et al. 2010) colonise the human body. These colonisers vary in diversity and abundance depending on the specific niche of the human body that they colonise. Based on the abundance, it is estimated that the gut bacterial microbiome (henceforth microbiome) represented by 3.8×10^{13} cells is numerically similar in abundance to the number of cells in the human body (3.0×10^{13}) (Sender et al. 2016). But the gut bacterial microbiome has several orders of magnitude more genes than the genes in the human body. The microbiome has approximately 3.3 million genes compared to the 23,000 genes in the human body (Qin et al. 2010). In contrast, the fungal community of the microbiome commonly referred to as the 'mycobiome' constituted only a minor component (0.1%) of the total microbes in the gut (Arumugam et al. 2011; Cui et al. 2013; Chin et al. 2020). Compared to the microbiome and mycobiome, the 'virome' which refers to all the viruses associated with the gut mucosal surface outnumber bacterial cells by 20:1 (Mukhopadhya et al. 2019) and a gram of human gut content contains at least 108–109 virus-like particles

S. Shivaji, *Human Ocular Microbiome*,
https://doi.org/10.1007/978-981-19-1754-7_1

(Kim et al. 2017; Guerin et al. 2018). Despite their numerical dominance compared to bacteria and viruses in the gut currently, less than 1% of the virome has been sequenced and characterised (Virgin 2014). Over the years it has been recognised that the microbiome, mycobiome and virome are likely to impact human health. This chapter would focus on the gut microbiome (bacteria) and its impact on human health.

1.2 Gut Microbiome Functions

The gut bacterial microbiome aids in digestion and in host immune system, protects against pathogenic bacteria and aids in synthesis of short-chain fatty acids (such as butyrate, acetate and propionate) and production of vitamins. The gut bacterial microbiome also preserves homeostasis of T helper 1 (Th1) and 17 (Th17) cells and regulatory T cells (Treg) which are vital in mounting an immune response against pathogens (Xue et al. 2021). The microbiome mediates these effects by secreting factors that modulate intestinal permeability, the mucus layer function, epithelial cell function, innate and adaptive immunity, intestinal motility and neurotransmission (Hollister et al. 2014). Gut microbiomes are known to be diverse between individuals (Livingston et al. 2010) and need to be clearly understood with respect to its potential influence on human health.

1.3 Core Gut Microbiome and 'Dysbiosis'

The two abundant bacterial phyla in the microbiome are the *Bacteroidetes* and *Firmicutes*, and they together constitute 70–90% of the microbiome. The genus *Clostridium* makes up 95% of the *Firmicutes* phylum, and the other genera include *Lactobacillus*, *Bacillus*, Enterococcus and *Ruminococcus*. Bacteroidetes is essentially represented by the genera *Bacteroides* and *Prevotella* (Rinninella et al. 2019). The phyla *Actinobacteria*, *Proteobacteria*, *Fusobacteria* and *Verrucomicrobia* make up the remaining 10–30% of the gut microbiome (Arumugam et al. 2011). The *Actinobacteria* phylum is mainly represented by the *Bifidobacterium* genus (Arumugam et al. 2011; Rinninella et al. 2019). The above phyla, namely, the *Bacteroidetes*, *Firmicutes*, *Proteobacteria*, *Actinobacteria*, *Verrucomicrobia* and *Fusobacteria* together represent the 'core' or 'substantial core' gut microbiome with most individuals sharing most phyla (Turnbaugh et al. 2007, 2009).The gut microbiome in a healthy individual exhibits a balance between the 'good' and 'bad' bacteria with the former including the bacteria with probiotic and anti-inflammatory properties and the latter including bacteria with pro-inflammatory and pathogenic properties (Wang et al. 2017). The balance in the microbiome tilts from the good or 'beneficial' to bad or 'harmful' bacteria (Nagpal et al. 2014) under certain conditions such as increased sugar intake, high-fat diet, sedentary lifestyle, diseased conditions,

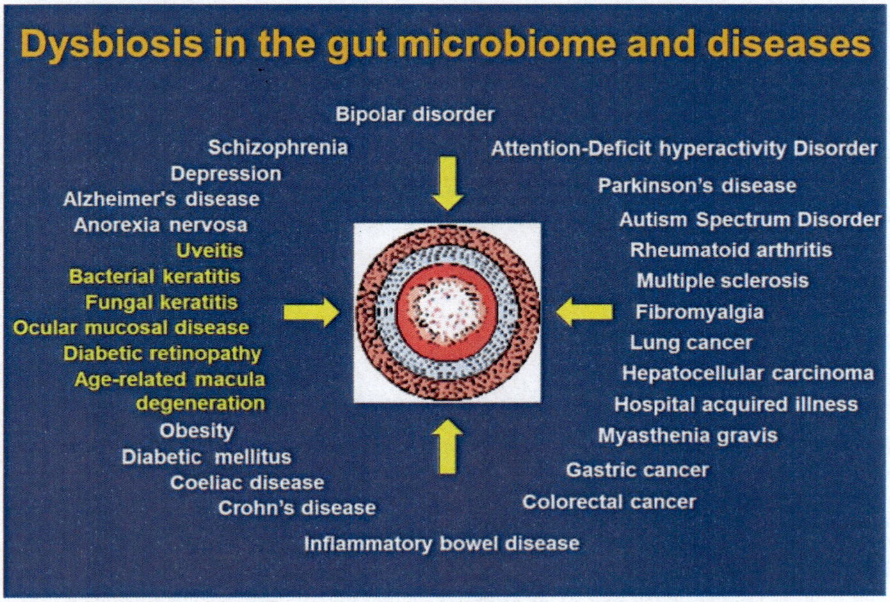

Fig. 1.1 Human diseases influenced by dysbiosis in the gut bacterial microbiome. Figure modified from Shivaji (2020)

increased uptake of antibiotics, etc. This imbalance is reflected in the gut microbiome, and distinct changes are observed at different taxonomic levels, from the phyla to the species level. These changes in diversity, abundance and function of the gut microbiome under the above conditions compared to that in the healthy human gut microbiome are referred to as 'dysbiosis'. The physiological implications of the gut microbiome dysbiosis is a hallmark of several intestinal and extra-intestinal diseases (Fig. 1.1) (Carding et al. 2015; Belizário and Faintuch 2018; Belizário et al. 2018).

1.4 Gut Microbiome Is Associated with Disease

Animal experiments have served as reliable model to study the role of dysbiosis of gut microbiomes and disease aetiology (Buttó and Haller 2016; Pizarro et al. 2000). It was observed that when a particular mouse strain TNF$^{\text{delta ARE}}$ was co-housed under specific pathogen-free (SPF) conditions, three phenotypes non-responder, low-responder and responder were developed (Fig. 1.2) (Buttó and Haller 2016). Using the above three phenotypes, it was demonstrated that only when oral gavage of dysbiotic microbiota from a responder was transferred to a susceptible host the disease developed. But, when microbiota from responder was transferred to

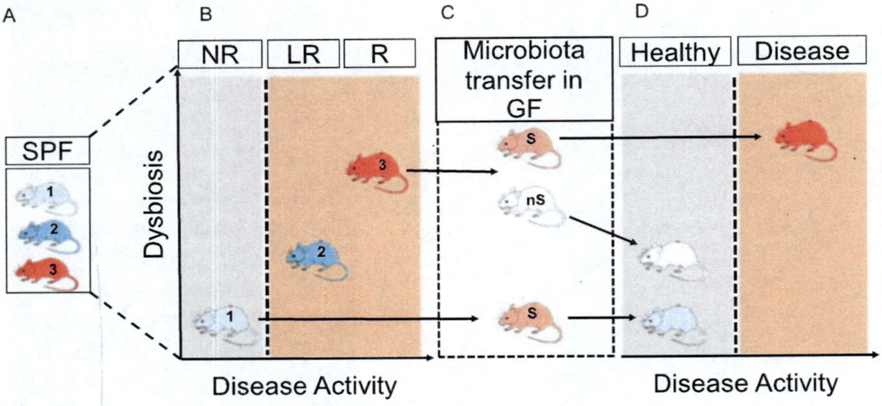

Fig. 1.2 Faecal gavage into germ-free mice demonstrating disease-causing nature of dysbiotic microbial communities. Mice marked 1, 2 and 3 are TNF$^{delta\ ARE}$ which when co-housed under specific pathogen-free (SPF) conditions developed as three phenotypes non-responder (NR, 1), low-responder (LR, 2) and responder (R, 3). Oral gavage of microbiota from NR mice in to germ-free (GF) susceptible mice (S, TNF$^{delta\ ARE}$) does not cause disease. In contrast, microbiota from R mice when transferred to S mice caused disease in the S host (S, TNF$^{delta\ ARE}$), but when microbiota from R mice was transferred into nS recipients, it did not induce disease. Figure modified from Buttó and Haller (2016)

non-susceptible host the disease did not develop. So also, when microbiota from a non-responder was transferred to a susceptible recipient, it did not induce the disease (Buttó and Haller 2016; Schaubeck et al. 2016) (Fig. 1.2). Other experiments that confirmed the role of dysbiotic gut microbiome in disease was the unequivocal demonstration that lean mice when administered with dysbiotic microbiota from obese mice or a human obese donor were transformed into obese mice (Turnbaugh et al. 2006; Ridaura et al. 2013). But if the faecal transplants from obese mice were supplemented with *Christensenella minuta* and administered to obese mice, the recipients became thinner, providing evidence that *C. minuta* regulates obesity (Goodrich et al. 2014). The other beneficial bacteria in the gut microbiome include *Akkermansia muciniphila*, which reversed obesity and produced endocannabinoids to decrease insulin resistance and *Faecalibacterium prausnitzii*, which is anti-inflammatory and protects against intestinal inflammation (Everard et al. 2013). Conversely, *Klebsiella pneumoniae* and *Proteus mirabilis* in the gut microbiome are harmful bacteria involved in colitis in mice (Garrett et al. 2010). Thus, just as a pathogen could cause a disease, a 'good' microbe could prevent a disease. However, to ascertain the beneficial and harmful effects of the gut microbiome, it is absolutely essential to normalize all the results with respect to several host factors (such as gender, age, genetics, region of origin and intrinsic factors of the gut like bile acids, pH, transit time and mucus), environmental factors (e.g. medication and nutrients) and microbial factors (e.g. bacterial enzymes, adhesion capability, metabolic strategies, bacteriophages, etc.) (Prakash et al. 2011) which are all recognized confounding factors.

1.5 Gut Microbiome Dysbiosis and Human Health

The formidable challenge in gut microbiome studies and human health is to establish a connection/involvement of a specific microbe or a group of microbes to a particular disease. Establishing such a connection is a challenging task since the bacteria that habitat the gut are extremely diverse and numerous. Nevertheless there have been several success stories as highlighted below.

1.5.1 Gut Microbiome Dysbiosis and Intestinal Diseases

Several intestinal diseases like obesity (Ignacio et al. 2016), Crohn's disease (Sokol et al. 2020a, b), Type 2 DM (T2DM), Type 1 diabetes mellitus (T1DM) (Qin et al. 2010; Murri et al. 2013), colorectal cancer (Flemer et al. 2017) and gastric cancer (Dicksved et al. 2009) have been associated with dysbiosis in the gut microbiome. In many of these diseases, candidate bacteria have been implicated based on the inferred functions of the bacteria (Table 1.1). For instance, in obese individuals pathogenic (*Enterobacter* spp.), probiotic (*Lactobacillus* spp.) and anti-inflammatory (*Roseburia* spp.) bacteria were increased (Ignacio et al. 2016; Fei and Zhao 2013; Santacruz et al. 2009), whereas *C. minuta* which controls obesity, *Bifidobacterium spp.* which is antibacterial, *Oscillospira* spp .which is probiotic, *Alistipes* spp. which is anti-inflammatory and *Succinivibrio* spp. and *Erwinia* spp. which are commensals were decreased in the gut microbiomes of obese individual (Goodrich et al. 2014; Murugesan et al. 2015). Thus, the microbiome results do indicate changes in diversity in obese individuals compared to the healthy individuals, but at the functional level, the results lacked consistency (Table 1.1). For instance, *C. minuta* (Goodrich et al. 2014) and *Akkermansia muciniphila* when present in abundance reversed obesity (Everard et al. 2013; Dao et al. 2016). In contrast, in inflammatory disease like Crohn's disease, pro-inflammatory bacteria (*Escherichia coli*) increased, whereas anti-inflammatory bacteria (*Faecalibacterium prausnitzii* and *Clostridium leptum*) decreased (Sokol et al. 2009, 2020a, b). Micro-array studies also showed that in addition to *F. prausnitzii*, other anti-inflammatory bacteria like *Ruminococcus albus*, *R. callidus* and *R. bromii* were decreased in CD patients, and several pathogenic bacteria (*Enterococcus* sp., *E. coli*, *Clostridium difficile* and *Shigella flexneri*) were more abundant (Kang et al. 2010). It was hypothesised that the anti-inflammatory bacterium *F. prausnitzii* could be used as probiotic in CD patients to counterbalance dysbiosis (Sokol et al. 2008). In a recent review, Gurung et al. (2020) summarised evidence from 42 human microbiome studies of T2DM and concluded that only a few genera with probiotic and anti-inflammatory properties like *Bifidobacterium* spp., *Bacteroides* spp. and a few only with anti-inflammatory properties like *Faecalibacterium*, *Akkermansia* and *Roseburia* were negatively associated with T2DM. It was also observed that pro-inflammatory *Fusobacterium* was positively associated with T2DM in all the

Table 1.1 Abundance changes (increase/decrease) in bacterial genera in the gut microbiomes of individuals with different intestinal diseases

Intestinal disease	Bacterial genus[a]	Increase/ decrease in disease condition	Function and reference
Obesity	*Lactobacillus* spp. (Pr), *Roseburia* spp. (Ai), *Enterobacter* spp. (Pa)	Increased	Probiotic, anti-inflammatory and pathogenic (Prakash et al. 2011; Fei and Zhao 2013; Santacruz et al. 2009)
Obesity	*Christensenella minutia and Oscillospira* spp. (Pr), *Akkermansia muciniphila and Alistipes* (Ai), *Succinivibrio and Erwinia (Co)*, *Bifidobacterium* spp. (*Ab*)	Decreased	Probiotic, anti-inflammatory and commensal reduced (Goodrich et al. 2014; Murugesan et al. 2015)
Crohn's disease	*Faecalibacterium prausnitzii* (Ai), *Clostridium leptum (Ai)*	Decreased	Anti-inflammatory (Sokol et al. 2009, 2020a, b)
Crohn's disease	*Escherichia coli* (Pi)	Increased	Pro-inflammatory Sokol et al. 2009; Sokol et al. 2020a, b)
Inflammatory bowel disease	*F. prausnitzii* (Ai), *Clostridium leptum (Ai)*	Decreased	Anti-inflammatory (Manichanh et al. 2006; Martinez-Medina et al. 2006)
Coeliac disease	*Bifidobacterium* sp. (Pr), *C. histolyticum* (Ai), *C. lituseburense* (Ai), *F. prausnitzii* (Ai)	Decreased	Probiotic and anti-inflammatory (De Palma et al. 2010)
Type 1 diabetes mellitus	*Akkermansia muciniphila* (Ai), *Bacteroidetes ovatus* (Pi), *B. thetaiotaomicron* (Co), *E. coli* (Pa)	Increased	Anti-inflammatory, pro-inflammatory, commensal and pathogenic. Increases gut permeability and endotoxemia (Qin et al. 2012a, b; Aviles-Jimenez et al. 2015; Ferreira et al. 2018)
Type 2 diabetes mellitus	*Bifidobacterium* (Pr and Ai), *Bacteroides* (Pr and Ai), *Faecalibacterium (*Ai), *Akkermansia* (Ai) *and Roseburia* (Ai*)*	Decreased	Probiotic and anti-inflammatory. Involved in onset and progression of disease (Gurung et al. 2020)
Type 2 diabetes mellitus	*Ruminococcus* (Pr), *Fusobacterium* (Pi) *and Blautia* (Pr)	Increased	Probiotic and pro-inflammatory. Involved in onset and progression of disease (Gurung et al. 2020; Wang et al. 2017; Li et al. 2017)
Colorectal cancer	*Bacteroides fragilis* (Ab), *Porphyromonas* spp. (Pi), *Prevotella* spp. (Pi), *Streptococcus gallolyticus* (Pi), *Peptostreptococcus anaerobius*	Increased	Antibacterial, pro-inflammatory, commensal, pathogenic and anti-inflammatory (De Palma et al.

(continued)

Table 1.1 (continued)

Intestinal disease	Bacterial genus[a]	Increase/ decrease in disease condition	Function and reference
	(Co), *Enterococcus faecalis* (Pa), *Streptococcus bovis* (Pi), *Fusobacterium nucleatum* (Pi), *Leptotrichia* spp. (Pi) *Campylobacter jejuni* (Pi), *Helicobacter hepaticus* (Pi), *E. coli* (Pa), *Odoribacter* (Ai), *Alistipes finegoldii* (Pi)		2010; Manichanh et al. 2006; Martinez-Medina et al. 2006)
Gastric cancer	*Helicobacter pylori* (Pi and Pa), *Peptostreptococcus stomatis* (Pi), *Parvimonas micra* (*Pa*) *Fusobacterium nucleatum* (Pi), *Streptococcus anginosus* (Pa)	Increased	Pro-inflammatory and pathogenic. Involved in disease progression (Aviles-Jimenez et al. 2015; Ferreira et al. 2018)
Colorectal cancer	*Bacteroides fragilis* (Ab), *Porphyromonas* spp. (Pi), *Prevotella* spp. (Pi), *Streptococcus gallolyticus* (Pi), *Peptostreptococcus anaerobius* (Co), *Enterococcus faecalis* (Pa), *Streptococcus bovis* (Pi), *Fusobacterium nucleatum* (Pi), *Leptotrichia* spp. (Pi), *Campylobacter jejuni* (Pi), *Helicobacter hepaticus* (Pi), *E. coli* (Pa), *Odoribacter* (Ai), *Alistipes finegoldii* (Pi)	Increased	Antibacterial, pro-inflammatory, commensal, pathogenic and anti-inflammatory (De Palma et al. 2010; Manichanh et al. 2006; Martinez-Medina et al. 2006)
Gastric cancer	*Helicobacter pylori* (Pi and Pa), *Peptostreptococcus stomatis* (Pi), *Parvimonas micra* (*Pa*) *Fusobacterium nucleatum* (Pi), *Streptococcus anginosus* (Pa)	Increased	Pro-inflammatory and pathogenic. Involved in disease progression (Aviles-Jimenez et al. 2015; Ferreira et al. 2018)

[a]*Pi* pro-inflammatory, *Ai* anti-inflammatory, *Ab* antibacterial, *Co* commensal, *Pr* probiotic, *Pa* pathogenic

42 studies (Li et al. 2017; Gurung et al. 2020). But it was intriguing that *a* few probiotic and anti-inflammatory bacteria like *Ruminococcus* spp. and probiotic bacteria like *Blautia* spp. were also positively associated withT2DM. Thus, it is difficult to understand the dual role of the probiotic and anti-inflammatory bacteria both in the healthy controls and T2DM patients. It was envisaged that the presence or absence of the disease would probably depend on the abundance of beneficial bacteria which was observed to be more dominant in healthy individuals. Further a dominance of pro-inflammatory and pathogenic bacteria appeared to be common for all other diseases of the intestine including gastric and colorectal cancer (Table 1.1). Microbiome studies also provided insight into how some of the bacteria may cause

the disease. It was proposed that obese persons have microbiota that increase carbohydrate metabolism and capacity to harvest energy and promote lipogenesis and fat accumulation (Turnbaugh et al. 2006; Després et al. 2008). Studies with *Fusobacterium nucleatum* and *Bacteroides fragilis* with respect to colorectal cancer (Tilg et al. 2018) demonstrated that the ability to induce tumours was dependent on several properties such as their ability to adhere to cancer cells, to modulate immune cells, to modify the microenvironment of the tumour, to activate the β-catenin pathway, etc. (Tilg et al. 2018).

1.5.2 Gut Microbiome Dysbiosis and Extra-intestinal Diseases

In contrast to intestinal diseases, it would be more challenging to implicate dysbiotic changes in the gut microbiome in extra-intestinal diseases like cancers, muscular dystrophy, cardiac, mental disorders, vaginosis, renal disease, hepatic diseases, etc. (Tang et al. 2017; Acharya and Bajaj 2017) (Table 1.2). Even more challenging is implicating dysbiosis in the gut microbiome in neurodevelopmental and neurodegenerative diseases like autism spectrum disorder (ASD) (Yim et al. 2017), major depressive disorder (MDD) (Zheng et al. 2016), Alzheimer's disease (AD) (Liu et al. 2019), Parkinson's disease (PD) (Wu et al. 2020) and myasthenia gravis (MG) (Qiu et al. 2018) (Tables 1.3). It was also observed that the specific taxa that were part of the dysbiotic change were not common between the diseases implying that the taxa may be disease-specific. It is also obvious that the dysbiotic changes observed are not always consistent with the functional attribute of the bacterium. For instance, in rheumatoid arthritis (RA) patients, the gut microbiomes exhibited increase in pro-inflammatory bacteria, pathogenic, probiotic and commensal bacteria (Scher et al. 2015; Liu et al. 2013), but in a few other studies, pro-inflammatory bacteria and probiotic were also decreased (Eerola et al. 1994; Drago 2019). This lack of consistency in the dysbiotic changes was also observed in MS patients who showed significant changes in the abundance of bacteria with anti-inflammatory and pro-inflammatory properties in the diseased condition (Kirby and Ochoa-Repáraz 2018; Mirza and Mao-Draayer 2017) and also in fibromyalgia patients which showed either increase or decrease only in anti-inflammatory bacteria (Nagy-Szakal et al. 2017; Minerbi et al. 2019). Results of this nature, where the bacteria with a particular function (such anti- or pro-inflammatory, etc.) were either increased or decreased in the diseased condition, make interpretation of the change difficult. In contrast, the observed decrease in anti-inflammatory *Faecalibacterium* and increase in inflammatory taxa such as *Salmonella*, *Enterobacter*, *Citrobacter*, *Erwinia*, *Serratia* and *Pantoea* in critically ill patients (Maddi et al. 2019) permits a clear-cut interpretation that increase in taxa with inflammatory properties and decrease in anti-inflammatory properties would support critical illness. Dysbiotic changes in patients with lung cancer (Zhuang et al. 2019), hepatocellular carcinoma (Fox et al.

Table 1.2 Abundance changes (increase/decrease) in bacterial genera in the gut microbiomes of individuals with different extra-intestinal diseases

Extra-intestinal disease	Bacterial genus[a]	Increase/ decrease in disease condition	Function and reference
Rheumatoid arthritis	*Clostridium perfringens* (Pi), *Anaeroglobus geminates* (Pi), *Leptotrichia* (Pa), *Lactobacillus mucosae* (Pr), *L. ruminis* (Pr), *Prevotella* (Pr), *Lactobacillus salivarius* (Co), *L. iners* (Co)	Increased	Pro-inflammatory, pathogenic, probiotic and commensal (Scher et al. 2015; Liu et al. 2013)
Rheumatoid arthritis	*Bifidobacteria* (Pr), Bacteroides (Pr), *Prevotella copri (Pi)*	Decreased	Pro-inflammatory and probiotic. Involved in immune barrier breakdown (Eerola et al. 1994; Drago 2019)
Multiple sclerosis	*Butyricimonas* (Ai), *F. prausnitzii* (Ai), *Prevotella* (Pi) *Bacteroides* (Pr), *Lactobacilli* (Pr), *Helicobacter pylori* (Pi and Pa), *Parabacteroides distasonis* (Co)	Decreased	Anti-inflammatory, pro-inflammatory, probiotic, pathogenic bacteria and commensal (Kirby and Ochoa-Repáraz 2018; Mirza and Mao-Draayer 2017)
Multiple sclerosis	*Akkermansia* (Ai), *Dorea* (AI), *Enterobacteriaceae* (Pi), *Akkermansia muciniphila* (Ai), *Acinetobacter calcoaceticus* (Pi)	Increased	Anti-inflammatory and pro-inflammatory bacteria (Kirby and Ochoa-Repáraz 2018; Mirza and Mao-Draayer 2017)
Fibromyalgia	*Faecalibacterium* (Ai), *Roseburia* (Ai), *Dorea* (Ai), *Coprococcus* (Ai), *Clostridium* (Ai), *Ruminococcus* (Ai), *Coprobacillus (Ai)*, *Clostridium scindens* (Ai), *Parabacteroides merdae* (Ai), *Alistipes* (Ai)	Increased	Anti-inflammatory (Nagy-Szakal et al. 2017)
Fibromyalgia	*Roseburia* (Ai), *Dialister* (Ai), *Prevotella copri* (Pi), *Bacteroides uniformis* (Ai), *Bifidobacterium* (Pr), *Eubacterium* (Ai) *and Clostridium* (Ai)	Decreased	Anti-inflammatory, proinflammatory and probiotic (Minerbi et al. 2019)
Hospital-acquired illness	*Faecalibacterium* (Ai)	Decreased	Anti-inflammatory (Maddi et al. 2019)
Hospital-acquired illness	*Salmonella* (Pa), *Enterobacter* (Pa), *Citrobacter* (Pa), *Erwinia* (Pi), *Serratia* (Pi) *Pantoea* (*Pi*)	Increased	Pathogenic and Pro-inflammatory (Maddi et al. 2019)
Lung cancer	*Enterococcus* (Pa)	Increased	Pathogenic (Zhuang et al. 2019)

(continued)

Table 1.2 (continued)

Extra-intestinal disease	Bacterial genus[a]	Increase/ decrease in disease condition	Function and reference
Hepatocellular carcinoma	*Helicobacter hepaticus* (*Pi*)	Increased	Pro-inflammatory (Fox et al. 2010)
Myasthenia gravis	*Clostridium* (Ai), *Eubacterium* (Ai), *Faecalibacterium prausnitzii* (Ai)	Decreased	Anti-inflammatory (Moris et al. 2018)
Myasthenia gravis	*Streptococcus* (Pr), *Parasutterella* (Pi)	Increased	Pro-inflammatory and probiotic (Moris et al. 2018)

[a]*Ai* anti-inflammatory, *Pi* pro-inflammatory, *Pa* pathogenic, *Co* commensal, *Pr* probiotic

2010) and myasthenia gravis (Moris et al. 2018; Qiu et al. 2018) (Table 1.2) also implicate a role for pathogenic, pro-inflammatory, anti-inflammatory and probiotic bacteria in the disease. Studies have also indicated that dysbiotic changes in specific bacteria residing in the gut are associated with various mental disorders (Table 1.3). But a careful analysis of the data indicated that implicating the candidate genera in the disease based on their inferred function is far from easy. For instance, in individuals with depression (Aizawa et al. 2016; Szczesniak et al. 2016, Schizophrenia (Shen et al. 2018), Parkinson's disease (Bedarf et al. 2017) and autism spectrum disorder (Sai Prashanthi et al. 2019), probiotic and anti-inflammatory bacteria are either decreased or increased thus making interpretation difficult. But in bipolar disorder (Painold et al. 2018) and attention-deficit hyperactivity disorder (Aarts et al. 2017), the dysbiotic changes were distinct with increase in probiotic and decrease in anti-inflammatory bacteria thus highlighting the important role of these two functional bacteria (Table 1.3). Dysbiotic changes have also been reported for Alzheimer's disease (Vogt et al. 2017) and anorexia nervosa (Schwensen et al. 2018).

1.5.3 Gut Microbiome Dysbiosis and Ocular Diseases

Systemic diseases also manifest as ocular diseases. A significant proportion of patients (10%) diagnosed with inflammatory bowel disease also exhibit ocular inflammatory diseases like uveitis, conjunctivitis and episcleritis and (Levine and Burakoff 2011; Vavricka et al. 2015). Apart from this, quite often ocular diseases like uveitis and keratitis are encountered in the absence of infections and occur under idiopathic conditions or Steven Johnson syndrome-induced keratitis or as autoimmune uveitis. Such diseases that are detected in the absence of infectious causative bacteria are likely to be due to dysbiosis in the gut microbiome. Gut microbiome dysbiosis has been implicated in ocular diseases like fungal keratitis (Kalyana Chakravarthy et al. 2018a), bacterial keratitis (Jayasudha et al. 2018), uveitis

Table 1.3 Abundance changes (increase/decrease) in bacterial genera in the gut microbiomes of individuals with various mental disorders

Mental disorder	Bacterial genus[a]	Increase/ decrease in disease condition	Function and reference
Depression	*Bacteroidetes* (Pr), *Alistipes* (Ai)	Increased	Probiotic and anti-inflammatory (Aizawa et al. 2016; Szczesniak et al. 2016)
Depression	*Faecalibacterium* (Ai), *Bifidobacterium* (Pr), *Lactobacillus* (Pr)	Decreased	Probiotic and anti-inflammatory (Aizawa et al. 2016; Szczesniak et al. 2016)
Schizophrenia	*Prevotella* (Pi), *Succinivibrio* (Ai), *Collinsella* (Pi), *Megasphaera* (Pi), *Klebsiella* (Pi), *Methanobrevibacter* (Pi), *Clostridium* (Ai)	Increased	Pro- and anti-inflammatory (Schwarz et al. 2018; Shen et al. 2018)
Schizophrenia	*Blautia* (Pr), *Coprococcus* (Ai), *Roseburia* (Ai)	Decreased	Probiotic - and anti-inflammatory (Schwarz et al. 2018; Shen et al. 2018)
Parkinson's disease	Bacteroides (Pr), *Roseburia* (Ai), *Akkermansia* (Ai), *Oscillospira* (Ai)	Increased	Pro- and anti-inflammatory (Keshavarzian et al. 2015; Bedarf et al. 2017)
Parkinson's disease	*Blautia* (Pr), *Coprococcus* (Ai), *Dorea* (Ai), *Oscillospira* (Ai), *Akkermansia* (Ai), *Roseburia* (Ai), *Ruminococcus* (Pr)	Decreased	Probiotic and anti-inflammatory (Keshavarzian et al. 2015; Bedarf et al. 2017)
Autism spectrum disorder	*Streptococcus* (Ai), *Akkermansia* (Ai), *Rhodococcus* (Pi), *Oscillospira* (Ai), *Desulvibrio* (Pi), *Burkholderia* (Pi), *Collinsella* (Ai), *Corynebacterium* (Pi), *Dorea* (Ai), *Lactobacillus* (Pr)	Increased	Pro- and anti-inflammatory (Kushak et al. 2017; Pulikkan et al. 2019)
Autism spectrum disorder	*Faecalibacterium (Ai)*, *Ruminococcus*, (Pr), *Fusobacteria* (Pi), *Verrucomicrobia* (Pi), *Bifidobacterium* (Pr), *Neisseria*, *Alistipes* (Ai), *Bilophila* (Pi), *Dialister* (Pr), *Parabacteroides* (Ai), *Veillonella* (Pi)	Decreased	Probiotic, pro- and anti-inflammatory (Kushak et al. 2017; Pulikkan et al. 2019)
Bipolar disorder	*Bacteroides* (Pr)	Increased	Probiotic (Evans et al. 2017; Painold et al. 2018)
Bipolar disorder	*Faecalibacterium* (Ai), *Roseburia* (Ai) *Alistipes* (Ai),	Decreased	Anti-inflammatory (Evans et al. 2017; Painold et al. 2018)

(continued)

Table 1.3 (continued)

Mental disorder	Bacterial genus[a]	Increase/ decrease in disease condition	Function and reference
Attention-deficit hyperactivity disorder	*Bifidobacterium* (Pr)	Increased	Probiotic (Aarts et al. 2017)
Attention-deficit hyperactivity disorder	*Clostridiales* (Ai)	Decreased	Anti-inflammatorry (Aarts et al. 2017)
Alzheimer's disease	*Blautia* (Pr), *Phascolarctobacterium (Pi)*, *Gemella* (Co), *E. coli (Pa)*, *Shigella* (Pi), *P. aeruginosa* (Pi)	Increased	Probiotic, pro-inflammatory and pathogenic (Cattaneo et al. 2017; Vogt et al. 2017)
Alzheimer's disease	*Genus SMB53* (Co), *Dialister* (Pr), *Clostridium* (Ai), *Turicibacter* (Co)	Decreased	Commensal, probiotic and anti-inflammatory (Cattaneo et al. 2017; Vogt et al. 2017)
Anorexia nervosa	*Methanobrevibacter smithii* (Pi)	Increased	Pro-inflammatory (Schwensen et al. 2018)

[a]*Pa* pathogenic, *Pi* pro-inflammatory, *Co* commensal, *Ai* anti-inflammatory, *Pr* probiotic

(Huang et al. 2018; Kalyana Chakravarthy et al. 2018b; Horai et al. 2015), age-related macular degeneration (Zinkernagel et al. 2017), ocular mucosal disease (de Paiva et al. 2016) and diabetic retinopathy (Das et al. 2021) thus providing evidence of a connection between gut microbiome dysbiosis and ocular diseases (Shivaji 2017, 2019, 2020). It was also demonstrated that the ocular fungal microbiome changed under conditions of fungal keratitis (Pulikkan et al. 2019; Shivaji et al. 2019). Though a connection between the gut microbiome and the diseases of the eye has been implied, the challenge is to establish the cellular and molecular basis of the gut microbiome-associated ocular disease.

1.5.3.1 Gut Microbiome Dysbiosis and Uveitis

Gut microbiome studies in uveitis (UVT) patients consistently showed an overall decrease in the diversity and abundance of bacterial communities compared to healthy control (HC) microbiomes thus implicating gut microbiome dysbiosis in UVT (Kalyana Chakravarthy et al. 2018b; Huang et al. 2018) (Fig. 1.3). Functional analyses of the dysbiotic genera in the gut microbiomes, based on inferred functions of the genera, revealed that several bacteria that contribute to anti-inflammatory response by producing butyric acid like *Dialister*, *Lachnospira*, *Blautia*, *Clostridium*, *Dorea*, *Odoribacter*, *Coprococcus*, *Faecalibacterium prausnitzii*, *Akkermansia muciniphila*, *Mitsuokella*, *Megasphaera* and *Roseburia* were decreased several folds in the gut microbiome of UVT patients compared to healthy controls (Table 1.4); simultaneously several others which include *Ruminococcus*, *Bacteroides*,

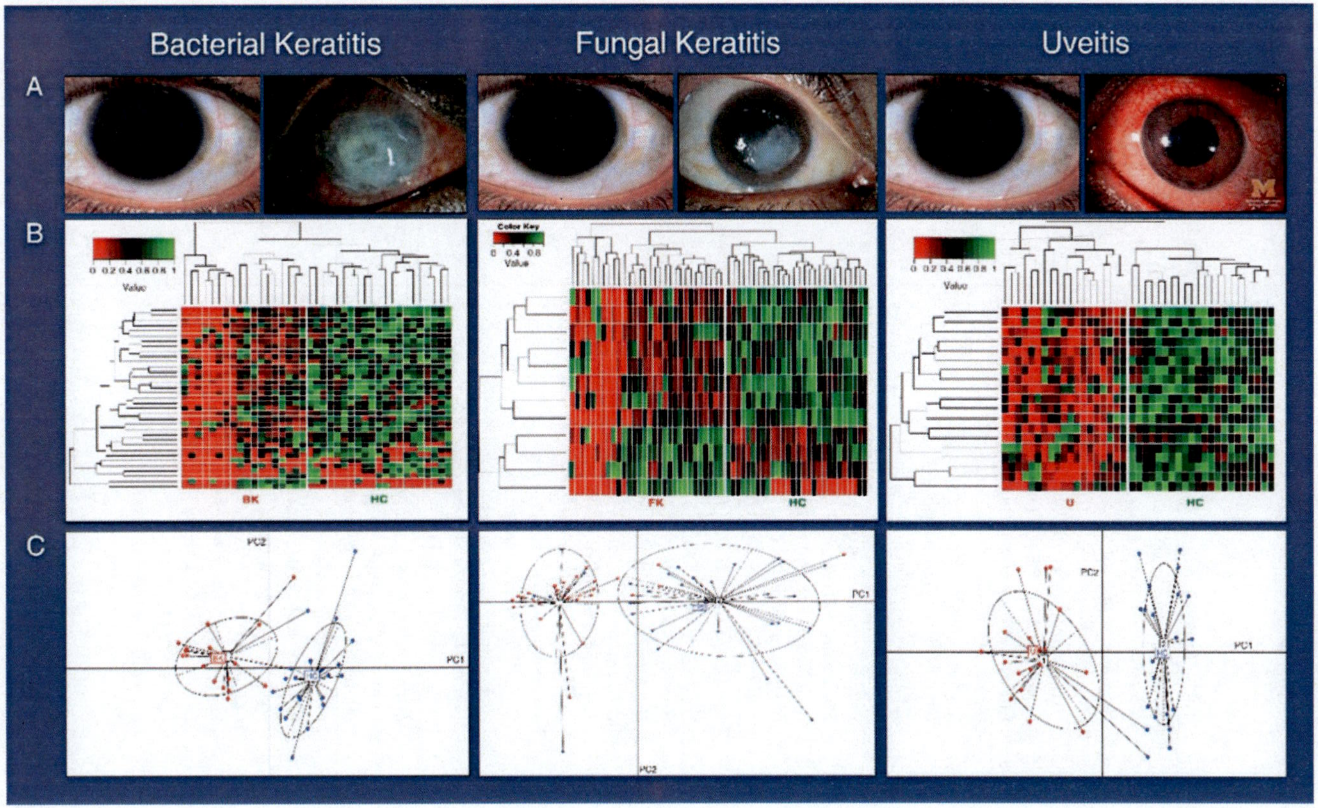

Fig. 1.3 Gut bacterial microbiome dysbiosis based on the abundance of the bacterial genera in healthy individuals and individuals with bacterial and fungal keratitis and uveitis (**a**) as analysed by heat map (**b**) and principal component analysis (**c**). Based on data from Kalyana Chakravarthy et al. (2018a, b) and Jayasudha et al. (2018)

Table 1.4 Abundance changes (increase/decrease) in bacterial genera in the gut microbiomes of individuals with ocular diseases

Ocular disease	Bacterial genus[a]	Increase/ decrease in disease condition	Function and reference
Bacterial keratitis	*Acidaminococcus* (Ai), *Bacteroides caccae* (Ai), *Bifidobacterium adolescentis* (Ai), *Blautia* (Ai), *Clostridium* (Ai), *Coprococcus eutactus* (Ai) *i Dialister* (Ai), *Dorea* (Ai), *Faecalibacterium prausnitzii* (Ai), *Lachnospira* (Ai), *Lactobacillus* (Ai/Pr), *Megamonas* (Ai), *Megasphaera* (Ai), *Mitsuokella multacida (Ai) Odoribacter* (Ai), *Parabacteroides (*Ai), *Phascolarctobacterium* (Ai), *Roseburia* (Ai), *Ruminococcus* (Ai), *Streptococcus* (Ai), *Veillonella dispar* (Ai)	Decreased	Anti-inflammatory (Jayasudha et al. 2018)
Bacterial keratitis	*Prevotella copri* (Pi), *Bilophila* (Pi)	Increased	Pro-inflammatory (Jayasudha et al. 2018)
Bacterial keratitis	*Bacteroides* (Pr), *Bifidobacterium adolescentis* (Pr), *Blautia*, *Dialister* (Pr), *Lactobacillus mucosae* (Pr), *Lactobacillus ruminis* (Pr), *Megamonas* (Pr), *Oscillospira* (Pr), *Phascolarctobacterium* (Pr), *Ruminococcus* (Pr), *Veillonella* (Pr), *Streptococcus* (Pr), *Turicibacter* (Pr)	Decreased	Probiotic (Jayasudha et al. 2018)
Bacterial keratitis	*Bacteroides* (Ab), *Lactobacillus ruminis* (Ab), *Bifidobacterium adolescentis* (Ab), *Blautia* (Ab)	Decreased	Antibacterial (Jayasudha et al. 2018)
Fungal keratitis	*Megasphaera* (Ai), *Ruminococcus* (Ai), *Roseburia* (Ai), *Lachnospira* (Ai), *Acidaminococcus* (Ai), *Clostridium* (Ai), *Dialister* (Ai), *Dorea* (Ai), *Mitsuokella multacida* (Ai), *Faecalibacterium prausnitzii* (Ai), *Lactobacillus* (Pr), *Bacteroides plebeius* (Pr), *Bifidobacterium adolescentis* (Pr), *Klebsiella* (Pi), *Sutterella* (Pi)	Decreased	Anti- and pro-inflammatory and probiotic (Kalyana Chakravarthy et al. 2018a)

(continued)

Table 1.4 (continued)

Ocular disease	Bacterial genus[a]	Increase/ decrease in disease condition	Function and reference
Fungal keratitis	*Bacteroides fragilis* (Ab), *Dorea* (Ai), *Shigella* (Pi), *Treponema* (Pa)	Increased	Antibacterial, anti-inflammatory, pro-inflammatory and pathogenic (Kalyana Chakravarthy et al. 2018a)
Uveitis	*Lachnospira* (Ai), *Ruminococcus* (Pr), *Bacteroides* (Pr), *Dialister* (Ai), *Dorea* (Ai), *Blautia* (Ai), *Clostridium* (Ai), *Coprococcus* (Ai), *Bifidobacterium adolescentis* (Pr), *Oscillospira* (Pr), *Odoribacter* (Ai), *Veillonella dispar* (Pr), *Faecalibacterium prausnitzii* (Ai), *Akkermansia muciniphila* (Ai), *Mitsuokella* (Ai), *Megasphaera* (Ai), *Roseburia* (Ai)	Decreased	Anti-inflammatory and probiotic (Kalyana Chakravarthy et al. 2018b; Huang et al. 2018; Horai et al. 2015)
Uveitis	*Bifidobacterium adolescentis* (Pr), *Bifidobacterium longum* (Pr)	Decreased	Probiotic (Kalyana Chakravarthy et al. 2018b)
Ocular mucosal disease	*Pseudobutyrivibrio* (Ai), *Escherichia* (Pa)/*Shigella* (Pi), *Blautia* (Ai), *Streptococcus* (Ai)	Increased	Anti- and pro-inflammatory and pathogenic (de Paiva et al. 2016)
Ocular mucosal disease	*Bacteroides* (Pr), *Parabacteroides* (Ai), *Faecalibacterium* (Ai), *Prevotella* (Pi)	Decreased	Probiotic, anti- and pro-inflammatory (de Paiva et al. 2016)
Age-related macular degeneration	*Anaerotruncus* (Ai), *Oscillibacter* (Pi), *Ruminococcus torques* (Pr, Ai), *Eubacterium ventriosum* (Ai)	Increased	Probiotic, anti- and pro-inflammatory (Zinkernagel et al. 2017)
Age-related macular degeneration	*Bacteroides eggerthii* (Co)	Decreased	Commensal (Zinkernagel et al. 2017)
Diabetic retinopathy	*Akkermansia* (Ai), *Parabacteroides* (Ai), *Megamonas* (Ai), *Acidaminococcus* (Ai), *Alistipes* (Ai), *Escherichia* (Pa), *Enterobacter* (Pa), *Cloacibacillus* (Pa), *Enterococcus* (Pa, Pr), *Oxalobacter* (Pr), *Shigella* (Pa, Pi)	Increased	Das et al. 2021

(continued)

Table 1.4 (continued)

Ocular disease	Bacterial genus[a]	Increase/ decrease in disease condition	Function and reference
Diabetic retinopathy	*Faecalibacterium* (Ai), *Roseburia* (Ai), *Lachnospira* (Ai), *Bifidobacterium* (Pr/Ai), *Mitsuokella* (Ai), *Streptococcus* (Pr, Ai), *Sutterella* (Pi), *Lacto-bacillus* (Pr, Ai), *Clostridium* (pathogen/Ai), *Haemophilus* (Pa), *Blautia* (Ai), *Erwinia* (Pa), *Desulfovibrio* (Pa), *Bulleidia*(-Pa), *Butyrivibrio* (Ai), *Asteroleplasma* (Nk), *Anaerovibrio* (Nk), *Comamonas* (Pa *Rothia* (Pa), *Turicibacter* (Nk)	Decreased	Das et al. 2021

[a]*Pa* pathogenic, *Pi* pro-inflammatory, *Pr* probiotic, *Ai* anti-inflammatory, *Co* commensal, *NK* not known

Bifidobacterium adolescentis, *Oscillospira* and *Veillonella dispar* and which are known to exhibit probiotic properties were also decreased in the UVT patients compared to healthy controls (Table 1.4). Thus, decrease in gut bacteria with anti-inflammatory and probiotic properties may contribute to the inflammatory reaction observed in UVT patients. Aarts et al. (2017) observed that a few short-chain fatty acid-producing bacteria, viz. *Faecalibacterium* and *Roseburia* which are anti-inflammatory, were also present in the guts of individuals with UVT, but their abundances were less than half when compared to HC implying that abundance is important.

Establishing a connection between the gut microbiome dysbiosis and UVT is a challenge. Convincing evidence is available to indicate that the gut commensal microbes are associated with both intestinal and extra-intestinal diseases like UVT (Heissigerova et al. 2016; Horai and Caspi 2019). Experimentally induced autoimmune uveitis (EAU) mice and in addition mice which are spontaneously uveitic (R161H mice) provide unequivocal evidence for the involvement of the gut microbiome in UVT. The first important observation was that in spontaneously uveitic R161H mice, it was observed that on depletion of microbiota with a cocktail of antibiotics resulted in attenuation of spontaneous uveitis. The second important finding was that in the same spontaneously uveitic R161H mice, it was also observed that the gut microbiome activates the TH17 uveitis-relevant cells, in the intestine, relevant to the onset of clinical uveitis (Horai and Caspi 2019) (Fig. 1.4). These studies supported the view that the gut microbiome triggered UVT, but the proof that the TH17 uveitis-relevant cells from the intestine enter the eye through systemic circulation and thus cause UVT is still lacking (Nakamura et al. 2016). This study also provided evidence that in the EAU mice, Foxp3+regulatory T cells (Tregs) both

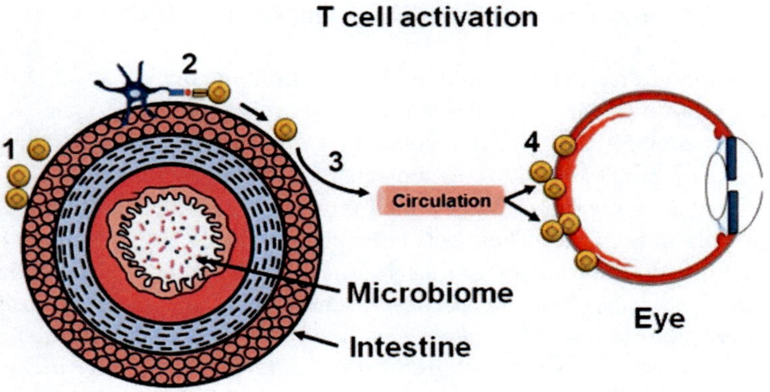

Fig. 1.4 Diagrammatic model supporting a causative role for the gut microbiota in triggering UVT. In this model autoreactive T cells in the gut of R161H mice (1) are presented by antigen presenting cells (2), which migrate to the eye (3), cross the blood retinal barrier and cause uveitis (4). Figure modified from Horai and Caspi (2019)

in the eye and lymphoid tissues acted as the trigger of UVT (Nakamura et al. 2016). Subsequently, it was also demonstrated that (Heissigerova et al. 2016) bacterial load in the gut before induction of inflammation was prerequisite for the susceptibility of mice to EAU induced by IRBP in complete Freund's adjuvant (CFA) (Heissigerova et al. 2016). Based on these studies, it was concluded that gut microbiota are crucial to the pathogenesis of uveitis in EAU mice and treatment with appropriate antibiotics could be used as an adjunct therapy for uveitis which could become sight-threatening (Nakamura et al. 2016).

1.5.3.2 Gut Microbiome Dysbiosis and Bacterial Keratitis (BK)

Comparison of the gut microbiome in the faeces of individuals diagnosed as having BK with that of HC individuals indicated dysbiosis in patients with BK (Jayasudha et al. 2018) (Fig. 1.3). Functional analysis of the bacteria in BK patients indicated decrease in antibacterial (4 nos.), anti-inflammatory (21 nos.) and probiotic (12 nos.) bacteria whereas increase in pro-inflammatory (2 nos.) bacteria (Table 1.4). Interestingly it was also observed that *Prevotella copri* which has earlier been implicated in rheumatoid arthritis and *Bilophila* were the only two pro-inflammatory bacteria that were increased in abundance in BK patients (Fisher and Phillips 2009; Lukiw 2016). In BK patients pathogens like *Bacteroides fragilis*, *Enterococcus*, *Dysgonomonas* and the genus CF231 (Lukiw 2016; Fisher and Phillips 2009) which cause gastroenteritis (Hatziioanou et al. 2017) were also enriched in the gut. Overall, in BK patients decrease in anti-inflammatory and probiotic bacteria is probably contributing to inflammation.

1.5.3.3 Gut Microbiome Dysbiosis and Fungal Keratitis (FK)

Analysis of the gut bacterial microbiomes in the healthy controls and FK patients indicated dysbiosis in the gut microbiome in the diseased individuals (Kalyana Chakravarthy et al. 2018a) (Fig. 1.3). As in the gut microbiomes of BK individuals, analysis of the faeces of FK subjects indicated a decrease in abundance of several anti-inflammatory bacteria (11 nos.), which promote human health as compared to the gut microbiomes of HC. These anti-inflammatory bacteria that decreased in the gut microbiome of BK compared to the HC individuals include those affiliated to *Lachnospiraceae* and *Ruminococcaceae* and to the *genera Roseburia, Megasphaera, Ruminococcus, Lachnospira, Acidaminococcus, Clostridium, Dorea* and *Dialister* and the species *Mitsuokella multacida and Faecalibacterium prausnitzii* (Table 1.4). Concomitantly a decrease in probiotic bacteria like *Bacteroides plebeius, Lactobacillus* and *Bifidobacterium adolescentis* was observed in the gut microbiome of BK patients. Further, increase in the pro-inflammatory bacteria *Shigella* and *Treponema* a pathogen would support FK (Kalyana Chakravarthy et al. 2018a). Thus, the observed specific decrease in the gut bacteria with anti-inflammatory and probiotic properties would exacerbate the inflammatory reaction in keratitis.

1.5.3.4 Gut Microbiome Dysbiosis and Ocular Mucosal Disease

In patients with Sjögren syndrome (SS), the mucosal tissues and secretory glands of the eye and mouth are affected. In the eye it causes severe dry eyes. Gut microbiome investigations revealed significantly altered oral and gut microbiomes in SS patients (Table 1.4) (de Paiva et al. 2016). The gut microbiome was altered in SS associated dry eye disease patients, and a careful analysis of the gut microbiome indicated ambiguous changes with respect to the functions of the dysbiotic changes observed. For instance, certain anti-inflammatory bacteria were increased (*Pseudobutyrivibrio, Blautia* and *Streptococcus*), whereas others like (*Parabacteroides* and *Faecalibacterium)* were decreased in SS*;* pro-inflammatory (*Shigella*) bacteria was increased, whereas *(Prevotella)* was decreased; pathogenic *Escherichia* was increased, whereas probiotic *Bacteroides* was reduced in abundance under dry eye condition. The involvement of dysbiosis in the gut microbiomes of patients with dry eye is evident, but the process of pathogenesis is not clearly understood.

1.5.3.5 Gut Microbiome Dysbiosis and Age-Related Macular Degeneration (AMD)

AMD is commonly encountered in elderly people and could cause blindness. Several factors have been implicated in AMD including inflammatory factors, genetic risk factors and nutrition. Andriessen et al. (2016) demonstrated that mice fed with high-fat diet show distinct changes in the gut microbiota, and this exacerbates choroidal neovascularisation, a feature of AMD. Additionally, high-glycaemic-index diet altered gut microbiota in mice and such animals developed AMD. But in these mice with AMD due to a diet of high-glycaemic-index when treated with a low-glycaemic-index diet, AMD was reverted. In human beings, dysbiosis in the gut microbiome is also associated with AMD (Zinkernagel et al. 2017) (Table 1.4). *Oscillibacter*, *Anaerotruncus*, *Eubacterium ventriosum* and *Ruminococcus torques* were relatively enriched in patients with AMD, whereas *Bacteroides eggerthii* was decreased in AMD patients (Table 1.4). Furthermore, the genes of glutamate degradation, the L-alanine fermentation and arginine biosynthesis pathways were enriched, whereas genes of the fatty acid elongation pathway decreased in the gut microbiomes of AMD patients thus confirming nutrition as an important variable in AMD manifestation. It has also been reported that the intestinal bacteria in AMD subjects possessed altered fatty acid metabolism and carotenoid biosynthetic pathways compared to control subjects, thus implicating these pathways in the pathogenesis of AMD (Lin 2019).

1.5.3.6 Gut Microbiome Dysbiosis and Diabetic Retinopathy

Das et al. (2021) confirmed gut microbiome dysbiosis in people with Type 2 diabetes mellitus (T2DM) and demonstrated dysbiosis in the gut microbiomes of people with diabetic retinopathy (DR), and differences were observed in the diversity and abundance of several different genera compared to healthy control individuals (HC) (Table 1.4). That the gut microbiome of the DR individuals was dysbiotic was also evident from the heat map and the NMDS plots, in which the two cohorts, the DR and HC groups, were separated distinctly. The gut microbiomes of T2DM and DR patients at the genera level were more discriminatory compared to the gut microbiomes of T2DM patients with HC. Since DR is considered, as an inflammatory disease, it was observed as anticipated that the gut microbiomes of DR individuals showed a decrease in abundance in probiotic and anti-inflammatory bacteria and in addition a few pathogenic bacteria were also decreased in abundance compared to HC (Das et al. 2021). Comparison of the interaction networks also indicated that the gut microbiomes of the three cohorts, namely, HC, T2DM and DR, were distinctly different and also reflected the inflammatory status of the microbiomes (Fig. 1.5). For instance, in HC, the hub genera (*Akkermansia* and *Barnesiella*) with anti-inflammatory properties probably suppressed the pathogenic hubs *Gardnerella*, *Anaerobiospirillum*, *Leptotrichia* and *Cloacibacillus* by interacting negatively with the pathogens. In contrasting T2DM network, majority of the genera (12 of 14) interacted positively with one another. Interestingly, 6 of these 12 hub genera (*Gardnerella, Gemella*, *Fusobacterium*, *Atopobium*, *Halomonas* and *Vagococcus*) were pathogens and would support the observation that T2DM is associated with chronic inflammation. In DR gut microbiomes, *Butyrivibrio* an anti-inflammatory hub genus negatively correlated with the anti-inflammatory hub genus *Megasphaera*. It was also observed that *Butyrivibrio* interacted negatively with

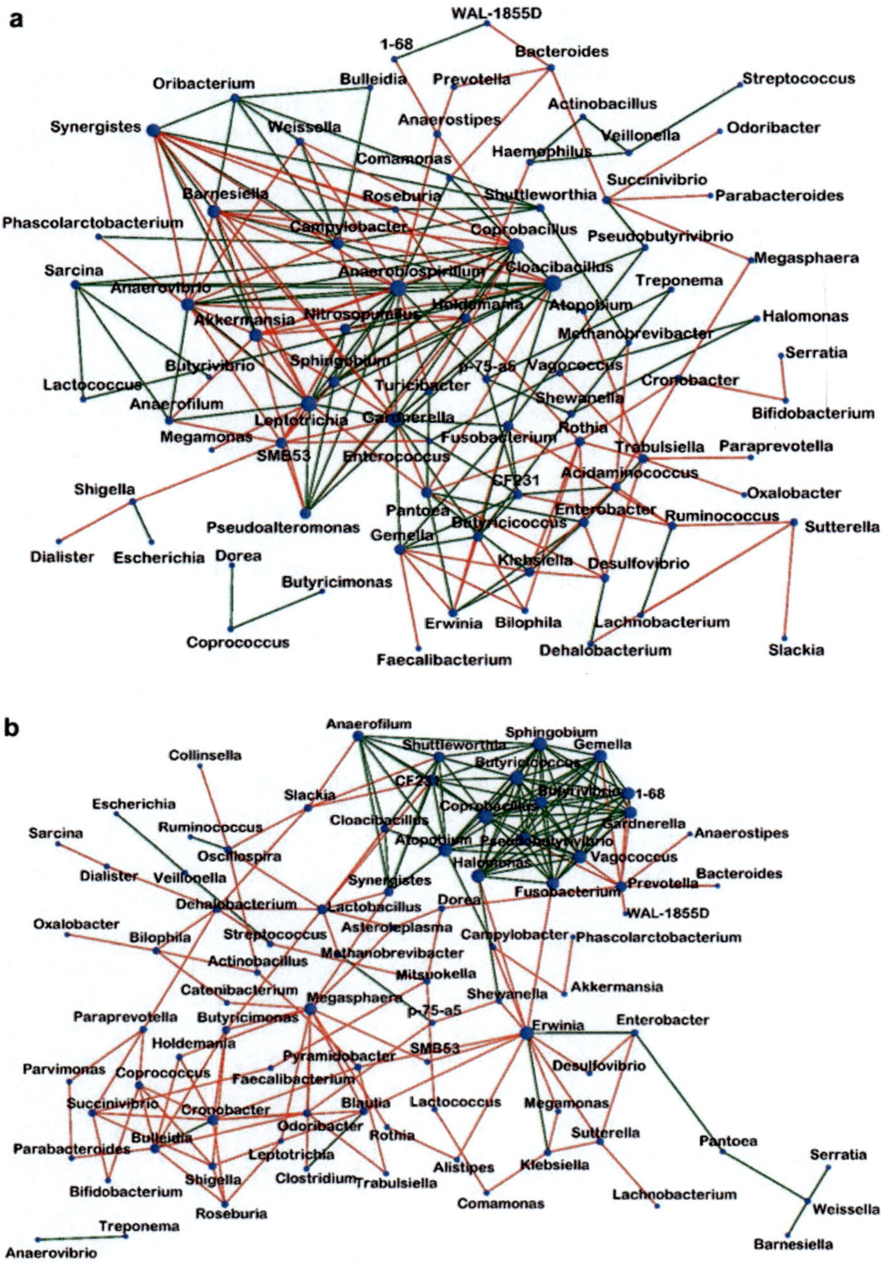

Fig. 1.5 Interaction networks (based on the abundance of bacterial genera) in the gut microbiomes of healthy controls (HC, $n = 30$) (**a**), Type 2 diabetes mellitus (T2DM, $n = 24$) (**b**) and diabetic retinopathy (DR, $n = 28$) (**c**) individuals. The size of the nodes is proportional to their degree of interaction. The edges (green and red) denote positive and negative correlations/interactions, respectively. Figure from Das et al. (2021)

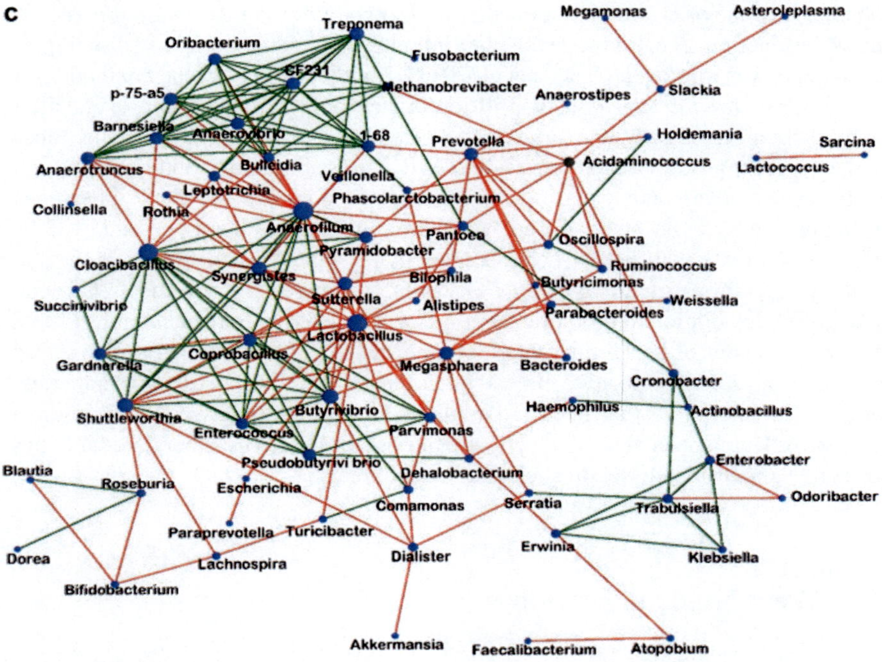

Fig. 1.5 (continued)

Lactobacillus, a probiotic, but interacted positively with *Synergistes* and *Cloacibacillus* which are two pathogenic hub genera. In contrast, *Megasphaera* and *Lactobacillus* showed negative correlation with the above two pathogenic hub genera *Synergistes* and *Cloacibacillus*. Thus, in DR patients, the gut microbiome influences DR, an inflammatory disease, by exhibiting a significant decrease in abundance in bacteria with probiotic anti-inflammatory and or possibly pathogenic bacteria compared to HC (Das et al. 2021).

1.6 Modulation of the Gut Microbiome as a Therapy

The gut microbiome composition is dependent on diet, and studies have demonstrated that the gut microbes are not passive partners and are known to change depending on several nutritional, genetic and environmental factors leading at times to dysbiosis in the gut microbiome. Reversal of the gut microbiome dysbiosis to restore normalcy is an attractive strategy which could be accomplished by the use of prebiotics, antibiotics and probiotics. In animal models of rheumatoid arthritis, probiotic therapy has been very successful but has not been replicated consistently in clinical settings. Rheumatoid arthritis patients who received *Lactobacillus rhamnosus* alone or in combination with *Bacillus coagulans* or *Lactobacillus reuteri* showed reduction in cytokine levels and inflammatory markers, and well-being of the patients was improved (Pineda et al. 2011). But the drawback was *Lactobacillus rhamnosus* or *Bifidobacterium* composition in the stool microbiota before, and after probiotic courses of *Bifidobacterium* and *Lactobacillus rhamnosus* were like those of the placebo group (Wieërs et al. 2020). It is possible that modulating the gut microbiome to overcome diseases may appear to be a straight forward strategy but probably needs to be standardised. And probiotics as a therapy could still be effective. Patients with vernal keratoconjunctivitis when treated for a month with probiotic eye drops of *Lactobacillus acidophilus* showed improvement (Lovieno et al. 2008). In addition to probiotic therapy, faecal microbial transplantation (FMT), involving transfer of faecal microbiota of a normal healthy individual to a diseased person, also resulted in positive clinical outcomes. FMT was successful as a therapy for *Clostridium difficile* diarrhoea and inflammatory bowel diseases (IBD) (Sandhya et al. 2016; Damman et al. 2012). In ocular biology FMT as a treatment modality has not yet been considered due to social and ethical reasons.

1.7 What Needs to Be Done

A possible connection or association between the gut microbiome and ocular diseases has been demonstrated and accepted (Shivaji 2017, 2019, 2020). This connection though yet to be understood at the cellular and molecular level has opened up several avenues for therapy. The importance of the gut microbiome in ocular health and disease would be better appreciated if the following are addressed:

1. Need to establish a connection between the gut and ocular microbiomes at the cellular and molecular level (metabolites, inflammatory molecules and cytokines).
2. Longitudinal studies should define changes in the ocular microbiome with the degree of severity of the ocular disease.
3. Undertake studies in close relatives to confirm that the changes in the microbiomes are not due to inter-individual variation.
4. Use mice and zebra fish to understand the cellular and molecular mechanisms underlying the disease (Sandhya et al. 2016).

However, it needs to be borne in mind that patient-driven microbiome research is bogged down with innumerable difficulties starting with identification of the patient, noncompliance of the patient, the variability between the patients, the time involved, etc. But, ultimately, microbiome research offers hope since there is light at the end of the tunnel and could lead to the development of a new treatment method involving gut microbes.

References

Aarts E, Ederveen THA, Naaijen J, Zwiers MP, Boekhorst J, Timmerman HM, Smeekens SP, Netea MG, Buitelaar JK, Franke B et al (2017) Gut microbiome in ADHD and its relation to neural reward anticipation. PLoS ONE 12:e0183509

Acharya C, Bajaj JS (2017) Gut microbiota and complications of liver disease. Gastroenterol Clin N Am 46:155–169

Aizawa E, Tsuji H, Asahara T, Takahashi T, Teraishi T, Yoshida S, Ota M, Koga N, Hattori K, Kunugi H (2016) Possible association of Bifidobacterium and Lactobacillus in the gut microbiota of patients with major depressive disorder. J Affect Disord 202:254–257

Andriessen EMMA, Wilson AM, Mawambo G, Dejda A, Miloudi K, Sennlaub F, Sapieha P (2016) Gut microbiota influences pathological angiogenesis in obesity-driven choroidal neovascularization. EMBO Mol Med 8:1366–1379

Arumugam M, Raes J, Pelletier E, Le Paslier D, Yamada T, Mende DR, Fernandes GR, Tap J, Bruls T, Batto JM et al (2011) Enterotypes of the human gut microbiome. Nature 473:174–180. https://doi.org/10.1038/nature09944

Aviles-Jimenez F, Vazquez-Jimenez F, Medrano-Guzman R et al (2015) Stomach microbiota composition varies between patients with non-atrophic gastritis and patients with intestinal type of gastric cancer. Sci Rep 4:4202. https://doi.org/10.1038/srep04202

Bedarf JR, Hildebrand F, Coelho LP et al (2017) Functional implications of microbial and viral gut metagenome changes in early-stage L-DOPA-naïve Parkinson's disease patients. Genome Med 9:39. https://doi.org/10.1186/s13073-017-0428-y

Belizário JE, Faintuch J (2018) Microbiome and gut dysbiosis. Experientia Suppl 109:459–476. https://doi.org/10.1007/978-3-319-74932-7_13

Belizário JE, Faintuch J, Garay-Malpartida M (2018) Gut microbiome dysbiosis and immunometabolism: new frontiers for treatment of metabolic diseases. Mediat Inflamm 2018: 2037838

Buttó LF, Haller D (2016) Dysbiosis in intestinal inflammation: cause or consequence. Int J Med Microbiol 306(5):302–309. https://doi.org/10.1016/j.ijmm.2016.02.010

Carding S, Verbeke K, Vipond DT, Corfe BM, Owen LJ (2015) Dysbiosis of the gut microbiota in disease. Microb Ecol Health Dis 26:26191. https://doi.org/10.3402/mehd.v26.26191

Cattaneo A, Cattane N, Galluzzi S, Provasi S, Lopizzo N, Festari C, Ferrari C, Guerra UP, Paghera B, Muscio C et al (2017) Association of brain amyloidosis with pro-inflammatory gut bacterial taxa and peripheral inflammation markers in cognitively impaired elderly. Neurobiol Aging 49:60–68

Chin VK, Yong VC, Chong PP, Nordin SA, Basir R, Abdullah M (2020) Mycobiome in the gut: a multiperspective review. Mediat Inflamm 2020:9560684. https://doi.org/10.1155/2020/9560684

Cui L, Morris A, Ghedin E (2013) The human mycobiome in health and disease. Genome Med 5(7): 63

Damman CJ, Miller SI, Surawicz CM, Zisman TL (2012) The microbiome and inflammatory bowel disease: is there a therapeutic role for fecal microbiota transplantation? Am J Gastroenterol 107: 1452–1459

Dao MC, Everard A, Aron-Wisnewsky J, Sokolovska N, Prifti E, Verger EO et al (2016) Akkermansia muciniphila and improved metabolic health during a dietary intervention in obesity: relationship with gut microbiome richness and ecology. Gut 65:426–436

Das T, Jayasudha R, Chakravarthy S et al (2021) Alterations in the gut bacterial microbiome in people with type 2 diabetes mellitus and diabetic retinopathy. Sci Rep 11:2738. https://doi.org/10.1038/s41598-021-82538-0

de Paiva CS, Jones DB, Stern ME et al (2016) Altered mucosal microbiome diversity and disease severity in Sjögren syndrome. Sci Rep 6:23561

De Palma G, Nadal I, Medina M, Donat E, Ribes-Koninckx C, Calabuig M et al (2010) Intestinal dysbiosis and reduced immunoglobulin-coated bacteria associated with coeliac disease in children. BMC Microbiol 10:63

Després JP, Lemieux I, Bergeron J, Pibarot P, Mathieu P, Larose E, Rodés-Cabau J, Bertrand OF, Poirier P (2008) Abdominal obesity and the metabolic syndrome: contribution to global cardiometabolic risk. Arterioscler Thromb Vasc Biol 28:1039–1049. https://doi.org/10.1161/ATVBAHA.107.159228

Dicksved J, Lindberg M, Rosenquist M, Enroth H, Jansson JK, Engstrand L (2009) Molecular characterization of the stomach microbiota in patients with gastric cancer and in controls. J Med Microbiol 58:509–516

Drago L (2019) Prevotella copri and microbiota in rheumatoid arthritis: fully convincing evidence? J Clin Med 8:1837. https://doi.org/10.3390/jcm8111837

Eerola E, Mottonen T, Hannonen P et al (1994) Intestinal flora in early rheumatoid arthritis. Br J Rheumatol 33:1030–1038

Evans SJ, Bassis CM, Hein R, Assari S, Flowers SA, Kelly MB, Young VB, Ellingrod VE, McInnis MG (2017) The gut microbiome composition associates with bipolar disorder and illness severity. J Psychiatr Res 87:23–29

Everard A, Belzer C, Geurts L, Ouwerkerk JP, Druart C, Bindels LB et al (2013) Cross-talk between Akkermansia muciniphila and intestinal epithelium controls diet-induced obesity. Proc Natl Acad Sci 110:9066–9071

Fei N, Zhao L (2013) An opportunistic pathogen isolated from the gut of an obese human causes obesity in germfree mice. ISME J 7:880–884

Ferreira RM, Pereira-Marques J, Pinto-Ribeiro I, Costa JL, Carneiro F, Machado JC, Figueiredo C (2018) Gastric microbial community profiling reveals a dysbiotic cancer-associated microbiota. Gut 67:226–236. https://doi.org/10.1136/gutjnl-2017-314205

Fisher K, Phillips C (2009) 2009 The ecology, epidemiology and virulence of Enterococcus. Microbiology 155:1749–1757

Flemer B, Lynch DB, Brown JMR, Jeffery IB, Ryan FJ, Claesson MJ et al (2017) Tumour-associated and non-tumour-associated microbiota in colorectal cancer. Gut 66:633–643

Fox JG, Feng Y, Theve EJ et al (2010) Gut microbes define liver cancer risk in mice exposed to chemical and viral transgenic hepatocarcinogens. Gut 59(1):88–97. https://doi.org/10.1136/gut.2009.183749

Garrett WS, Gallini CA, Yatsunenko T, Michaud M, DuBois A, Delaney ML et al (2010) Enterobacteriaceae act in concert with the gut microbiota to induce spontaneous and maternally transmitted colitis. Cell Host Microbe 8:292–300

Goodrich JK, Waters JL, Poole AC, Sutter JL, Koren O, Blekhman R et al (2014) Human genetics shape the gut microbiome. Cell 159:789–799

Guerin E, Shkoporov A, Stockdale S et al (2018) Biology and taxonomy of crAss-like bacteriophages, the most abundant virus in the human gut. Cell Host Microbe 24:653–664

Gurung M, Li Z, You H, Rodrigues R, Jump DB et al (2020) Role of gut microbiota in type 2 diabetes pathophysiology. EBioMedicine 51:102590. https://doi.org/10.1016/j.ebiom.2019.11.051

Hatziioanou D, Gherghisan-Filip C, Saalbach G, Horn N, Wegmann U, Duncan SH, Flint HJ, Mayer MJ et al (2017) Discovery of a novel l antibiotic nisin O from Blautiaobeum A2-162, isolated from the human gastrointestinal tract. Microbiology 163:1292–1305

Heissigerova J, Seidler Stangova P, Klimova A, Svozilkova P, Hrncir T, Stepankova R et al (2016) The microbiota determines susceptibility to experimental autoimmune uveoretinitis. J Immunol Res 2016:5065703. https://doi.org/10.1155/2016/5065703

Hollister EB, Gao C, Versalovic J (2014) Compositional and functional features of the gastrointestinal microbiome and their effects on human health. Gastroenterology 146(6):1449–1458. https://doi.org/10.1053/j.gastro.2014.01.052

Horai R, Caspi RR (2019) Microbiome and autoimmune uveitis. Front Immunol 10:232

Horai R, Zarate-Blades CR, Dillenburg-Pilla P, Chen J, Kielczewski JL, Silver PB et al (2015) Microbiota-dependent activation of an autoreactive T cell receptor provokes autoimmunity in an immunologically privileged site. Immunity 43:343–353. https://doi.org/10.1016/j.immuni.2015.07.0146

Huang X, Ye Z, Cao Q, Su G, Wang Q, Deng J, Zhou C, Kijlstra A, Yang P (2018) Gut microbiota composition and fecal metabolic phenotype in patients with acute anterior uveitis. Invest Ophthalmol Vis Sci 59:1523–1531. https://doi.org/10.1167/iovs.17-22677

Ignacio A, Fernandes MR, Rodrigues VA, Groppo FC, Cardoso AL, Avila-Campos MJ et al (2016) Correlation between body mass index and faecal microbiota from children. Clin Microbiol Infect 22:251–258

Jayasudha R, Chakravarthy SK, Prashanthi GS, Sharma S, Garg P, Murthy SI, Shivaji S (2018) Alterations in gut bacterial and fungal microbiomes are associated with bacterial Keratitis, an inflammatory disease of the human eye. J Biosci 43(67):1–22. https://doi.org/10.1007/s12038-018-9798-6

Kalyana Chakravarthy S, Jayasudha R, Ranjith K et al (2018a) Alterations in the gut bacterial microbiome in fungal Keratitis patients. PLoS ONE 13(6):e0199640. https://doi.org/10.1371/journal.pone.0199640

Kalyana Chakravarthy S, Jayasudha R, Sai Prashanthi G, Ali MH, Sharma S, Tyagi M, Shivaji S (2018b) Dysbiosis in the gut bacterial microbiome of patients with uveitis, an inflammatory disease of the eye. Indian J Microbiol. https://doi.org/10.1007/s12088-018-0746-9

Kang S, Denman SE, Morrison M, Yu Z, Dore J, Leclerc M, McSweeney CS (2010) Dysbiosis of fecal microbiota in Crohn's disease patients as revealed by a custom phylogenetic microarray. Inflamm Bowel Dis 16(12):2034–2042. https://doi.org/10.1002/ibd.21319

Keshavarzian A, Green SJ, Engen PA, Voigt RM, Naqib A, Forsyth CB, Mutlu E, Shannon KM (2015) Colonic bacterial composition in Parkinson's disease. Mov Disord 30:1351–1360

Kim S, Kim H, Yim YS, Ha S, Atarashi K, Tan TG et al (2017) Maternal gut bacteria promote neurodevelopmental abnormalities in mouse offspring. Nature 549:528

Kirby TO, Ochoa-Repáraz J (2018) The gut microbiome in multiple sclerosis: a potential therapeutic avenue. Med Sci 6:69. https://doi.org/10.3390/medsci6030069

Kushak RI, Winter HS, Buie TM, Cox SB, Phillips CD, Ward NL (2017) Analysis of the duodenal microbiome in autistic individuals: association with carbohydrate digestion. J Pediatr Gastroenterol Nutr 64:110–116

Levine JS, Burakoff R (2011) Extraintestinal manifestations of inflammatory bowel disease. Gastroenterol Hepatol 7:235–241

Li X, Wang E, Yin B, Fang D, Chen P, Wang G, Zhao J, Zhang H, Chen W (2017) Effects of lactobacillus casei CCFM419 on insulin resistance and gut microbiota in type 2 diabetic mice. Benefic Microbes 8:421–432. https://doi.org/10.3920/BM2016.0167

Lin P (2019) Importance of the intestinal microbiota in ocular inflammatory diseases: a review. Clin Exp Ophthalmol 47:418–422. https://doi.org/10.1111/ceo.13493

Liu X, Zou Q, Zeng B, Fang Y, Wei H (2013) Analysis of fecal Lactobacillus community structure in patients with early rheumatoid arthritis. Curr Microbiol 67:170–176

Liu P, Wu L, Peng G, Han Y, Tang R, Ge J et al (2019) Altered microbiomes distinguish Alzheimer's disease from amnestic mild cognitive impairment and health in a Chinese cohort. Brain Behav Immun 80:633–643

Livingston M, Loach D, Wilson M, Tannock GW, Baird M (2010) Gut commensal Lactobacillus reuteri 100-23 stimulates an immunoregulatory response. Immunol Cell Biol 88(1):99–102. https://doi.org/10.1038/icb.2009.71

Lovieno A, Lambiase A, Sacchetti M, Stampachiacchiere B, Micera A, Bonini S (2008) Preliminary evidence of the efficacy of probiotic eye-drop treatment in patients with vernal keratoconjunctivitis. Graefes Arch Clin Exp Ophthalmol 246(3):435–441

Lukiw WJ (2016) Bacteroides fragilis Lipopolysaccharide and Inflammatory Signaling in Alzheimer's disease. Front Microbiol 7:1544

Maddi A, Sabharwal A, Violante T et al (2019) The microbiome and lung cancer. J Thorac Dis 11(1):280–291. https://doi.org/10.21037/jtd.2018.12.88

Manichanh C et al (2006) Reduced diversity of faecal microbiota in Crohn's disease revealed by a metagenomic approach. Gut 55:205–211

Martinez-Medina M, Aldeguer X, Gonzalez-Huix F, Acero D, Garcia-Gil LJ (2006) Abnormal microbiota composition in the ileocolonic mucosa of Crohn's disease patients as revealed by polymerase chain reaction-denaturing gradient gel electrophoresis. Inflamm Bowel Dis 12: 1136–1145

Minerbi A, Gonzalez E, Brereton NJB, Anjarkouchian A, Dewar K, Fitzcharles M-A, Chevalier S, Shir Y (2019) Altered microbiome composition in individuals with fibromyalgia. Pain. https://doi.org/10.1097/j.pain.0000000000001640

Mirza A, Mao-Draayer Y (2017) The gut microbiome and microbial translocation in multiple sclerosis. Clin Immunol 183:213–224

Moris G, Arboleya S, Mancabelli L et al (2018) Fecal microbiota profile in a group of myasthenia gravis patients. Sci Rep 8:14384. https://doi.org/10.1038/s41598-018-32700-y

Mukhopadhya I, Segal JP, Carding SR, Hart AL, Hold GL (2019) The gut virome: the 'missing link' between gut bacteria and host immunity? Ther Adv Gastroenterol 12:1756284819836620

Murri M, Leiva I, Gomez-Zumaquero JM, Tinahones FJ, Cardona F, Soriguer F et al (2013) Gut microbiota in children with type 1 diabetes differs from that in healthy children: a case-control study. BMC Med 11:46

Murugesan S, Ulloa-Martinez M, Martinez-Rojano H, Galvan-Rodriguez FM, Miranda-Brito C, Romano MC et al (2015) Study of the diversity and short-chain fatty acids production by the bacterial community in overweight and obese Mexican children. Eur J Clin Microbiol Infect Dis 34:1337–1346

Nagpal R, Hariom Y, Francesco M (2014) Gut microbiota: the next-gen frontier in preventive and therapeutic medicine? Front Med 1:15. https://doi.org/10.3389/fmed.2014.00015

Nagy-Szakal D, Williams BL, Mishra N, Che X, Lee B, Bateman L, Klimas NG, Komaroff AL, Levine S, Montoya JG, Peterson DL, Ramanan D, Jain K, Eddy ML, Hornig M, Lipkin WI (2017) Fecal metagenomic profiles in subgroups of patients with myalgic encephalomyelitis/ chronic fatigue syndrome. Microbiome 5(1):44

Nakamura YK, Metea C, Karstens L, Asquith M, Gruner H, Moscibrocki C et al (2016) Gut microbial alterations associated with protection from autoimmune uveitis. Invest Ophthalmol Vis Sci 57:3747–3758. https://doi.org/10.1167/iovs.16-197338

Painold A, Mörkl S, Kashofer K, Halwachs B, Dalkner N, Bengesser S, Birner A, Fellendorf F, Platzer M, Queissner R et al (2018) A step ahead: Exploring the gut microbiota in inpatients with bipolar disorder during a depressive episode. Bipolar Disord 21(1):40–49. https://doi.org/10.1111/bdi.12682

Pineda ML, Thompson SF, Summers K, de Leon F, Pope J, Reid G (2011) A randomized, double-blinded, placebo-controlled pilot study of probiotics in active rheumatoid arthritis. Med Sci Monit 17:347–354

Pizarro TT, Arseneau KO, Cominelli F (2000) Lessons from genetically engineered animal models XI. Novel mouse models to study pathogenic mechanisms of Crohn's disease. Am J Physiol Gastrointest Liver Physiol 278:665–669

Prakash S, Rodes L, Coussa-Charley M, Tomaro-Duchesneau C (2011) Gut microbiota: Next frontier in understanding human health and development of biotherapeutics. Biol Theory 5: 71–86

Pulikkan J, Mazumder A, Grace T (2019) Role of the gut microbiome in autism spectrum disorders. Adv Exp Med Biol 1118:253–269. https://doi.org/10.1007/978-3-030-05542-4_13

Qin J, Li R, Raes J, Arumugam M, Burgdorf KS, Manichanh C et al (2010) A human gut microbial gene catalogue established by metagenomic sequencing. Nature 464(7285):59–65. https://doi.org/10.1038/nature08821

Qin J, Li Y, Cai Z, Li S, Zhu J, Zhang F et al (2012a) A metagenome-wide association study of gut microbiota in type 2 diabetes. Nature 490:55–60

Qin J, Li Y, Cai Z et al (2012b) A metagenome-wide association study of gut microbiota in type 2 diabetes. Nature 490:55–60

Qiu D, Xia Z, Jiao X, Deng J, Zhang L, Li J (2018) Altered gut microbiota in myasthenia gravis. Front Microbiol 9:2627

Ridaura VK, Faith JJ, Rey FE, Cheng J, Duncan AE, Kau AL et al (2013) Gut microbiota from twins discordant for obesity modulate metabolism in mice. Science 341:1241214

Rinninella E, Raoul P, Cintoni M et al (2019) What is the healthy gut microbiota composition? A changing ecosystem across age, environment, diet, and diseases. Microorganisms 7(1):14

Sai Prashanthi GS, Jayasudha R, Chakravarthy SK, Padakandla SR, SaiAbhilash CR, Sharma S, Bagga B, Murthy SI, Garg P, Shivaji S (2019) Alterations in the ocular surface fungal microbiome in fungal keratitis patients. Microorganisms 7:309. https://doi.org/10.3390/microorganisms7090309

Sandhya P, Debashish D, Disha and S Vinod. (2016) Does the buck stop with the bugs? An overview of microbial dysbiosis in rheumatoid arthritis. Int J Rheum Dis 19:8–20

Santacruz A, Marcos A, Warnberg J, Marti A, Martin-Matillas M, Campoy C et al (2009) Interplay between weight loss and gut microbiota composition in overweight adolescents. Obesity 17: 1906–1915

Schaubeck M, Clavel T, Calasan J et al (2016) Dysbiotic gut microbiota causes transmissible Crohn's disease-like ileitis independent of failure in antimicrobial defence. Gut 65:225–237

Scher JU, Ubeda C, Artacho A et al (2015) Decreased bacterial diversity characterizes the altered gut microbiota in patients with psoriatic arthritis, resembling dysbiosis in inflammatory bowel disease. Arthritis Rheumatol 67:128–139. https://doi.org/10.1002/art.38892

Schwarz E, Maukonen J, Hyytiäinen T, Kieseppä T, Orešiˇc M, Sabunciyan S, Mantere O, Saarela M, Yolken R, Suvisaari J (2018) Analysis of microbiota in first episode psychosis identifies preliminary associations with symptom severity and treatment response. Schizophr Res 192:398–403

Schwensen HF, Kan C, Treasure J, Høiby N, Sjögren M (2018) A systematic review of studies on the faecal microbiota in anorexia nervosa: Future research may need to include microbiota from the small intestine. Eat Weight Disord 23:399–418

Sender R, Fuchs S, Milo R (2016) Revised estimates for the number of human and bacteria cells in the body. PLoS Biol 14(8):e1002533. https://doi.org/10.1371/journal.pbio.1002533

Shen Y, Xu J, Li Z, Huang Y, Yuan Y, Wang J, Zhang M, Hu S, Liang Y (2018) Analysis of gut microbiota diversity and auxiliary diagnosis as a biomarker in patients with schizophrenia: a cross-sectional study. Schizophr Res 197:470–477. https://doi.org/10.1016/j.schres.2018.01.002

Shivaji S (2017) We are not alone: a case for the human microbiome in extra intestinal diseases. Gut Pathogens 9:13

Shivaji S (2019) Connect between gut microbiome and diseases of the human eye. J Biosci 44:110. https://doi.org/10.1007/s12038-019-9931-1

Shivaji S (2020) A systematic review on gut microbiome and ocular inflammatory diseases: are they associated? Indian J Ophthalmol 69:535–542

Shivaji S, Jayasudha R, Prashanthi GS, Chakravarthy SK, Sharma S (2019) The human ocular surface fungal microbiome. Invest Ophthalmol Vis Sci 60:451–459

Sokol H, Pigneur B, Watterlot L, Lakhdari O, Bermúdez-Humarán LG et al (2008) Faecalibacterium prausnitzii is an anti-inflammatory commensal bacterium identified by gut microbiota analysis of Crohn disease patients. PNAS 105:16731–16736. https://doi.org/10.1073/pnas.0804812105

Sokol H, Seksik P, Furet JP, Firmesse O, Nion-Larmurier I, Beaugerie L et al (2009) Low counts of Faecalibacterium prausnitzii in colitis microbiota. Inflamm Bowel Dis 15:1183–1189

Sokol H, Landman C, Seksik P, Berard L, Montil M, Nion-Larmurier I et al (2020a) Fecal microbiota transplantation to maintain remission in Crohn's disease: a pilot randomized controlled study? Microbiome 8:12. https://doi.org/10.1186/s40168-020-0792-5

Sokol H, Landman C, Seksik P, Berard L, Montil M, Nion-Larmurier I, Bourrier A, Le Gall G, Lalande V, De Rougemont A, Kirchgesner J, Daguenel A, Cachanado M, Rousseau A, Drouet É, Rosenzwajg M, Hagege H, Dray X, Klatzman D, Marteau P, Beaugerie L, Simon T (2020b) Fecal microbiota transplantation to maintain remission in Crohn's disease: a pilot randomized controlled study. Microbiome 8:12. https://doi.org/10.1186/s40168-020-0792-5

Szczesniak O, Hestad KA, Hanssen JF, Rudi K (2016) Isovaleric acid in stool correlates with human depression. Nutr Neurosci 19:279–283

Tang WH, Kitai T, Hazen SL (2017) Gut microbiota in cardiovascular health and disease. Circ Res 120:1183–1196

Tilg H, Adolph TE, Gerner RR, Moschen AR (2018) The intestinal microbiota in colorectal cancer. Cancer Cell 33(6):954–964. https://doi.org/10.1016/j.ccell.2018.03.004

Turnbaugh PJ, Ley RE, Mahowald MA, Magrini V, Mardis ER, Gordon JI (2006) An obesity-associated gut microbiome with increased capacity for energy harvest. Nature 444:1027–1031

Turnbaugh PJ, Ley RE, Hamady M, Fraser-Liggett CM, Knight R, Gordon JI (2007) The human microbiome project. Nature 449:804–810

Turnbaugh PJ, Hamady M, Yatsunenko T, Cantarel BL, Duncan A, Ley RE, Sogin ML, Jones WJ, Roe BA, Affourtit JP et al (2009) A core gut microbiome in obese and lean twins. Nature 457: 480–484

Vavricka SR, Schoepfer A, Scharl M, Lakatos PL, Navarini A, Rogler G (2015) Extraintestinal manifestations of inflammatory bowel disease. Inflamm Bowel Dis 21:1982–1992. https://doi.org/10.1097/MIB.0000000000000392

Virgin HW (2014) The virome in mammalian physiology and disease. Cell 157:142–150

Vogt NM, Kerby RL, Dill-McFarland KA, Harding SJ, Merluzzi AP, Johnson SC, Carlsson CM, Asthana S, Zetterberg H, Blennow K et al (2017) Gut microbiome alterations in Alzheimer's disease. Sci Rep 7:13537

Wang G, Li X, Zhao J, Chen W (2017) Lactobacillus casei CCFM419 attenuates type 2 diabetes via a gut microbiota dependent mechanism. Food Funct 8:3155–3164. https://doi.org/10.1039/c7fo00593h

Wieërs G, Belkhir L, Enaud R, Leclercq S, Philippart de Foy J-M, Dequenne I, de Timary P, Cani PD (2020) How probiotics affect the microbiota. Front Cell Infect Microbiol 9:454. https://doi.org/10.3389/fcimb.2019.00454

Wu W, Kong Q, Tian P et al (2020) Targeting gut microbiota dysbiosis: potential intervention strategies for neurological disorders. Engineering. https://doi.org/10.1016/j.eng.2019.07.026

Xue W, Jing LJ, Yanli Z, Bin Z, Lai W (2021) Microbiota and ocular diseases. Front Cell Infect Microbiol 11:1047

Yim YS, Park A, Berrios J, Lafourcade M, Pascual LM, Soares N et al (2017) Reversing behavioural abnormalities in mice exposed to maternal inflammation. Nature 549:482–487

Zheng P, Zeng B, Zhou C, Liu M, Fang Z, Xu X et al (2016) Gut microbiome remodeling induces depressive-like behaviors through a pathway mediated by the host's metabolism. Mol Psychiatry 21:786–796

Zhuang H, Cheng L, Wang Y, Zhang YK, Zhao MF, Liang GD, Zhang MC, Li YG, Zhao JB, Gao YN, Zhou YJ, Liu SL (2019) Dysbiosis of the gut microbiome in lung cancer. Front Cell Infect Microbiol 19:112. https://doi.org/10.3389/fcimb.2019.00112

Zinkernagel M, Zysset-Burri D, Keller I et al (2017) Association of the intestinal microbiome with the development of neovascular age-related macular degeneration. Sci Rep 7:40826. https://doi.org/10.1038/srep40826

Chapter 2
The Cultivable Bacteria of the Human Eye

Abstract This chapter highlights the cultivable bacteria of the human eye based on microscopic examination of the ocular sample and cultivating the bacteria on suitable growth media. This time-tested method has demonstrated that eye has its own component of bacteria which varied between individuals depending on their ethnicity, geographical origin, etc. Over the years, this microscopic examination and cultivable method of identifying ocular bacteria has become the gold standard for aetiological diagnosis and subsequent treatment of ocular diseases.

Keywords Ocular surface bacteria · Bacterial load · Commensals · Cultivable bacteria · Ocular disease

2.1 Introduction

Antonie Van Leeuwenhoek, a Dutch scientist, was the first to discover bacteria in the year 1676. What followed was the discovery of bacteria from the environment including the biotic (plants, animals, etc.) and abiotic components (water, air soil, etc.). The major challenge was to isolate, cultivate and identify bacteria taxonomically and establish their attributes (beneficial or harmful). Bacteria of the eye (ocular surface) are numerically less compared to bacteria reported from other parts of the human body. Thus the eye is less conducive to the colonisation of bacteria, and this is probably due to factors in the tears which exhibit antimicrobial activity (McDermott 2013) and also due to the mechanical and relentless movement of the eyelids (Shovlin et al. 2013).

2.2 Commensal Microbes of the Ocular Surface

Despite the various strategies that prevent bacterial colonisation of the ocular surface (McDermott 2013; Shovlin et al. 2013), it is possible that an environmental organism could become a part of the ocular surface flora. But it is still not clear whether these environmental visitors are transient or permanent residents of the ocular surface. The

S. Shivaji, *Human Ocular Microbiome*,
https://doi.org/10.1007/978-981-19-1754-7_2

commensal nature of these ocular microbes was confirmed by in vitro experiments which indicated that when commensal bacteria of *Propionibacterium acnes* and *Staphylococcus epidermidis* were used to challenge epithelial cells of the cornea and conjunctiva in culture, the host cells did not mount an anti-inflammatory response. Further, in 2017, it was observed that *Corynebacterium mastitidis*, a candidate commensal, elicited a commensal-specific interleukin 17 response from γδ T cells in the ocular mucosa (St Leger et al. 2017). In contrast, when epithelial cells of the eye were exposed to *Pseudomonas aeruginosa*, a pathogen, the ocular cells produced inflammatory cytokines (IL-6 and IL-8) (Cavuoto et al. 2019; Ueta and Kinoshita 2010).

2.2.1 Bacterial Load on the Ocular Surface

The ocular mucosal surface harbours bacteria just as other mucosal surfaces like the buccal, vaginal and intestinal mucosa. But the fact that cultures are not consistently isolated from conjunctival swabs may also imply that the bacterial load of the ocular surface is very low. Graham et al. (2007) observed that the bacterial load in the conjunctival swabs and conjunctival impression cytology samples when monitored using conventional broad-range 16S rDNA PCR was detectable in 40 of 59 of the swabs and in 28 of 51 impression cytology samples. Estimates indicated that the numbers of cfu varied depending on the region of the eye sampled and followed the order 0–1000 cfu for eyelid swab (Stapleton et al. 1997), 0–85 cfu per conjunctival swab (Stapleton et al. 1997; McNatt et al. 1978), 0–64 cfu for tear film (Larkin and Leeming 1991) and 1.6 cfu from tears (Hart et al. 1996) thus indicating that the eyelid has the greatest bacterial load (Table 2.1). Oguz et al. (2000) identified a total of 460 cfu of bacteria on the ocular surface including the lids and the conjunctiva, and the predominant taxa were *S. epidermidis* (178 cfu), diphtheroids (166 cfu), *S. aureus* (77 cfu), *Haemophilus* spp. (32 cfu), *Corynebacterium* spp. (4 cfu) and *Streptococcus* spp. (2 cfu). A study from Australia characterised the normal microbiota of the ocular surface and found that the lower lid margins contained several bacteria including *Propionibacterium* spp. (5–340 cfu/swab), coagulase-negative staphylococci (5–100 cfu/swab), *S. aureus* (5–75 cfu/swab), *Micrococcus*

Table 2.1 Bacterial load of the different regions of the eye

	Colony forming units	Reference
Eyelid swab	0–100	Stapleton et al. (1997), Larkin and Leeming (1991), McNatt et al. (1978), Hart et al. (1996)
Conjunctival swab	0–85	Stapleton et al. (1997), Larkin and Leeming 1991), Hart et al. (1996), McNatt et al. (1978)
Tear film	0–64	Larkin and Leeming (1991)
Tears	1.6	Hart et al. (1996)

spp. (15 cfu/swab) and *Bacillus* spp. (10 cfu/swab) (Ozkan et al. 2012). These studies indicated that in the lids and conjunctiva, the cultivable bacteria are limited in diversity and number. In contrast to the above the oral and gastrointestinal mucosae contain a greater bacterial load (between 10^7 and 10^8 cfu/mL) (Abelson et al. 2016; Willcox 2013). In a subsequent study, using qPCR of 16S rDNA gene qPCR (Doan et al. 2016), it was demonstrated that the normalised bacterial load for conjunctival samples was 0.14 ± 0.03 16S rDNA copies/human actin copy. As compared to the facial skin and buccal mucosa, the bacterial load on the conjunctiva was approximately 150- to 200-fold less. A human cell in the buccal mucosa and skin contained approximately 12 and 16 bacteria, respectively, which was considerably high in number compared to approximately 0.06 bacterium per human conjunctival cell. The results also indicated more bacterial DNA in the lower compared to the upper conjunctiva, but the bacterial load was not influenced by sex of the individual and remained similar in the two eyes within a subject; but the bacterial load showed a statistically significant difference in subjects older than 60 years when compared to individuals less than 30 years (Doan et al. 2016). Willcox (2013) showed microbial growth was slightly higher in the lids compared to the conjunctiva but detected the same species. These findings indicated that the healthy conjunctiva has a paucibacterial microbiome (Doan et al. 2016).

2.2.2 Viable Community of Bacteria on the Ocular Surface

Several media such as chocolate agar or blood agar plates, broth culture, etc. have been used successfully to culture bacteria from conjunctival swabs (10–90%) (Perkins et al. 1975; Willcox 2013; de Kaspar et al. 2004; Doan et al. 2016; Graham et al. 2007; Kittipibul et al. 2020). But the positivity of cultivability varied (10–90%) and was attributed to several variables like collection, transportation and culture conditions. For instance, conjunctival swabs in thioglycolate broth yielded predominantly coagulase-negative staphylococci, whereas blood agar plates yielded more of *Corynebacterium* spp. and agar plates without blood more of *Propionibacterium* spp. (de Kaspar et al. 2008). It was also observed that the cultivability of *Staphylococcus aureus* and *S. epidermidis* approximately doubled when the conjunctiva were sampled after 8 h of sleep (Ramachandran et al. 1995). Among the various bacteria isolated from the ocular surface, haemolytic *Staphylococcus* were among the first to be cultivated using conjunctival swabs of individuals without ocular infection (Keilty 1930). In a subsequent study, Khorazo and Thompson (1935) studied the normal conjunctiva in 1122 individuals and confirmed that staphylococci were the most predominant bacteria present in 64% of the sampled individuals. Among the staphylococci, coagulase-negative staphylococci (CoNS) that colonise the mucosa and lid margins were the most commonly found bacteria (45%) (Graham et al. 2007; Doan et al. 2016; Sthapit 2014). At the species level, *Staphylococcus epidermidis* was the predominant species which was detected as early as 1954 and after (Smith 1954; Perkins et al. 1975; Suto et al. 2012; Cason and

Table 2.2 Cultivable bacteria from the conjunctivae of normal healthy individuals

Bacteria detected	Cultivable bacteria
Gram-positive	Coagulase-negative staphylococcus, *Staphylococcus epidermidis*, *S. aureus*, methicillin-resistant *Staphylococcus aureus*, *S. lentus*, *S. xylosus*, *S. citrius*, *S. hominis*, *S. haemolyticus*, *S. capitis*, *S. warneri*, *S. lugdunensis*, *S. hyicus*, *S. intermedius*, *S. schleiferi*, *Propionibacterium* spp., *P. acnes*, *P. lymphophilum*, *Propionibacterium acidipropionici*, *Diphtheroids* spp., *Corynebacterium* spp., *Streptococcus* spp., *Streptococcus pyogenes*, *S. viridians*, *Micrococcus* spp., *Bacillus* spp., *Peptostreptococcus* spp., *Lactobacillus* spp., *Peptococcus niger*, *Enterococcus* spp., *Stomatococcus mucilaginosus*, *Pneumococcus* spp., *Haemophilus* spp., *Haemophilus influenza*, *Rothia mucilaginosa*, *Rothia* spp., *Eubacterium* spp., *Clostridium* spp., *Sarcinae* spp.
Gram-negative	*Gram-negative rods*, *Pseudomonas* spp., *P. aeruginosa*, Enterobacter spp., *Escherichia coli*, *Proteus* spp., *Acinetobacter* spp., Citrobacter spp., unidentified Gram-negative bacteria, *Neisseria* spp., *Moraxella* spp., *Bifidobacterium* spp., *Megasphaera elsdenii*, *Bacteroides ureolyticus*, *B. pneumosintes*, *Klebsiella pneumonia*

Data from Keilty (1930), Rajvanshi (1968), Khorazo and Thompson (1935), Perkins et al. (1975), Hovding (1981), Larkin and Leeming (1991), Leitch et al. (1998), Thiel and Schumacher (1994), Hofling-Lima et al. (2002), Moeller et al. (2005), Graham et al. (2007), Grzybowski et al. (2017), Suto et al. (2012), Sharma (2012), Sharma et al. (2013a, b), Willcox (2013), Sthapit (2014), Kittipibul et al. (2020), Ranjith et al. (2017), Doan et al. (2016), Jensen and Gluud (1985), Capriotti et al. (2009)

Winkler 1954; Graham et al. 2007; Khorazo and Gutierrez 1972; Soudakoff 1954; Sharma et al. 2013b; Kittipibul et al. 2020). *Staphylococcus epidermidis* has a global presence on the ocular surface and has been reported from healthy individuals from several countries (Suto et al. 2012; Olson et al. 2010; Rajvanshi 1968; Capriotti et al. 2009; Khorazo and Thompson 1935; Graham et al. 2007). The other common staphylococcal species from the ocular surface included *S. aureus*, *S. haemolyticus*, *S. citrius*, *S. xylosus*, *S. capitis*, *S. lentus*, *S. warneri*, *S. intermedius*, *S. lugdunensis*, *S. hyicus*, *S. hominis* and *S. schleiferi* (Sharma et al. 2013b; Hovding 1981; Willcox 2013; Larkin and Leeming 1991; Rajvanshi 1968; Graham et al. 2007; Khorazo and Thompson 1935; Leitch et al. 1998). Several studies have also indicated that coagulase-positive staphylococci are the predominant flora of conjunctiva (Keilty 1930; Khorazo and Thompson 1935; Rajvanshi 1968; Sthapit 2014). In 1935 it was also reported that apart from staphylococci which were predominant (64%) in conjunctival swabs, the other bacteria identified were diphtheroids in 36%, alpha-haemolytic streptococci in 3–4%, *Streptococcus pyogenes* in 1.2%, *Pneumococci* in 2.6% and sarcinae in 1% (Khorazo and Thompson 1935). With time several other bacteria were also identified on the ocular surface (Table 2.2) (Sharma et al. 2013b; Willcox 2013; Doan et al. 2016; Keilty 1930; Sharma 2012; Suto et al. 2012; Grzybowski et al. 2017; Hofling-Lima et al. 2002; Moeller et al. 2005; Graham et al. 2007; Kittipibul et al. 2020; Perkins et al. 1975). Methicillin-resistant *S. aureus* (MRSA) has also been observed on the ocular surface with a preponderance in older patients (Olson et al. 2010; Hsu et al. 2015). Many of the above bacteria detected on the ocular surface are commonly detected associated with the skin and mucus membrane and are likely acquired in conjunctiva from the adjacent eyelid skin or from the hand (Sthapit 2014). Overall, the studies indicated that normal ocular bacterial flora is diverse (Grzybowski et al. 2017).

2.2.3 *Impact of Age on the Ocular Surface Cultivable Microbiome*

The ocular surface microbiome is dynamic and changes with the age of the individual. In the normal conjunctiva, the incidence of bacteria increases with age (Rao and Rao 1972). Singer et al. (1988) studied bacteria isolated from conjunctival swabs from 229 eyes of 144 unaffected individuals and observed that the number of species isolated per eye was greater in number in adults than in younger subjects (1.47 versus 1.13: $p < 0.05$). The incidence of *S. epidermidis* was minimum (30.8%) in elderly (51 years age) and nearly same (45.2–46.7%) in middle (26–50 years age) and young age groups (13–25 years age). In contrast, *Staphylococcus aureus* was found to increase significantly from 8.9% in young to 16.7% in middle age and further to 23.1% in elderly subjects (Sharma et al. 2013a). However, this age-dependent trend in the abundance of *S. epidermidis* and *S. aureus* was inconsistent between studies. For instance, Liu et al. (2011) reported greater incidence of *S. epidermidis* in elderly (94.64%) which decreased in middle-aged and was considerably low in children (33.33%). Sthapit (2014) attributed the above shift in floral incidence to aqueous tear deficiency with age, goblet cell changes and lipid deregulatory states that accumulate over time (Sthapit 2014). Sharma et al. (2013a, b) also found *S. viridian* and *Haemophilus* spp. were more prevalent in younger than middle-aged individuals and were conspicuous by their absence in elderly. Conversely, diphtheroids were more prevalent in the elderly tapering to zero in the younger age group. *E. coli* was not found in middle-aged individuals, and incidence was equal in the other two age groups. Ocular surface isolates have been identified immediately after birth (Eder et al. 2005) and up to 90 years of age (Thiel and Schumacher 1994). Studies indicated that the bacterial flora on the conjunctiva of the newborns was identical to that associated with the vagina and cervix of pregnant woman (Grunberger and Kofler 1954; Hong 1974), and *S. aureus*, *S. epidermidis*, *Streptococcus* spp. and *Escherichia coli* were identified indicating that the conjunctiva was contaminated by the vaginal and cervical bacterial flora of the mother (Fig. 2.1). Hong (1974) found that the conjunctiva of babies was dominated by coagulase-negative *Staphylococcus* spp. and *Propionibacterium* spp. and several other bacteria which are common to the human cervix and included predominantly *S. viridans*, *S. epidermidis*, *Haemophilus vaginalis*, *Micrococcus* spp., diphtheroids, *Bacillus* spp., *Bacteroides* spp., *Peptostreptococcus* spp., *Propionibacterium acnes*, *Gardnerella* spp., *Peptococcus* spp., *Lactobacillus* spp., *Escherichia coli*, *Bifidobacteria* spp. and *S. aureus* (Brook et al. 1979; Isenberg et al.

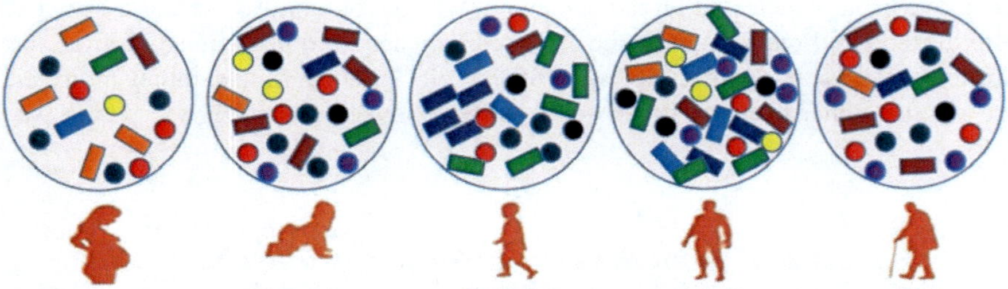

Fig. 2.1 Changes in the ocular surface cultivable microbiome with age. The differently coloured filled circles and rectangles indicate the diversity of cocci and rods in the ocular surface microbiome as detailed in Sect. 2.3. (Modified from Kowalski and Roat (www.oculist.net/downaton502/prof/ebook/duanes/pages/v8/v8c041) and data from Weiss et al. 1993, Singer et al. 1988, McNatt et al. 1978, Brook et al. 1979 and Brown 1978)

1986; Eder et al. 2005) (Fig. 2.1). Similar microbes were also identified in babies delivered following caesarean section, but the bacteria were numerically reduced (Eder et al. 2005). Two days after birth, fewer microbes (including pathogenic bacteria) were isolated from the conjunctiva (Eder et al. 2005; Brook et al. 1979; Ozkan and Willcox 2019), and the dominating bacteria were coagulase-negative *Staphylococcus* spp., *S. aureus* and *Propionibacterium acnes*. In addition, several other bacteria that were not consistently detected included *Micrococcus* spp., *Corynebacterium* spp., *Streptococcus viridans* group, *Enterococcus* spp., *Haemophilus* spp., *Peptostreptococcus* spp., *Bacillus* spp., *Lactobacillus* spp., *Eubacterium* spp., *Clostridium* spp., *E. coli* and *Klebsiella* spp. (Eder et al. 2005; Brook et al. 1979; Ozkan and Willcox 2019) (Fig. 2.1). Subsequently, the conjunctival flora continued to evolve up to pediatric age and stabilised as in the adults such that the microorganisms from the conjunctiva were similar to those observed in adults and the most representative were coagulase-negative staphylococci, *Propionibacterium* spp., *Streptococcus* spp., *Micrococcus* spp., *S. aureus*, *Enterococcus* spp., *Bacillus* spp. and *Corynebacterium* spp. (Sankaridurg et al. 2009; Eder et al. 2005; Thiel and Schumacher 1994; Brook et al. 1979; Petrillo et al. 2020) (Fig. 2.1). Thiel and Schumacher (1994) studied human conjunctival flora in individuals varying in age from 3 to 90 years and observed characteristic changes at different stages of life. *Megasphaera elsdenii*, *Bacteroides ureolyticus*, *Bacteroides pneumosintes*, *Stomatococcus mucilaginosus* and group ANF *Corynebacterium* were isolated for the first time in the eye by this study. Further, *Propionibacterium* spp., and aerobic cocci are common in the younger age group and decrease with age. *Corynebacterium* spp. and anaerobic cocci become more frequent with age (Thiel and Schumacher 1994). Sthapit (2014) observed that *Streptococcus* spp. were remarkably more in early adults (15–45 years). It was also observed that in adults >17 years up to 60 years, coagulase-negative staphylococci, *S. aureus*, *Corynebacterium* spp., *Streptococcus* spp., *Proteus* spp., *Micrococcus* spp., *Propionibacterium acnes*., *Haemophilus* spp. and *Peptostreptococcus* spp. were dominant. Further, between

26 and 99 years, coagulase-negative staphylococci, *Staphylococcus aureus*, *Streptococcus* spp., *Corynebacterium* spp., *Proteus* spp., *Enterococcus* spp., *Escherichia coli* and *Klebsiella* spp., were dominant (Thiel and Schumacher 1994) Sthapit (2014). Thus the bacterial diversity is dynamic and changes with age. Whether this change has a functional relevance or whether it is influenced by biotic factors is a question that remains to be addressed.

2.2.4 Impact of Sex on Ocular Surface Cultivable Microbiome

Rao and Rao (1972) observed that incidence of conjunctival bacterial flora is similar in either sex (22.5% males 18.2% females). But Javed et al. (2008) showed that male patients had higher percent of growth of normal flora (68.1%) than in the females (31.9%), and the increase was attributed to the outdoor agricultural activities of the males. But Sthapit (2014) found no significant difference in the prevalence of normal flora between male (93%) and female (85%) individuals ($p = 0.317$) and between the two eyes.

Coagulase-negative staphylococci in both eyes of male and females were statistically not significant ($p > 0.05$), but *Streptococcus* spp. were found significantly higher in females. Despite the discrepancy in the results on the possibility of gender influencing the ocular microbiota, an attractive hypothesis for the changes observed by some between the sexes could be attributed to a sex-steroid imbalance after menopause which is known to initiate profound changes in the micro-environment of the ocular surface predisposing women to many inflammatory, autoimmune and allergic ocular diseases, including dry eye syndrome (Wen et al. 2017; Versura et al. 2015; Truong et al. 2014; Sacchetti et al. 2015; Mantelli et al. 2016). This aspect, that sex impacts the ocular surface microbiome is also evident from the fact that women are more prone to certain ocular diseases than men (Mantelli et al. 2016) thus establishing a relationship between ocular surface microbiome, age and sex. Versura et al. (2015) reviewed the influence of hormones (androgens and oestrogens) on glands (lachrymal and meibomian glands) that influence the ocular surface and on ocular surface epithelia including conjunctival goblet cells and concluded that these hormones may have an impact on the ocular surface microbiota.

2.2.5 Impact of Geographical Location on Ocular Surface Cultivable Microbiome

Differences in the normal microbiota in individuals from different climatic and geographical locations have been studied. Doan et al. (2016) analysed 428 conjunctivae from a total of 107 healthy adult subjects which included 85 Caucasian, five

African-American, 14 Asian and three self-classified as 'others' for their microbial diversity and observed that in a good majority of the samples (78.5%) at least 1 organism could be grown. The most frequently isolated organism (45.3%) were coagulase-negative *Staphylococcus* spp., *Propionibacterium* (33.9%) and diphtheroids (15.4%) (Table 2.2). In addition, *Streptococcus viridans*, *Neisseria* spp., *Bacillus* spp., *Micrococcus* spp., unidentified Gram-positive rods, unidentified Gram-positive cocci, *Rothia mucilaginosa*, *Mycelia sterilia*, *Haemophilus* spp., *Rhodotorula* spp., *E. coli*, *Lactobacillus* spp., *Klebsiella pneumonia*, *and Moraxella* spp. were also identified in one individual. Earlier studies had indicated that diphtheroids (*Corynebacterium* sp.) are common in Japanese eyes whereas *S. aureus* in Paraguayan eyes (Willcox 2013; Jensen and Gluud 1985). Capriotti et al. (2009) analysed the conjunctival swabs of healthy eyes from a rural population in Sierra Leone, and the isolated organisms were coagulase-negative *Staphylococcus* (28.6%) and *S. aureus* (19.9%) followed by Gram-negative bacteria like *Pseudomonas* spp., *Haemophilus* spp. (9.8%), *Nocardia* spp./*Actinomyces* spp. (6.5%) and *Pseudomonas aeruginosa* (6.2%) differing from that seen in the urban population (Table 2.2). But some studies were unable to demonstrate that the microbiota varied in different regions like Korea, Brazil, the United States, Japan, Finland, Uganda and Sierra Leone (Grzybowski et al. 2017; Hofling-Lima et al. 2002; Dong et al. 2011; Moeller et al. 2005; Kittipibul et al. 2020; Gomes et al. 2020). A few of the variations in the above studies in the ocular surface microbiota included *Pseudomonas aeruginosa*, *Haemophilus influenzae* (Grzybowski et al. 2017), *Megasphaera elsdenii*, *Bacteroides pneumosintes*, *Bacteroides ureolyticus* and *Stomatococcus* spp. (Thiel and Schumacher 1994). In an earlier study, Rao and Rao (1972) analysed 115 patients from India, with no clinical evidence of extraocular infections and observed that in the summer (April to October), characterised by high temperature >31 °C and high humidity, there is a predominance of bacterial flora compared to the slightly cooler season (November to March).

2.3 Cultivable Microbiome of Eyelids and Tears

Very few studies have attempted to study the viable bacteria of different regions of the eye such as the eyelids, tears, tear film, etc. The conjunctiva is known to have a predominance of the coagulase-negative staphylococci and *Propionibacterium* spp. and also a low occurrence of *Rhodococcus erythropolis*, *S. aureus*, *Klebsiella oxytoca*, *Micrococcus* spp. *and Erwinia* spp. (Graham et al. 2007, see also Table 2.2). Further the same species were isolated from both the lids and conjunctiva, the difference being that higher cfu was recovered from the conjunctiva compared to the lids (Willcox 2013). Further, in the lids the predominant organisms which were isolated were coagulase-negative *Staphylococcus*, *S. aureus* and diphtheroids/*Corynebacterium* (Table 2.3) (Groden et al. 1991; Taylor et al. 1988; Au et al. 1993; Doyle et al. 1995). 16S rRNA gene sequencing confirmed the presence of coagulase-negative staphylococci *Rhodococcus erythropolis*, *Klebsiella oxytoca*

Table 2.3 Cultivable bacteria from the eyelids and tears of normal healthy individuals

Bacteria detected	Eyelids	Tears
Gram-positive	Coagulase-negative	Coagulase-negative
	Staphylococci; *S. epidermidis Propionibacterium* spp. *Propionibacterium* acnes	Staphylococci; *S. epidermidis*
	Diphtheroids spp.	*Propionibacterium* sp.
	S. aureus	*Diphtheroids* spp.
	Corynebacterium spp.	*S. aureus*
	Micrococcus spp.	
	Streptococcus spp., *S. viridans*, *Eubacterium* spp., *Clostridium* spp., *Bacillus* spp., *Enterococcus* spp.	
Gram-negative	*Moraxella* spp.	
	Branhamella catarrhalis	
	Pseudomonas spp., Neisseria spp., *Proteus* spp., *Acinetobacter* spp.	

Modified from Willcox (2013), Weiss et al. (1993), Groden et al. (1991), Au et al. (1993), Taylor et al. (1988), Doyle et al. (1995)

and Erwinia spp. (Graham et al. 2007). The study also indicated that similar bacteria are present in the eyelids and tears and they were all common to the conjunctiva.

2.4 Ocular Microbiota Changes in the Diseased Eye

Studies have demonstrated differences in ocular surface microbiota between patients with ocular diseases and healthy controls (Suto et al. 2012; Watters et al. 2017; Graham et al. 2007; Hori et al. 2008; Frizon et al. 2014; Venugopal et al. 2016). What follows is a description of the various taxa that are associated with the eye impacted by ocular diseases.

2.4.1 Keratitis

The cornea is inflamed in individuals with keratitis and requires immediate attention to avoid complications of the eye and even loss of vision. The common aetiological agents are bacteria and fungi, and at times viruses and parasites cause keratitis (Fig. 2.2a, b). Though keratitis is normally caused by the above infectious agents, non-infectious keratitis is also seen and commonly occurs in individuals with an eye injury, or following contact lenses use or by the introduction an extraneous foreign body in the eye, etc. The symptoms of keratitis include redness of eye, pain, blurred vision, photophobia, excess tearing and discharge from the eye (Fig. 2.2a, b). The

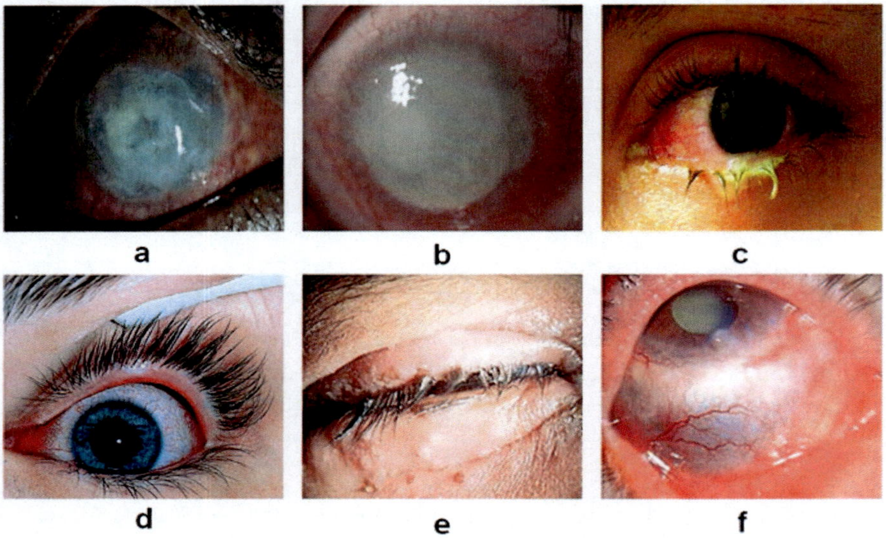

Fig. 2.2 Disease of the eye. (**a**) Bacterial keratitis; (**b**) *Acanthamoeba* keratitis; (**c**) bacterial conjunctivitis; (**d**) viral conjunctivitis; (**e**) herpetic blepharitis; (**f**) scleritis. (Except (**a**) which was obtained courtesy LVPEI, Hyderabad the remaining photographs were retrieved either from Creative Commons Attribution-Share Alike 4.0 (**b**, **d**, **f**) or 3.0 (**d**, **e**))

incidence of keratitis in India follows the trend: bacteria (32.77%) > fungi (34.4%) > bacteria and fungi (2.39%) > *Acanthamoeba* (1.04%) (Bharathi et al. 2007). Studies have also indicated that the incidence of keratitis depended on several factors. For instance, keratitis was higher for rigid gas permeable lenses and daily wear soft contact lenses compared to daily wear, and it was also increased in patients using extended wear (overnight wear) compared to users of soft contact lenses (Dart et al. 1991), daily-use silicone hydrogel contact lenses (Stapleton et al. 2008; Dart et al. 2008) and also in patients with poor hygiene and continuous use of lens storage cases instead of regular replacement (Stapleton et al. 2012). But good hygiene and cleaning routines did not ensure protection from keratitis (McLaughlin-Borlace et al. 1998). In individuals with bacterial keratitis, the bacteria cultured from swabs include several Gram-positive and Gram-negative bacteria with a preponderance of *Staphylococcus epidermidis*, *Staphylococcus* spp., *S. aureus*, *Pseudomonas aeruginosa*, *Propionibacterium acnes*, *Propionibacterium* spp., *Streptococcus pneumoniae*, *Bradyrhizobium* spp. and several other species (Table 2.4) (Bourcier et al. 2003; Egrilmez and Yildirim-Theveny 2020; Ormerod et al. 1987; Neumann and Sjostrand 1993; Rau et al. 2008; Turner and Stinson 1965; Zhai et al. 2018; Van Tyne et al. 2016; Das et al. 2019). Pathogens like *Staphylococcus* spp. and *P. aeruginosa* are also dominant on the corneal surface of bacterial keratitis patients (Al-Najjar 2019).

Table 2.4 Bacteria associated with human ocular diseases

Ocular disease	Associated bacteria	Reference
Keratitis	*Actinomyces* spp., *Bradyrhizobium* spp., *Corynebacterium* spp., *Escherichia coli*, *Klebsiella* spp., *Moraxella* spp., *Mycobacterium* spp., *Nocardia* spp., *Propionibacterium acnes*, *Propionibacterium* spp., *Pseudomonas aeruginosa*, *Proteus* spp., *Staphylococcus aureus*, *Staphylococcus epidermidis*, *Staphylococcal* spp., *Serratia* spp., *Streptococcus pneumoniae* and *Stenotrophomonas maltophilia*	Bourcier et al. (2003), Egrilmez and Yildirim-Theveny (2020), Neumann and Sjostrand (1993), Ormerod et al. (1987), Turner and Stinson (1965), Van Tyne et al. (2016), Zhai et al. (2018), Das et al. (2019)
Conjunctivitis	*Chlamydia trachomatis*, *Haemophilus influenzae*, *Escherichia coli*, *Neisseria gonorrhoeae*, *Klebsiella* spp., *Moraxella catarrhalis*, *Staphylococcus epidermidis*, *Staphylococcus aureus* and *Streptococcus pneumoniae*	Alfonso et al. (2015), Ghosh et al. (1995), Quinet et al. (2010), Rietveld et al. (2004), Cavuoto et al. (2008)
Blepharitis	*Bacillus oleronius*, *Corynebacterium* spp., *Enhydrobacter* spp., *Streptococcus pyogenes*, *Staphylococcus aureus* and *Streptophyta* spp.	Lee et al. (2012), Szkaradkiewicz et al. (2012)
Scleritis	*Corynebacterium diphtheriae*, *Mycobacterium tuberculosis*, *Mycobacterium* spp., *Nocardia* spp., *Pseudomonas aeruginosa*, *Staphylococcus pneumoniae*, *Streptococcus* spp. and *Stenotrophomonas maltophilia*	Bloomfield et al. (1976), Jain et al. (2009), Reddy et al. (2015)
Endophthalmitis	*Bacillus cereus*, *Haemophilus influenza*, *Enterococcus faecalis*, *Escherichia coli*, *Klebsiella* spp., *Moraxella* spp., *Micrococcus* spp., *Neisseria meningitidis*, *Propionibacterium acnes*, *Proteus* spp., *Pseudomonas* spp., *Serratia* spp., *Staphylococcus aureus*, *Staphylococcus epidermidis*, *Staphylococcus pyogenes*, *Streptococcus viridians* and *Streptococcus* spp.	Bode et al. (1985), Callegan et al. (2002), Driebe et al. (1986), Maitray et al. (2019), Tseng et al. (1996), Wong (2000), Rishi et al. (2009)
Orbital cellulitis	*Arcanobacterium haemolyticum*, *Bacteroides* spp., *Eikenella corrodens*, *Escherichia coli*, *Fusobacterium necrophorum*, *Fusobacterium lentum*, *Haemophilus influenzae*, *Moraxella catarrhalis*, *Peptostreptococcus* spp., *Prevotella* spp., *Propionibacterium acnes*,	Malik et al. (2004), McKinley et al. (2007), Partamian et al. (1983), Sharma et al. (2013a, b), Weiss et al. (1983), Youssef et al. (2008)

(continued)

Table 2.4 (continued)

Ocular disease	Associated bacteria	Reference
	Pseudomonas spp., *Staphylococcus aureus*, *Staphylococcus* spp., *Streptococcus* spp. and *Veillonella* spp.	
Dacryo-cystitis	*Citrobacter* spp., *Escherichia coli*, *Enterobacter* spp., *Haemophilus influenzae*, *Klebsiella pneumoniae*, *Pseudomonas* spp., *P. vulgaris*, *P. mirabilis*, *P. aeruginosa*, *Peptostreptococcus* spp., *Propionibacterium acnes*, *Staphylococcus* spp., *Staphylococcus epidermidis*, *S. aureus*, *Streptococcus pyogenes* and *Streptococcus pneumoniae*	Briscoe et al. (2005), Chaudhry et al. (2005), Eslami et al. (2018), Mills et al. (2007), Bharathi et al. (2008)
Retinitis	*Bartonella henselae*, *Borrelia burgdorferi*, *Toxoplasma gondii*, *Mycobacterium tuberculosis* and *Treponema pallidum*	Chi et al. (2012), Gupta et al. (2007), Hadfield and Guy (2017), Moradi et al. (2015), Sathiamoorthi and Smith (2016)
Post-fever retinitis	*Leptospira* spp., *Mycobacterium tuberculosis*, *Salmonella typhi* and *Rickettsia* spp.	Mahendradas et al. (2020), Prabhushanker et al. (2017), Relhan et al. (2014)
Uveitis	*Borrelia burgdorferi*, *Chlamydia trachomatis*, *Mycobacterium tuberculosis* and *Treponema pallidum* spp.	Barisani-Asenbauer et al. (2012)
Dry eye disease	*Acinetobacter lwoffii*, *Klebsiella oxytoca*, *Corynebacterium* spp., *Erwinia* spp., *Escherichia coli*, *Kocuria roseus*, *Lactococcus raffinolactis*, *Proteus mirabilis*, *Micrococcus luteus*, *Micrococcus lylae*, *Pseudomonas* spp., *Pseudomonas aeruginosa*, *Propionibacterium* spp., *Propionibacterium acnes*, *Rhodococcus erythropolis*, *Streptococcus angina*, *Staphylococcus aureus*, *Staphylococcus haemolyticus* and coagulase-negative *Staphylococci*	Albietz and Lenton (2006), Graham et al. (2007), Kohanim et al. (2016), Seal et al. (1986), Hori et al. (2008), Zhang et al. (2013), Suto et al. (2012)
Meibomian gland dysfunction	*Acinetobacter calcoaceticus*, *Actinomyces* spp., *Bacillus* spp., *Cladosporium* spp., *Corynebacterium* spp., *Corynebacterium* spp., *Corynebacterium macginleyi*, *Moraxella osloensis*, *Haemophilus influenzae*, *Lysinibacillus* spp., *Microbacteriaceae*, *Paenibacillus* spp., *Pantoea agglomerans*, *Staphylococcus* spp., *S. aureus*, *Staphylococcus lugdunensis*,	Jiang et al. (2018), Watters et al. (2017)

(continued)

Table 2.4 (continued)

Ocular disease	Associated bacteria	Reference
	Stenotrophomonas maltophilia and *Xanthomonadaceae*	
Chronic Stevens-Johnson syndrome	*Acinetobacter* spp., *Bacillus amyloliquefaciens*, *Cellulomonas* spp., *Corynebacterium* spp., *Enterobacter* spp., *Escherichia coli*, *Haemophilus* spp., *Morganella morganii*, *Proteus mirabilis*, *Pseudomonas* spp., *Serratia nonliquefaciens*, *Staphylococcus aureus*, *S. epidermidis*, *Staphylococcus* spp., *S. capitis*, *Streptococcus* spp., *Streptococcus agalactiae*, *S. milleri* and *S. pneumoniae*	Frizon et al. (2014), Venugopal et al. (2016), Moeller et al. (2005), Sotozono et al. (2002), Kittipibul et al. (2020)
Sjogren's syndrome	*Actinobacteria*, *Bacteroidetes*, *Firmicutes* and *Proteobacteria*	de Paiva et al. (2016)
Ocular graft vs. host disease	*Corynebacterium* spp., *Enterobacteriaceae*, *Haemophilus influenza*, *Pasteurellaceae*, *Propionibacterium acnes*, *Selenomonas* spp., *Staphylococcus* species, haemolytic *Streptococcus* spp., *Lactobacillus/Streptococcus* spp. and aerobic Gram-positive cocci	Shimizu et al. (2019)
Prior to cataract surgery	*Acinetobacter* spp., *Alcaligenes* spp., *Corynebacterium* spp., *E. coli*, *Enterobacter aerogenes*, *Klebsiella oxytoca*, *Moraxella catarrhalis*, *Pseudomonas aeruginosa*, *Proteus vulgaris*, *Proteus mirabilis Serratia marcescens*, *Staphylococcus aureus* and coagulase-negative *Staphylococci*	Suto et al. (2012), Paul et al. (2021)
Diabetes mellitus	Diphtheroids, *Moraxella* spp., *Neisseria* spp., *Staphylococcus aureus*, *S. epidermidis*, non-haemolytic *Streptococcus*, coagulase-negative staphylococci	Arbab et al. (2010)
Contact lens keratitis	*Corynebacterium* spp. *Klebsiella oxygenate*, *Pseudomonas aeruginosa*, *Serratia* spp., *Nocardia* spp., *Streptococcus pneumoniae*, *Staphylococcus aureus*, coagulase-negative *Staphylococcus*, and other Gram-positive and Gram-negative bacteria	Willcox (2007), Stapleton et al. (2007), Galentine et al. (1984), Alfonso et al. (1986)
Alcoholism	*Acinetobacter* spp., *Corynebacterium* spp., *Bacillus* spp., *Haemophilus* spp. *Neisseria* spp. and *Micrococcus* spp. *Moraxella* spp., *Staphylococcus*	Gunduz et al. (2016)

(continued)

Table 2.4 (continued)

Ocular disease	Associated bacteria	Reference
	aureus, coagulase-negative *Staphylococcus* and *Streptococcus* spp.	
Limbal stem cell deficiency (LSCD)	*Staphylococcus epidermidis*, *S. aureus* and *P. aeruginosa*	Gunasekaran et al. (2019)
Behçet's disease	*Bacillus* spp., *Neisseria* spp. *Moraxella* spp., *Streptococcus* spp. coagulase-negative *Staphylococcus* and *S. aureus*	Gunduz et al. (2008)

2.4.2 Conjunctivitis

In patients with conjunctivitis, the eye is swollen, inflamed and pink to an extent that the white of the eye is not visible. Conjunctivitis could be contagious and may effect both the eyes producing a sticky discharge (Azari and Barney 2013). Conjunctivitis occurs following infection with bacteria or viruses (Azari and Barney 2013) or as an allergic reaction (Fig. 2.2c, d). The prevalence of conjunctivitis is 12.22% (Kahol et al. 2019) and the aetiological agents include several and Gram-positive and Gram-negative bacteria which include *Chlamydia trachomatis*, *Haemophilus influenzae*, *Neisseria gonorrhoeae*, etc. In neonatal conjunctivitis *E. coli* is the most common pathogen (Quinet et al. 2010; Ghosh et al. 1995) (Table 2.4).

2.4.3 Blepharitis

Patients present with conjunctival hyperaemia, inflamed eyelid margin (Putnam 2016) discomfort and epiphora leading to light sensitivity and vision loss (Lemp and Nichols 2009). The incidence of blepharitis in India is 12.33% (Lipa and Banu 2016), whereas in the United States, it is higher at 37–47% of patients with all eye diseases (Lemp and Nichols 2009). The aetiological factors of blepharitis include viral, bacterial and parasitic infections (Fig. 2.2e), dermatoses, trauma, immunologic conditions and benign and malignant eyelid tumours. The bacteria cultivated from blepharitis patients include *Streptophyta* spp., *Staphylococcus aureus*, *Corynebacterium* spp., *Streptococcus pyogenes*, *Bacillus oleronius and Enhydrobacter* spp. (Table 2.4) (Lee et al. 2012; Szkaradkiewicz et al. 2012).

2.4.4 Scleritis

Individuals with scleritis have inflammation of the sclera and (or) adjacent tissues which is painful and chronic, and the situation at times could lead to vision impairment (Fig. 2.2f). Inflammation of the sclera is rare in children (Azari and Barney 2013). Scleritis could be infectious, idiopathic or due to autoimmune conditions (Jain et al. 2009; Pradhan and Jacob 2013). Infectious aetiology causes 7–8% of all scleritis cases in adults, and tuberculosis and herpes zoster virus have been implicated as the most common cause of infectious scleritis (Jabs et al. 2000). Tuberculous scleritis is due to direct invasion of *Mycobacterium tuberculosis* or mediated by inflammatory reaction against proteins of *M. tuberculosis* (Bloomfield et al. 1976). The other bacteria that have been reported to be associated with scleritis include *Corynebacterium diphtheriae*, *Pseudomonas aeruginosa*, *Staphylococcus pneumoniae*, *Stenotrophomonas maltophilia*, *Mycobacterium* spp., *Streptococcus* spp. and *Nocardia* spp. (Table 2.4) (Reddy et al. 2015; Jain et al. 2009).

2.4.5 Orbital Cellulitis

In individuals with orbital cellulitis, a membrane in front of the orbit known as the orbital septum is inflamed (Fig. 2.3a). The symptoms include oedema, decreased ocular motility, pain, proptosis and chemosis (Chaudhry et al. 2012). The predominant cause of orbital cellulitis is sinusitis (58%), and the other causes include infection of the lid or face (28%), foreign body presence (11%) and haematogenous reasons (4%) and could be odontogenic (2–5%). Depending on the specific region that is affected, orbital cellulitis is classified as abscess (5.6%), orbital cellulitis (22.2%) and pre-septal cellulitis (72.2%). Orbital cellulitis is more prevalent in children as compared to adults (Nageswaran et al. 2006; Al-Madani et al. 2013). *Staphylococcus* spp., *Staphylococcus aureus* and *Streptococcus* spp. are the common causative organisms in adults, whereas in children *Haemophilus influenzae* is common. The less common organisms include *E. coli*, *Arcanobacterium haemolyticum*, *Moraxella catarrhalis*, *Pseudomonas* spp., *Peptostreptococcus* spp. and *Eikenella corrodens*. Anaerobes are less frequently cultivated, and these include *Fusobacterium necrophorum*, *Peptostreptococcus* spp. and *Bacteroides* spp. (Partamian et al. 1983; Malik et al. 2004; Youssef et al. 2008; Sharma et al. 2013a) (Table 2.4).

2.4.6 Endophthalmitis

Endophthalmitis is a disease of the eye and occurs when tissues or fluids inside the eyeball are infected either with bacteria or fungi (Fig. 2.3b). Normally it occurs

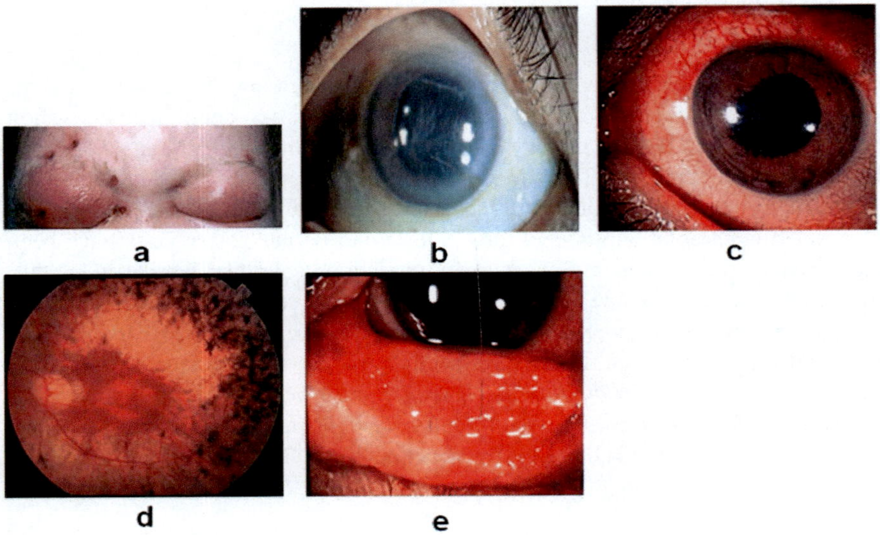

Fig. 2.3 Disease of the eye continued. (**a**) Orbital cellulitis; (**b**) endophthalmitis; (**c**) anterior uveitis; (**d**) retinitis and (**e**) Stevens-Johnson syndrome conjunctivitis. (**a**), (**c**) and (**e**) were retrieved from Creative Commons Attribution-Share Alike 3.0, (**d**) is a fundus photograph retrieved from Creative Commons Attribution-Share Alike 2.0 and (**b**) was obtained courtesy LVPEI, Hyderabad)

following surgery of the eye or due to an eye injury, and this is known as exogenous endophthalmitis (Durand 2013). Quite often, endophthalmitis is also seen in the absence of an injury to the eye, and the infection spreads from the blood or urinary tract infection (Ramakrishnan et al. 2009; Chee and Jap 2001; Durand 2013). Endophthalmitis is said to be exogenous when the organism is from an external source, but when the causative organism is from within the individual, it is known as endogenous endophthalmitis. Further, endophthalmitis is said to be acute endophthalmitis when symptoms show up a few days after surgery or following trauma, whereas in chronic endophthalmitis symptoms occur after a prolonged period. The symptoms of endophthalmitis include red eye, eye pain, swollen eyelids, white or yellow pus or discharge and visual impairment (Durand 2013). The incidence of endophthalmitis in India is between 0.04% and 0.15% (Lalitha et al. 2017). Significant decline in the incidence of post-operative endophthalmitis has been seen owing to multitude of factors like modern surgical techniques and instrumentation, better sterility protocols and use of prophylactic antibiotics (Sunaric-Megevand and Pournaras 1997). The incidence of endophthalmitis depends on the prior surgical procedure and ranges from a minimum of 0.01–0.3% in post-intravitreal injection, post-vitrectomy, post-trabeculectomy and post-cataract surgery to a maximum of 0–12.5% in post-kerato-prostheses (Levison et al. 2014). In several other surgical procedures like phacoemulsification, extracapsular lens extraction, IOL implantation, etc., it ranged between 0.07% and 0.3% (Mandelbaum et al. 1985; D'Amico et al. 1985; Wood 1996; Levison et al. 2014). In most cases the endogenous ocular flora of the patient such as *S. aureus*, *S. epidermidis* and various streptococci spp. were identified (Speaker et al. 1991; Bannerman et al. 1997) (Table 2.4). The contaminant organisms on the ocular surface could also be traced to the lens surface and contaminated air (Vafidis et al. 1984) and solutions used for irrigation in the operating room. The Gram-positive bacteria constitute the major portion of the bacterial infectious agents (75–90%) of endophthalmitis, and the predominant species are *S. aureus*, *S. epidermidis*, *Streptococcus* spp., *Streptococcus viridans*, *Bacillus cereus* and *Propionibacterium acnes;* the Gram-negative bacteria constituted only 9–10% of the infected bacteria and included *Proteus* spp., *Pseudomonas* spp., *Klebsiella* spp., *Haemophilus influenza* and coliform bacteria (Driebe et al. 1986; Wong 2000; Callegan et al. 2002; Tseng et al. 1996; Bode et al. 1985; Maitray et al. 2019) (Table 2.4). The ocular surface bacterial contamination of post-cataract surgery endophthalmitis patients was studied and the predominant organisms were coagulase-negative *Staphylococcus*, *Staphylococcus aureus* (1.8–72.31%), *Streptococcus* spp. (1.8–8%), *Pseudomonas* spp. (0.56–2%), *Corynebacterium* spp. (1–46.79%), *Proteus* spp. (0.8–2%) *Propionibacterium* spp. (5.4–33.00%) and *Enterococcus* spp. (1.8–8%) (Table 2.4). The disparities found between these studies were attributed to factors such as the number of patients included in the studies, geographical origin of the patients and the mean age of patients (de Kaspar et al. 2009; Walker and Claoue 1986; Arantes et al. 2006). According to Callegan et al. (2002), endophthalmitis is a stepwise process initiated by the bacteria gaining access to the eye, replicating, producing inflammation and toxins and finally culminating in death of retinal photoreceptor cells and disruption of retinal architecture thus inhibiting function of the retina.

2.4.7 Uveitis

In uveitis (UVT) patients, the uveal tract which includes the ciliary body, iris and choroid are inflamed (Fig. 2.3c). It could affect one or more of the above regions. Trauma, infection, autoimmunity and systemic diseases are the common causes of UVT. Symptoms are usually bilateral and include pain, photophobia and blurred vision (Azari and Barney 2013). The prevalence of UVT varies depending on the country with 730 per 100,000 in India, 200 per 100,000 in the United States and around 38 per 100,000 in France (Sangwan 2010). Viruses, bacteria, fungi and parasites cause infectious uveitis (Lin 2015). A few of the bacteria that have been identified in UVT patients include *Treponema pallidum*, *Mycobacterium tuberculosis*, *Borrelia burgdorferi* and *Chlamydia trachomatis* (Barisani-Asenbauer et al. 2012) (Table 2.4). Neither the ocular surface nor posterior chamber bacterial microbiomes have been reported so far in the case of uveitis patients. But several studies provide evidence that the gut bacterial microbiome is associated with idiopathic and autoimmune uveitis (Horai and Caspi 2019; Kalyana Chakravarthy et al. 2018; Fu et al. 2021; Jayasudha et al. 2019).

2.4.8 Retinitis

Retinitis or retinitis pigmentosa is the leading cause of blindness in 20- to 60-year-old patients. In these patients the retina is inflamed, and at times it is permanently damaged (Fig. 2.3d). The prevalence of retinitis is just about 1 in every 2500–7000 people. Viruses, bacteria, fungi, viruses, ticks and parasites are the aetiological agents (Williamson and Haynes 2018) that cause retinitis, and retinitis is commonly manifested in immunocompromised patients (Gupta et al. 2013). The common viruses that cause retinitis include cytomegalovirus (CMV), varicella zoster virus (VZV) and herpes simplex virus (HSV) (Guex-crosier et al. 1997), and *Treponema pallidum* (Moradi et al. 2015), *Toxoplasma gondii* (Hadfield and Guy 2017), *Borrelia burgdorferi* (Sathiamoorthi and Smith 2016), *Mycobacterium tuberculosis* (Gupta et al. 2007) and *Bartonella henselae* (Chi et al. 2012) are the common bacteria that cause the disease (Table 2.4).

2.4.9 Post-fever Retinitis

Post-fever retinitis (PFR) is an ocular disease caused by bacteria, protozoans and viruses (Kawali et al. 2019; Vishwanath et al. 2014; Arunasri et al. 2020a, b; Mahendradas et al. 2020) normally visible several days to weeks after the onset of febrile illness. Patients with PFR present with macular oedema, focal or multifocal, unilateral or multifocal patches of retinitis with possible optic nerve involvement and serous detachment at the macula, and the walls of retinal vessels appear beaded and exhibit tortuosity and perivascular sheathing. The aetiology of PFR in many cases remains unknown and may compromised immune status, and the patients present with sudden onset of diminishing vision which is painless (Vishwanath et al. 2014; Panchal 2018). Infectious aetiology has been reported for a small proportion of PFR cases caused by bacteria like *Salmonella typhi*, *Mycobacterium tuberculosis* (Prabhushanker et al. 2017; Relhan et al. 2014), *Leptospira* spp. and *Rickettsia* spp. A protozoan, *Toxoplasma gondii*, also is a causative agent (Table 2.4). Viruses like Zika virus *Flavivirus* serotypes (DENV-1, DENV-2, DENV-3 and DENV-4) and West Nile chikungunya (Zhao et al. 2017; Chan et al. 2006; Chawla et al. 2017; Duke-Elder and Perkins 1958; Shah et al. 1988; Xia et al. 2014) have been also identified as causing PFR. Kawali et al. (2019) studied 119 PFR patients and reported dengue IgM, chikungunya IgM and Weil-Felix virus in the patients and suggested the term 'epidemic retinitis' due to its seasonal variation and to differentiate it from other sporadic forms of retinitis. In retinitis caused by rickettsia, retinitis could be an immune-driven process or due to direct invasion in to the retinal tissue. Despite the above clinical manifestations, it is a challenge to identify causative agents associated with PFR.

2.4.10 Dacryocystitis

In dacryocystitis the nasolacrimal sac is inflamed due to infection, and it results in obstruction of the nasolacrimal duct. Therefore, the patients present with redness of the eye, pain and swelling of the lower eyelid. Bacteria, fungi and viruses are the common aetiological factors (Chaudhary et al. 2010), and the prevalence of dacryocystitis is 0.67% (Lipa and Banu 2016). The commonly detected bacteria in dacryocystitis include *Haemophilus influenzae*, *Escherichia coli*, *Staphylococcus aureus*, *S. epidermidis*, *Streptococcus pyogenes*, *S. pneumoniae*, *Proteus vulgaris*, etc. (Table 2.4). The bacteria detected in the chronic and the cute dacryocystitis patients are different. In the former chronic dacryocystitis, the Gram-positive bacteria are predominant and include *Staphylococcus* spp., *S. aureus*, coagulase-negative staphylococci, *H. influenzae and Streptococcus pneumoniae* were the most predominant Gram-negative bacteria. In acute dacryocystitis *Pseudomonas aeruginosa* and *S. aureus* were dominant (Bharathi et al. 2008). Furthermore, the anaerobic bacteria *Propionibacterium acnes* and *Peptostreptococcus* spp. were also identified as the causative agents (Chaudhry et al. 2005).

2.4.11 Meibomian Gland Dysfunction

The meibomian glands (MG) of the eye regulate tear composition by secreting meibum which along with mucus and water forms the tears. Meibum prevents the water in the tears from evaporating. Thus, the secretion of the MG is very important in tear formation. When conditions block the duct of the MG or alter the secretion of the gland, less oil reaches the surface of the eye, and as a consequence it becomes a leading cause of evaporative dry eye disease (DED) (Groden et al. 1991). Jiang et al. (2018) recruited 70 subjects with dry eye and isolated bacterial species from the conjunctival sac and meibomian gland (MG). The Gram-positive bacteria identified were *Staphylococcus* spp., *Lysinibacillus* spp., *Bacillus* spp. and bacteria-affiliated *Microbacteriaceae*. The Gram-negative bacteria included taxa affiliated to *Xanthomonadaceae* and *Moraxella osloensis* both in the control and moderate MGD groups (Table 2.4). In contrast in the severe MGD group, in addition to the above bacteria, the following Gram-positive bacteria *Corynebacterium* spp. *Corynebacterium macginleyi* and *Paenibacillus* spp. were also identified (Jiang et al. 2018) (Table 2.4). It was also observed that *S. epidermidis* was the frequently isolated bacterium in MGD patients, and the Gram-negative bacteria affiliated to *Xanthomonadaceae* were the only bacteria associated with the surface layer of MG secretions in the moderate MGD group (Jiang et al. 2018). Nattis et al. (2019) reported the presence of *Staphylococcus aureus* and *S. lugdunensis* in mild MGD patients and *Acinetobacter calcoaceticus*, *Corynebacterium* spp. and *Haemophilus influenzae* in individuals with mild to moderate MGD and *Corynebacterium* spp., *Actinomyces* spp., *Stenotrophomonas maltophilia*, *Cladosporium* spp. and *Pantoea agglomerans* in individuals with clinically significant MGD (Table 2.4) thus indicating that their findings were only partly in agreement with Jiang et al. (2018). Watters et al. (2017) confirmed that coagulase-negative staphylococcus was the most prevalent organism isolated from MGD patients. Further, alpha- and beta-haemolytic streptococci (<5%) and other Gram-positive bacteria (<10%) were also detected in MGD groups.

2.4.12 *Dry Eye Disease*

Unlike MGD where the gland is blocked leading to dry eye condition, the dry eye disease (DED) is caused by multiple factors such as decrease in the production of tears, increase in the evaporation of tears resulting in a decrease in the stability of the tear film, surface inflammation and ocular irritation (Helmick et al. 2008; Bron et al. 2014). Symptoms also include redness, photophobia, foreign body sensation, ocular burning and reduced visual acuity (Kassan and Moutsopoulos 2004). DED prevalence rates are higher among women and the elderly population and is known to affect 5–31% of the population (Brito-Zeron et al. 2014). DED patients presented with a greater bacterial load compared to healthy patients (Albietz and Lenton 2006) and coagulase-negative staphylococcal species (*Staphylococcus epidermidis*, *S. xylosus*, *S. sauri* and *S. lentus*) were most frequently isolated (Seal et al. 1986; Graham et al. 2007). The other bacteria identified in DED patients included coagulase-negative *Staphylococcus aureus*, *Staphylococcus haemolyticus*, *Propionibacterium* spp., *Propionibacterium acnes*, *Corynebacterium* spp., *Micrococcus luteus*, *Pseudomonas* spp., *Pseudomonas aeruginosa*, etc. (Seal et al. 1986; Suto et al. 2012; Graham et al. 2007; Zhang et al. 2013; Hori et al. 2008) (Table 2.4) and are more likely to also have fluoroquinolone-resistant conjunctival bacteria compared to the controls (Hori et al. 2008). Human herpes virus-6 (HHV-6) has also been detected in the tear fluid of dry eye patients (Robert et al. 2002). It is not clear, as yet, whether ocular surface microbiota contributes to the disease process in dry eye (Akpek et al. 2019).

2.4.13 *Stevens-Johnson Syndrome*

Stevens-Johnson syndrome (SJS) is an abnormal immune-mediated response, and patients present with exfoliation of the mucosal tissue and epidermis, and the mortality rate is 1–5%. SJS could be caused by systemic infections and due to prolonged use of sulphonamides, allopurinol, paracetamol, etc. Incidence of SJS is 1.2–6 out of 1,000,000 persons per year, and in a good number of SJS patients (25–75%), ocular involvement has been involved (Jain et al. 2016), and patients present with pseudomembrane formation, conjunctivitis and corneal abrasion (Albietz and Lenton 2006) (Fig. 2.3e). In SJS patients the ocular bacterial diversity had a predominance of Gram-positive cocci (55%) followed by Gram-negative bacilli (25%) and Gram-positive bacilli (19%) compared to the healthy controls (Moeller et al. 2005). In individuals with SJS, the conjunctiva revealed growth of several bacteria such as coagulase-negative *Staphylococcus*, *Streptococcus* spp., *Staphylococcus aureus*, *S. capitis*, *Proteus mirabilis*, *Corynebacterium* spp., *Morganella morganii*, *Enterobacter* spp., *Bacillus amyloliquefaciens*, *E. coli*, *Haemophilus* spp. and *Serratia nonliquefaciens* and commensals like *Streptococcus* spp. and *Micrococcus* spp. compared to the healthy controls (Kittipibul et al. 2020; Venugopal et al. 2016; Frizon et al. 2014) (Table 2.4). It was also observed that in SJS patients using either soft or hard scleral contact lenses, the bacteria encountered were coagulase-negative *Staphylococcus*, *Proteus mirabilis* and *Corynebacterium* spp. in agreement with the results of Erdogan et al. (2002). Sotozono et al. (2002) reported that SJS patients show a significant number of methicillin-resistant *Staphylococcus epidermidis* (MRSE) and *Staphylococcus aureus* (MRSA) keratitis. Thus, in SJS patients the normal ocular microbiota residing in the conjunctival sac is altered compared to the controls.

2.4.14 Sjogren's Syndrome

Sjogren's syndrome (SS) is an autoimmune disease causing ocular and oral dryness due to effects on the moisture-producing lacrimal and salivary glands, respectively. The symptoms include vaginal and skin dryness, muscle and joint pains, a chronic cough, thyroid complications, numbness in the arms and legs and fatigue. SS is more common in females and could develop at any age (Singh et al. 2016; Brito-Zeron et al. 2016). It affects about 0.1–4% of people in the United States. Attempts to characterise the viable ocular surface microbiota from SS patients have not yielded any results. Zaheer et al. (2018) have analysed the microbiome of the ocular surface of SS patients. Emerging evidence also suggested that dysbiosis in the gut microbiota is associated with autoimmunity in Sjögren's syndrome (Horai et al. 2017; Zarate-Blades et al. 2016).

2.4.15 Lax Eyelid Syndrome

Patients with lax eyelid syndrome (LES) present with conjunctivitis and 'floppy' eyelids. In contrast, patients with lax eyelid condition (LEC) present with lax eyelids but do not have conjunctivitis (Culbertson and Ostler 1981). Both LEC and LES are clubbed under floppy eyelid syndrome (FES) (van den Bosch and Lemij 1994; Culbertson and Ostler 1981). LES is observed in patients of all weight and ages (van den Bosch and Lemij 1994), and LES is associated with systemic increase in inflammatory markers like matrix metalloproteinases (Sward et al. 2018) and patients experience obstructive sleep apnoea (Peppard et al. 2013). The viable microbiota of the ocular surface of LEC and LES patients has not yet been characterised though microbiome changes have been reported (Zilliox et al. 2020).

2.4.16 Ocular Graft vs. Host Disease

Ocular graft-versus-host disease (oGVHD) is an immune-mediated complication observed in patients after haematopoietic stem cell transplantation. oGVHD causes mortality and morbidity (Ferrara et al. 2009; Mirza et al. 2016). oGVHD occurs in more than 60% of GVHD patients (Herretes et al. 2015), and the symptoms include keratoconjunctivitis sicca, blepharitis, cataract, filamentary keratitis and ocular hypertension (Lin and Cavanagh 2015). But these symptoms vary depending on whether the patients have acute or chronic oGVHD. Keratoconjunctivitis sicca, dry eye with conjunctivitis, increased eye discharge, tearing, dryness, red eye and punctate keratopathy are the common symptoms in acute oGVHD patients (Ogawa et al. 1999; Shikari et al. 2013). In contrast, in chronic oGVHD patients, the symptoms often reported were increased eye discharge, eye dryness, photophobia and alacrimia (Ogawa et al. 2013). *Staphylococcus* spp., *Streptococcus* spp., *Propionibacterium acnes*, *Corynebacterium* spp., aerobic Gram-positive rods, aerobic Gram-positive cocci and *Haemophilus influenzae* are the commensal microflora in the conjunctiva of oGVHD patients (Shikari et al. 2013; Ogawa et al. 1999). In contrast, only a few species were detected in the non-GVHD patients.

2.4.17 Prior to Cataract Surgery

Cataract contributes to 48% of world blindness and patients recover after surgery. But endophthalmitis is a severe complication occurring after cataract surgery, and endophthalmitis has been attributed to sterility of the operative field, compromised aseptic status of the operation theatre, surgical manipulations and surgical instruments and to the indigenous bacteria from the ocular surface (Speaker et al. 1991). Suto et al. (2012) isolated bacteria from conjunctival sac of 227 of the 579 eyes of cataract patients prior to cataract surgery and reported Gram-positive cocci (191 strains), methicillin-resistant coagulase-negative Staphylococci spp. (37 strains), methicillin-sensitive coagulase-negative Staphylococci spp. (127 strains), methicillin-sensitive *Staphylococcus aureus* (eight strains) and methicillin-resistant *S. aureus* (three strains). In addition, they isolated 76 Gram-positive bacilli of the genus *Corynebacterium* spp., and *Moraxella catarrhalis* was the only one Gram-negative coccus that was identified. There were 16 Gram-negative bacilli including *Acinetobacter* spp., three *E. coli*, two *Pseudomonas aeruginosa*, non-glucose-fermenting bacteria and, in addition, one isolate each of *Proteus vulgaris*, *P. mirabilis*, *Enterobacter aerogenes*, *Serratia marcescens*, *Klebsiella oxytoca* and *Alcaligenes* spp. Paul et al. (2021) also confirmed that out of 500 cases, *Staphylococcus aureus* was found in 72.13% cases, followed by coagulase-negative *Staphylococcus* (CoNS) from 18.03% of the cases. The study by Suto et al. (2012) also indicated higher bacterial detection rate in elderly patients. But the bacterial abundance was not significantly influence by factors such as steroid therapy, diabetic retinopathy, diabetes, haemoglobin A1c or glycosuria. The study indicated that the bacterial flora in the conjunctival swabs of cataract patients prior to cataract surgery is different from that of the conjunctival swab of normal individuals (Suto et al. 2012).

2.4.18 Diabetes

Diabetes mellitus is a multifactorial disease which is known to cause retinopathy and its complications and which could eventually lead to blindness (Armaly and Baloglou 1976; Sadun 1999). Post-operative endophthalmitis is common in diabetic patients (Phillips and Tasman 1994; Cohen et al. 1995). Bacterial colonisation on the conjunctiva of diabetic patients was observed to be higher, and this observation led to the prediction that a higher incidence of bacterial pathogens on the conjunctiva might be a predictor of the severity of diabetes mellitus (Arbab et al. 2010). Data from culture-dependent studies have not been consistent: some studies found higher culture-positive rates for diabetic patients compared to nondiabetic patients (Miah et al. 2014), whereas others found that the bacterial detection rate was not significantly different between the cohorts (Adam et al. 2015; Grzybowski et al. 2017), and the prevalence of bacterial load among individuals was not dependent on diabetes status, diabetic retinopathy, haemoglobin A1c levels or glycosuria (Suto et al. 2012). It was also observed that culture positivity was significantly increased in diabetic retinopathy patients (75%) compared to those without retinopathy (20%) (Arbab et al. 2010; Martins et al. 2004; Karimsab and Razak 2013). The most common isolated bacteria in diabetic and nondiabetic groups were *Staphylococcus epidermidis*, and this was the only bacterium with a significantly higher proportion in diabetic patients (80.2%) as compared to nondiabetic (61.1%) patients. The other isolates included diphtheroids, *Staphylococcus aureus*, *Proteus mirabilis*, *E. coli*, etc. (Tables 2.3 and 2.4) (Moreno et al. 2014; Cason and Winkler 1954; Smith 1954; Arbab et al. 2010; Walker and Claoue 1986; Miah et al. 2014). Further, coagulase-negative *Staphylococcus* and *S. epidermidis* were more frequent in patients with retinopathy than in patients without diabetic retinopathy (Arbab et al. 2010; Karimsab and Razak 2013). A higher frequency of coagulase-negative *Staphylococcus* was detected in diabetic patients with endophthalmitis when compared with nondiabetic subjects (Johnson et al. 1997; Rowsey et al. 1982; Puliafito et al. 1982; Martins et al. 2004; Driebe et al. 1986).

2.4.19 *Contact Lens Use*

The cultivable microbiota from the conjunctiva or lids (Willcox 2013) was increased in abundance (50%) in lens wearers (Larkin and Leeming 1991) compared to non-lens wearers (Callender et al. 1986). A review of the available literature indicated that the microbiota recovered either from the conjunctiva or the lids of lens users was dependent on the lens used. The lens has had no effect on the conjunctival or lid microbiota as in the case of use of HEMA-based lenses which are long-term extended wear (Stapleton et al. 1995) or when daily wear lenses were used (Erdogan et al. 2002). But the extended lens wearers showed a propensity of potential pathogenic microbes (such as Gram-negative bacilli). Elander et al. (1992) also observed that lens wearers swabbed prior to or after 1 week of either daily or extended lens wear did not show any significant change in the conjunctival microbiota. In fact the abundance of microbes cultured from the conjunctiva remained unchanged even after 3 months of use of rigid gas permeable lenses for orthokeratology (Boost and Cho 2005). The second trend observed in the microbiota of lens wearers was an increase in number in the conjunctiva and lids as in in daily wear HEMA-based lenses which showed increase in colonisation by coagulase-negative staphylococci (Stapleton et al. 1995). In contrast, in individuals who used extended rigid gas permeable lenses for 2 months or silicone hydrogel lens potential pathogenic microbes (*Aeromonas hydrophila*, *S. aureus* and '*viridans'* streptococci) increased in abundance, whereas concomitantly a decrease in coagulase-negative staphylococci from the conjunctiva (Fleiszig and Efron 1992; Iskeleli et al. 2005) was observed. Iskeleli et al. (2005) observed that diphtheroid rods and coagulase-negative staphylococci in conjunctival swabs significantly increased in individuals after use of silicone hydrogel lenses (Iskeleli et al. 2005).

Contact lens wear is a recognised predisposing factor in microbial keratitis (Stapleton and Carnt 2012), and *Pseudomonas aeruginosa* is the most common bacterium isolated from such patients, and this preponderance of *P. aeruginosa* has been attributed to the virulence factors and the ability of the bacterium to survive in the ocular environment, on the contact lens and on the surface of the lens storage case (Willcox 2007). The other bacteria detected in contact lens microbial keratitis are *Serratia* spp., *Klebsiella oxygenate*, coagulase-negative *Staphylococcus*, *Staphylococcus aureus*, *Streptococcus pneumoniae*, *Nocardia* spp., *Corynebacterium* spp. and other Gram-positive bacteria, fungus and *Acanthamoeba* (Galentine et al. 1984; Alfonso et al. 1986; Stapleton et al. 2007). Several of these bacteria like *Pseudomonas* spp., *Staphylococcus* spp., *Serratia* spp. and *Nocardia* spp. are important pathogens in microbial keratitis (Varaprasathan et al. 2004; Parment 1997). Contact lens-related microbial keratitis is more likely to occur in summer when it is warmer and humid compared with cooler conditions in winter (Stapleton et al. 2007).

2.4.20 Alcoholism

The conjunctival flora was compared between a group of individuals who were diagnosed with alcohol abuse and the control group, without a history of alcohol abuse (Gunduz et al. 2016). In the alcohol abuse group coagulase-negative *Staphylococcus* (CNS), *Staphylococcus aureus*, *Streptococcus* spp., *Bacillus* spp., *Corynebacterium* spp., *Moraxella* spp., *Acinetobacter* spp., *Haemophilus* spp., *Neisseria* spp. and *Micrococcus* spp. were cultured. Of these in the control group only CNS, *Bacillus* spp. *S. aureus* and *Corynebacterium* spp. were cultured, but *Streptococcus* spp., *Moraxella* spp., *Haemophilus* spp., *Neisseria* spp., *Acinetobacter* spp., and *Micrococcus* spp. grew specifically only in the conjunctival flora samples of the study group but not in the control group (Gunduz et al. 2016). Further, *S. aureus* colonisation in the study was significantly higher compared to control group. Disruption of hygiene in alcoholic patients and effects of alcohol on the immune system may be responsible for the observed changes in the conjunctival flora (Gunduz et al. 2016).

2.4.21 Chronic Limbal Stem Cell Deficiency

In patients with corneal and ocular surface diseases such as ocular cicatricial pemphigoid (OCP) and Stevens-Johnson syndrome (SJS), limbal stem cell transplantation is the method of choice. In patients with bilateral total limbal stem cell deficiency (LSCD) due to SJS and OCP *S. aureus*, *S. epidermidis* and *P. aeruginosa* were the common organisms isolated (Gunasekaran et al. 2019). It was suggested that the conjunctival flora could serve as an important determinant for post-operative infections like endophthalmitis (Gunasekaran et al. 2019).

2.4.22 Behçet's Disease

Behçet's disease is a systemic, autoimmune and inflammatory disease characterised by vasculitis. It is a multifactorial disease caused by viral and bacterial infectious agents, due to genetic predisposition, endothelial dysfunction and immune-related events (Bonfioli and Orefice 2005; Evereklioglu 2005; Gunduz et al. 2008) in Behçet's disease. Bacterial growth in the conjunctivae was observed in majority of the Behçet's patients (92%) compared to the control group (56%). Analysis of the conjunctival flora indicated that in the Behçet's group colonisation was significantly higher compared to the control group and *Moraxella* spp. (16% vs. 4%), *S. aureus* (24% vs. 4% in the control) and *Streptococcus* spp. (16% vs. 2%) were detected. Further, the abundance of coagulase-negative *Staphylococcus*, *Bacillus* spp. and *Neisseria* spp. remained unchanged (Gunduz et al. 2008).

2.5 Treatment of Ocular Infectious Diseases

Antibiotics are the first line of treatment for bacterial infections of the cornea and conjunctiva. Eye drops absorbed by the conjunctiva, sclera and choroid and reach the retina across the vitreous (Weijtens et al. 2002). Alternatively, the eye drops could also reach the intraocular tissue through the aqueous humour. At times the antibiotics are delivered directly in the anterior chamber or into the vitreous cavity to treat intraocular infections (Weng et al. 2017). The emergence of antimicrobial resistance (AMR) in ocular bacteria has further complicated the treatment of ocular infections.

2.6 Conclusions

The emphasis has been on the characterisation of the ocular surface bacteria of the eye to identify the pathogenic bacteria and develop strategies for treatment. The ocular surface cultivable microbiota is dynamic, and changes with the age of the individual, with respect to different climatic and geographical locations, with the diseased status of the eye, in different regions of the eye may be influenced by gender. Several ocular bacteria are yet to be identified to the species level. Physiological attributes of these ocular bacteria with respect to resistance to drugs, the reasons for the resistance and strategies to overcome resistance and to resolve the infection need to be addressed separately.

References

Abelson MB, Lane K, Slocum C (2016) The secrets of ocular microbiomes. http://www.reviewofophthalmology.com/content/t/ocular_disease/c/55178/

Adam M, Balci M, Bayhan HA, Inkaya AC, Uyar M, Gurdal C (2015) Conjunctival flora in diabetic and nondiabetic individuals. Turk J Ophthalmol 45:193–196. https://doi.org/10.4274/tjo.33230

Akpek EK, Bunya VY, Saldanha IJ (2019) Sjogren's syndrome: more than just dry eye. Cornea 38: 658–661. https://doi.org/10.1097/ICO.0000000000001865

Albietz JM, Lenton LM (2006) Effect of antibacterial honey on the ocular flora in tear deficiency and meibomian gland disease. Cornea 25:1012–1019. https://doi.org/10.1097/01.ico.0000225716.85382.7b

Alfonso E, Mandelbaum S, Fox MJ, Forster RK (1986) Ulcerative keratitis associated with contact lens wear. Am J Ophthalmol 101:429–433. https://doi.org/10.1016/0002-9394(86)90641-0

Alfonso SA, Fawley JD, Alexa LX (2015) Conjunctivitis. Prim Care 42:325–345. https://doi.org/10.1016/j.pop.2015.05.001

Al-Madani MV, Khatatbeh AE, Rawashdeh RZ, Al-Khtoum NF, Shawagfeh NR (2013) The prevalence of orbital complications among children and adults with acute rhinosinusitis. Braz J Otorhinolaryngol 79:716–719. https://doi.org/10.5935/1808-8694.20130131

Al-Najjar MAA (2019) An overview of ocular microbiology: ocular microbiota, the effect of contact lenses and ocular disease. Archiv Pharm Pharmacol Res 1:1. https://doi.org/10.33552/appr.2019.01.000522

Arantes TE, Cavalcanti RF, Diniz Mde F, Severo MS, Lins Neto J, Castro CM (2006) Conjunctival bacterial flora and antibiotic resistance pattern in patients undergoing cataract surgery. Arq Bras Oftalmol 69:33–36. https://doi.org/10.1590/s0004-27492006000100007

Arbab TM, Qadeer S, Iqbal S, Mirza MA (2010) Aerobic bacterial conjunctival flora in diabetic patients. Pak J Ophthalmol 26:177–181

Armaly M, Baloglou P (1976) Diabetes mellitus and eye changes in anterior segment. Arch Ophthalmol 77:485–492

Arunasri K, Mahesh M, Sai Prashanthi G et al (2020a) Comparison of the vitreous fluid bacterial microbiomes between individuals with post fever retinitis and healthy controls. Microorganisms 8:751. https://doi.org/10.3390/microorganisms8050751

Arunasri K, Mahesh M, Sai Prashanthi G et al (2020b) Mycobiome changes in the vitreous of post fever retinitis patients. PLoS One 15:e0242138. https://doi.org/10.1371/journal.pone.0242138

Au YK, Jensen HG, Rowsey J, Reynolds M (1993) Coagulase-negative staphylococci in conjunctivitis and blepharitis. Yan Ke Xue Bao 9:129–135

Azari AA, Barney NP (2013) Conjunctivitis: a systematic review of diagnosis and treatment. JAMA 310:1721–1729. https://doi.org/10.1001/jama.2013.280318

Bannerman TL, Rhoden DL, McAllister SK, Miller JM, Wilson LA (1997) The source of coagulase-negative staphylococci in the Endophthalmitis Vitrectomy Study. A comparison of eyelid and intraocular isolates using pulsed-field gel electrophoresis. Arch Ophthalmol 115: 357–361. https://doi.org/10.1001/archopht.1997.01100150359008

Barisani-Asenbauer T, Maca SM, Mejdoubi L, Emminger W, Machold K, Auer H (2012) Uveitis- a rare disease often associated with systemic diseases and infections- a systematic review of 2619 patients. Orphanet J Rare Dis 7:57. https://doi.org/10.1186/1750-1172-7-57

Bharathi MJ, Ramakrishnan R, Meenakshi R, Padmavathy S, Shivakumar C, Srinivasan M (2007) Microbial keratitis in South India: influence of risk factors, climate, and geographical variation. Ophthalmic Epidemiol 14:61–69. https://doi.org/10.1080/09286580601001347

Bharathi MJ, Ramakrishnan R, Maneksha V, Shivakumar C, Nithya V, Mittal S (2008) Comparative bacteriology of acute and chronic dacryocystitis. Eye (Lond) 22:953–960. https://doi.org/10.1038/sj.eye.6702918

Bloomfield SE, Mondino B, Gray GF (1976) Scleral tuberculosis. Arch Ophthalmol 94:954–956. https://doi.org/10.1001/archopht.1976.03910030482009

Bode DD Jr, Gelender H, Forster RK (1985) A retrospective review of endophthalmitis due to coagulase-negative staphylococci. Br J Ophthalmol 69:915–919. https://doi.org/10.1136/bjo.69.12.915

Bonfioli AA, Orefice F (2005) Behcet's disease. Semin Ophthalmol 20:199–206. https://doi.org/10.1080/08820530500231953

Boost MV, Cho P (2005) Microbial flora of tears of orthokeratology patients, and microbial contamination of contact lenses and contact lens accessories. Optom Vis Sci 82:451–458. https://doi.org/10.1097/01.opx.0000168587.72893.ec

van den Bosch WA, Lemij HG (1994) The lax eyelid syndrome. Br J Ophthalmol 78:666–670. https://doi.org/10.1136/bjo.78.9.666

Bourcier T, Thomas F, Borderie V, Chaumeil C, Laroche L (2003) Bacterial keratitis: predisposing factors, clinical and microbiological review of 300 cases. Br J Ophthalmol 87:834–838. https://doi.org/10.1136/bjo.87.7.834

Briscoe D, Rubowitz A, Assia E (2005) Changing bacterial isolates and antibiotic sensitivities of purulent dacryocystitis. Orbit 24:29–32. https://doi.org/10.1080/01676830590894897

Brito-Zeron P, Ramos-Casals M, Group E-Stf (2014) Advances in the understanding and treatment of systemic complications in Sjogren's syndrome. Curr Opin Rheumatol 26:520–527. https://doi.org/10.1097/BOR.0000000000000096

Brito-Zeron P, Baldini C, Bootsma H et al (2016) Sjogren syndrome. Nat Rev Dis Primers 2:16047. https://doi.org/10.1038/nrdp.2016.47

Bron AJ, Tomlinson A, Foulks GN et al (2014) Rethinking dry eye disease: a perspective on clinical implications. Ocul Surf 12:S1–S31. https://doi.org/10.1016/j.jtos.2014.02.002

Brook I, Barrett CT, Brinkman CR III, Martin WJ, Finegold SM (1979) Aerobic and anaerobic bacterial flora of the maternal cervix and newborn gastric fluid and conjunctiva: a prospective study. Pediatrics 63:451–455

Brown DH (1978) The conjunctival flora of nursing home patients and staff. Ann Ophthalmol 10: 333–334

Callegan MC, Engelbert M, Parke DW II, Jett BD, Gilmore MS (2002) Bacterial endophthalmitis: epidemiology, therapeutics, and bacterium-host interactions. Clin Microbiol Rev 15:111–124. https://doi.org/10.1128/CMR.15.1.111-124.2002

Callender MG, Tse LS, Charles AM, Lutzi D (1986) Bacterial flora of the eye and contact lens. Cases during hydrogel lens wear. Am J Optom Physiol Optic 63:177–180. https://doi.org/10.1097/00006324-198603000-00002

Capriotti JA, Pelletier JS, Shah M, Caivano DM, Ritterband DC (2009) Normal ocular flora in healthy eyes from a rural population in Sierra Leone. Int Ophthalmol 29:81–84. https://doi.org/10.1007/s10792-008-9196-4

Cason L, Winkler CH Jr (1954) Bacteriology of the eye. I. Normal flora. AMA Arch Ophthalmol 51:196–198. https://doi.org/10.1001/archopht.1954.00920040198005

Cavuoto K, Zutshi D, Karp CL, Miller D, Feuer W (2008) Update on bacterial conjunctivitis in South Florida. Ophthalmology 115:51–56. https://doi.org/10.1016/j.ophtha.2007.03.076

Cavuoto KM, Banerjee S, Galor A (2019) Relationship between the microbiome and ocular health. Ocul Surf 17:384–392. https://doi.org/10.1016/j.jtos.2019.05.006

Chan DP, Teoh SC, Tan CS et al (2006) Ophthalmic complications of dengue. Emerg Infect Dis 12: 285–289. https://doi.org/10.3201/eid1202.050274

Chaudhary M, Bhattarai A, Adhikari SK, Bhatta DR (2010) Bacteriology and antimicrobial susceptibility of adult chronic dacryocystitis. Nepal J Ophthalmol 2:105–113. https://doi.org/10.3126/nepjoph.v2i2.3716

Chaudhry IA, Shamsi FA, Al-Rashed W (2005) Bacteriology of chronic dacryocystitis in a tertiary eye care center. Ophthalmic Plast Reconstr Surg 21:207–210. https://doi.org/10.1097/01.iop.0000161718.54275.7d

Chaudhry IA, Al-Rashed W, Arat YO (2012) The hot orbit: orbital cellulitis. Mid E Afr J Ophthalmol 19:34–42. https://doi.org/10.4103/0974-9233.92114

Chawla R, Pundlik GA, Chaudhry R, Thakur C (2017) Rickettsial retinitis: direct bacterial infection or an immune-mediated response? Indian J Ophthalmol 65:1038–1041. https://doi.org/10.4103/ijo.IJO_369_17

Chee SP, Jap A (2001) Endogenous endophthalmitis. Curr Opin Ophthalmol 12:464–470. https://doi.org/10.1097/00055735-200112000-00012

Chi SL, Stinnett S, Eggenberger E, Foroozan R, Golnik K, Lee MS, Bhatti MT (2012) Clinical characteristics in 53 patients with cat scratch optic neuropathy. Ophthalmology 119:183–187. https://doi.org/10.1016/j.ophtha.2011.06.042

Cohen SM, Flynn HW Jr, Murray TG, Smiddy WE (1995) Endophthalmitis after pars plana vitrectomy. The Postvitrectomy Endophthalmitis Study Group. Ophthalmology 102:705–712. https://doi.org/10.1016/s0161-6420(95)30965-7

Culbertson WW, Ostler HB (1981) The floppy eyelid syndrome. Am J Ophthalmol 92:568–575. https://doi.org/10.1016/0002-9394(81)90652-8

D'Amico DJ, Caspers-Velu L, Libert J, Shanks E, Schrooyen M, Hanninen LA, Kenyon KR (1985) Comparative toxicity of intravitreal aminoglycoside antibiotics. Am J Ophthalmol 100:264–275. https://doi.org/10.1016/0002-9394(85)90792-5

Dart JK, Stapleton F, Minassian D (1991) Contact lenses and other risk factors in microbial keratitis. Lancet 338:650–653. https://doi.org/10.1016/0140-6736(91)91231-i

Dart JK, Radford CF, Minassian D, Verma S, Stapleton F (2008) Risk factors for microbial keratitis with contemporary contact lenses: a case-control study. Ophthalmology 115:1647–1654. https://doi.org/10.1016/j.ophtha.2008.05.003. 1654.e1641–1643

Das S, Samantaray R, Mallick A, Sahu SK, Sharma S (2019) Types of organisms and in-vitro susceptibility of bacterial isolates from patients with microbial keratitis: a trend analysis of 8 years. Indian J Ophthalmol 67:49–53. https://doi.org/10.4103/ijo.IJO_500_18

Doan T, Akileswaran L, Andersen D et al (2016) Paucibacterial microbiome and resident DNA virome of the healthy conjunctiva. Invest Ophthalmol Vis Sci 57:5116–5126. https://doi.org/10.1167/iovs.16-19803

Dong Q, Brulc JM, Iovieno A et al (2011) Diversity of bacteria at healthy human conjunctiva. Invest Ophthalmol Vis Sci 52:5408–5413. https://doi.org/10.1167/iovs.10-6939

Doyle A, Beigi B, Early A, Blake A, Eustace P, Hone R (1995) Adherence of bacteria to intraocular lenses: a prospective study. Br J Ophthalmol 79:347–349. https://doi.org/10.1136/bjo.79.4.347

Driebe WT Jr, Mandelbaum S, Forster RK, Schwartz LK, Culbertson WW (1986) Pseudophakic endophthalmitis. Diagnosis and management. Ophthalmology 93:442–448. https://doi.org/10.1016/s0161-6420(86)33722-9

Duke-Elder S, Perkins ES (1958) Diseases of the uveal tract. In: System of ophthalmology, vol 9. Henry Kimpton, London. https://nla.gov.au/nla.cat-vn1995859

Durand ML (2013) Endophthalmitis. Clin Microbiol Infect 19:227–234. https://doi.org/10.1111/1469-0691.12118

Eder M, Farina N, Sanabria RR et al (2005) Normal ocular flora in newborns delivered in two hospital centers in Argentina and Paraguay. Graefes Arch Clin Exp Ophthalmol 243:1098–1107. https://doi.org/10.1007/s00417-004-1096-3

Egrilmez S, Yildirim-Theveny S (2020) Treatment-resistant bacterial keratitis: challenges and solutions. Clin Ophthalmol 14:287–297. https://doi.org/10.2147/OPTH.S181997

Elander TR, Goldberg MA, Salinger CL, Tan JR, Levy B, Abbott RL (1992) Microbial changes in the ocular environment with contact lens wear. CLAO J 18:53–55

Erdogan H, Kemal M, Toker MI, Topalkara A, Bakici Z (2002) Effect of frequent-replacement contact lenses on normal conjunctival flora. CLAO J 28:94–95

Eslami F, Ghasemi Basir HR, Moradi A, Heidari Farah S (2018) Microbiological study of dacryocystitis in northwest of Iran. Clin Ophthalmol 12:1859–1864. https://doi.org/10.2147/OPTH.S175463

Evereklioglu C (2005) Current concepts in the etiology and treatment of Behcet disease. Surv Ophthalmol 50:297–350. https://doi.org/10.1016/j.survophthal.2005.04.009

Ferrara JL, Levine JE, Reddy P, Holler E (2009) Graft-versus-host disease. Lancet 373:1550–1561. https://doi.org/10.1016/S0140-6736(09)60237-3

Fleiszig SM, Efron N (1992) Conjunctival flora in extended wear of rigid gas permeable contact lenses. Optom Vis Sci 69:354–357. https://doi.org/10.1097/00006324-199205000-00004

Frizon L, Araujo MC, Andrade L, Yu MC, Wakamatsu TH, Hofling-Lima AL, Gomes JA (2014) Evaluation of conjunctival bacterial flora in patients with Stevens-Johnson Syndrome. Clinics (Sao Paulo) 69:168–172. https://doi.org/10.6061/clinics/2014(03)04

Fu X, Chen Y, Chen D (2021) The role of gut microbiome in autoimmune uveitis. Ophthalmic Res 64:168–177. https://doi.org/10.1159/000510212

Galentine PG, Cohen EJ, Laibson PR, Adams CP, Michaud R, Arentsen JJ (1984) Corneal ulcers associated with contact lens wear. Arch Ophthalmol 102:891–894. https://doi.org/10.1001/archopht.1984.01040030711025

Ghosh S, Chatterjee BD, Chakraborty CK, Chakravarty A, Khatua SP (1995) Bacteria in surface infections of neonates. J Indian Med Assoc 93:132–135

Gomes JAP, Frizon L, Demeda VF (2020) Ocular surface microbiome in health and disease. Asia Pac J Ophthalmol (Phila) 9:505–511. https://doi.org/10.1097/APO.0000000000000330

Graham JE, Moore JE, Jiru X et al (2007) Ocular pathogen or commensal: a PCR-based study of surface bacterial flora in normal and dry eyes. Invest Ophthalmol Vis Sci 48:5616–5623. https://doi.org/10.1167/iovs.07-0588

Groden LR, Murphy B, Rodnite J, Genvert GI (1991) Lid flora in blepharitis. Cornea 10:50–53

Grunberger V, Kofler E (1954) Neonatal bacterial flora of the conjunctiva. Gynaecologia 137:329–334

Grzybowski A, Brona P, Kim SJ (2017) Microbial flora and resistance in ophthalmology: a review. Graefes Arch Clin Exp Ophthalmol 255:851–862. https://doi.org/10.1007/s00417-017-3608-y

Guex-Crosier Y, Rochat C, Herbort CP (1997) Necrotizing herpetic retinopathies. A spectrum of herpes virus-induced diseases determined by the immune state of the host. Ocul Immunol Inflamm 5:259–265. https://doi.org/10.3109/09273949709085066

Gunasekaran S, Dhiman R, Vanathi M, Mohanty S, Satpathy G, Tandon R (2019) Ocular surface microbial flora in patients with chronic limbal stem cell deficiency undergoing cultivated oral mucosal epithelial transplantation. Mid E Afr J Ophthalmol 26:23–26. https://doi.org/10.4103/meajo.MEAJO_172_16

Gunduz A, Gunduz A, Cumurcu T, Seyrek A (2008) Conjunctival flora in Behcet patients. Can J Ophthalmol 43:476–479. https://doi.org/10.3129/i08-089

Gunduz G, Gunduz A, Polat N, Cumurcu BE, Yakupogullari Y (2016) The effect of chronic alcoholism on the conjunctival flora. Curr Eye Res 41:734–739. https://doi.org/10.3109/02713683.2015.1056805

Gupta V, Gupta A, Rao NA (2007) Intraocular tuberculosis--an update. Surv Ophthalmol 52:561–587. https://doi.org/10.1016/j.survophthal.2007.08.015

Gupta S, Vemulakonda GA, Suhler EB, Yeh S, Albini TA, Mandelcorn E, Flaxel CJ (2013) Cytomegalovirus retinitis in the absence of AIDS. Can J Ophthalmol 48:126–129. https://doi.org/10.1016/j.jcjo.2012.12.002

Hadfield SJ, Guy EC (2017) Toxoplasmosis. Medicine 45:763–766. https://doi.org/10.1016/j.mpmed.2017.09.013

Hart DE, Hosmer M, Georgescu M, Farris RL (1996) Bacterial assay of contact lens wearers. Optom Vis Sci 73:204–207. https://doi.org/10.1097/00006324-199603000-00014

Helmick CG, Felson DT, Lawrence RC et al (2008) Estimates of the prevalence of arthritis and other rheumatic conditions in the United States. Part I. Arthritis Rheum 58:15–25. https://doi.org/10.1002/art.23177

Herretes S, Ross DB, Duffort S et al (2015) Recruitment of donor T cells to the eyes during ocular GVHD in recipients of MHC-matched allogeneic hematopoietic stem cell transplants. Invest Ophthalmol Vis Sci 56:2348–2357. https://doi.org/10.1167/iovs.14-15630

Hofling-lima AL, Eid F, Montenegro L, Alvarenga L, Chalita M, You M (2002) Changes in conjunctival and lid flora after topical use of lomefloxacin and tobramycin in cataract and refractive surgery. Arq Bras Oftalmol 65:21–29

Hong YJ (1974) Studies on microbial flora of the conjunctival sac in new-born infants. J Korean Ophthalmol Soc 15:57–64

Horai R, Caspi RR (2019) Microbiome and autoimmune uveitis. Front Immunol 10:232. https://doi.org/10.3389/fimmu.2019.00232

Horai R, Sen HN, Caspi RR (2017) Commensal microbiota as a potential trigger of autoimmune uveitis. Expert Rev Clin Immunol 13:291–293. https://doi.org/10.1080/1744666X.2017.1288098

Hori Y, Maeda N, Sakamoto M, Koh S, Inoue T, Tano Y (2008) Bacteriologic profile of the conjunctiva in the patients with dry eye. Am J Ophthalmol 146:729–734. https://doi.org/10.1016/j.ajo.2008.06.003

Hovding G (1981) The conjunctival and contact lens bacterial flora during lens wear. Acta Ophthalmol 59:387–401. https://doi.org/10.1111/j.1755-3768.1981.tb03004.x

Hsu HY, Lind JT, Miller D, Tseng L (2015) Assessment of risk factors for oxacillin-resistant ocular flora in eyes having cataract surgery. J Cataract Refract Surg 41:387–392. https://doi.org/10.1016/j.jcrs.2014.05.050

Isenberg SJ, Apt L, Yoshimori R, Alvarez SR (1986) Bacterial flora of the conjunctiva at birth. J Pediatr Ophthalmol Strabismus 23:284–286

Iskeleli G, Bahar H, Eroglu E, Torun MM, Ozkan S (2005) Microbial changes in conjunctival flora with 30-day continuous-wear silicone hydrogel contact lenses. Eye Cont Lens 31:124–126. https://doi.org/10.1097/01.icl.0000141923.63458.df

Jabs DA, Mudun A, Dunn JP, Marsh MJ (2000) Episcleritis and scleritis: clinical features and treatment results. Am J Ophthalmol 130:469–476. https://doi.org/10.1016/s0002-9394(00)00710-8

Jain V, Garg P, Sharma S (2009) Microbial scleritis-experience from a developing country. Eye (Lond) 23:255–261. https://doi.org/10.1038/sj.eye.6703099

Jain R, Sharma N, Basu S et al (2016) Stevens-Johnson syndrome: the role of an ophthalmologist. Surv Ophthalmol 61:369–399. https://doi.org/10.1016/j.survophthal.2016.01.004

Javed E, Sultan M, Mehmood R (2008) Normal conjunctival flora. Ann Punjab Med Coll 2:17–20. https://doi.org/10.29054/apmc/2008.622

Jayasudha R, Kalyana Chakravarthy S, Sai Prashanthi G, Sharma S, Tyagi M, Shivaji S (2019) Implicating dysbiosis of the gut fungal microbiome in uveitis, an inflammatory disease of the eye. Invest Ophthalmol Vis Sci 60:1384–1393. https://doi.org/10.1167/iovs.18-26426

Jensen OL, Gluud BS (1985) Bacterial growth in the conjunctival sac and the local defense of the outer eye. Acta Ophthalmol Suppl 173:80–82. https://doi.org/10.1111/j.1755-3768.1985.tb06849.x

Jiang X, Deng A, Yang J et al (2018) Pathogens in the Meibomian gland and conjunctival sac: microbiome of normal subjects and patients with Meibomian gland dysfunction. Infect Drug Resist 11:1729–1740. https://doi.org/10.2147/IDR.S162135

Johnson MW, Doft BH, Kelsey SF, Barza M, Wilson LA, Barr CC, Wisniewski SR (1997) The endophthalmitis vitrectomy study. Relationship between clinical presentation and microbiologic spectrum. Ophthalmology 104:261–272. https://doi.org/10.1016/s0161-6420(97)30326-1

Kahol P, Thakur M, Gupta A, Saini SK (2019) Prevalence, morbidity and treatment seeking behavior for allergic conjunctivitis in children in a North Indian community. Clin Epidemiol Glob Health 7:239–245. https://doi.org/10.1016/j.cegh.2018.07.001

Kalyana Chakravarthy S, Jayasudha R, Sai Prashanthi G, Ali MH, Sharma S, Tyagi M, Shivaji S (2018) Dysbiosis in the gut bacterial microbiome of patients with uveitis, an inflammatory disease of the eye. Indian J Microbiol 58:457–469. https://doi.org/10.1007/s12088-018-0746-9

Karimsab D, Razak SK (2013) Study of aerobic bacterial conjunctival flora in patients with diabetes mellitus. Nepal J Ophthalmol 5:28–32. https://doi.org/10.3126/nepjoph.v5i1.7818

de Kaspar HM, Kreidl KO, Singh K, Ta CN (2004) Comparison of preoperative conjunctival bacterial flora in patients undergoing glaucoma or cataract surgery. J Glaucoma 13:507–509. https://doi.org/10.1097/01.ijg.0000137872.19942.cf

de Kaspar HM, Kreutzer TC, Aguirre-Romo I et al (2008) A prospective randomized study to determine the efficacy of preoperative topical levofloxacin in reducing conjunctival bacterial flora. Am J Ophthalmol 145:136–142. https://doi.org/10.1016/j.ajo.2007.08.031

de Kaspar HM, Ta CN, Froehlich SJ, Schaller UC, Engelbert M, Klauss V, Kampik A (2009) Prospective study of risk factors for conjunctival bacterial contamination in patients undergoing intraocular surgery. Eur J Ophthalmol 19:717–722. https://doi.org/10.1177/112067210901900505

Kassan SS, Moutsopoulos HM (2004) Clinical manifestations and early diagnosis of Sjogren syndrome. Arch Intern Med 164:1275–1284. https://doi.org/10.1001/archinte.164.12.1275

Kawali A, Mahendradas P, Mohan A, Mallavarapu M, Shetty B (2019) Epidemic retinitis. Ocul Immunol Inflamm 27:571–577. https://doi.org/10.1080/09273948.2017.1421670

Keilty RA (1930) The bacterial flora of the normal conjunctiva with comparative nasal culture study. Am J Ophthalmol 13:876–879. https://doi.org/10.1016/s0002-9394(30)92437-3

Khorazo DL, Gutierrez EH (1972) The bacterial flora of the healthy eye. In: Locather-Khorazo D, Seegal BC (eds) Microbiology of the eye. Mosby Press, St. Louis, MO, pp 13–23

Khorazo D, Thompson R (1935) The bacterial flora of the normal conjunctiva. Am J Opthalmol 18: 1114–1116

Kittipibul T, Puangsricharern V, Chatsuwan T (2020) Comparison of the ocular microbiome between chronic Stevens-Johnson syndrome patients and healthy subjects. Sci Rep 10:4353. https://doi.org/10.1038/s41598-020-60794-w

Kohanim S, Palioura S, Saeed HN et al (2016) Acute and chronic ophthalmic involvement in Stevens-Johnson syndrome/toxic epidermal necrolysis - a comprehensive review and guide to therapy. II. Ophthalmic disease. Ocul Surf 14:168–188. https://doi.org/10.1016/j.jtos.2016.02.001

Lalitha P, Sengupta S, Ravindran RD, Sharma S, Joseph J, Ambiya V, Das T (2017) A literature review and update on the incidence and microbiology spectrum of postcataract surgery endophthalmitis over past two decades in India. Indian J Ophthalmol 65:673–677. https://doi.org/10.4103/ijo.IJO_509_17

Larkin DF, Leeming JP (1991) Quantitative alterations of the commensal eye bacteria in contact lens wear. Eye (Lond) 5(Pt 1):70–74. https://doi.org/10.1038/eye.1991.14

Lee SH, Oh DH, Jung JY, Kim JC, Jeon CO (2012) Comparative ocular microbial communities in humans with and without blepharitis. Invest Ophthalmol Vis Sci 53:5585–5593. https://doi.org/10.1167/iovs.12-9922

Leitch EC, Harmis NY, Corrigan KM, Willcox MD (1998) Identification and enumeration of staphylococci from the eye during soft contact lens wear. Optom Vis Sci 75:258–265. https://doi.org/10.1097/00006324-199804000-00022

Lemp MA, Nichols KK (2009) Blepharitis in the United States 2009: a survey-based perspective on prevalence and treatment. Ocul Surf 7:S1–S14. https://doi.org/10.1016/s1542-0124(12)70620-1

Levison AL, Mendes TS, Bhisitkul R (2014) Postprocedural endophthalmitis: a review. Expert Rev Ophthalmol 8:45–62. https://doi.org/10.1586/eop.12.77

Lin P (2015) Infectious uveitis. Curr Ophthalmol Rep 3:170–183. https://doi.org/10.1007/s40135-015-0076-6

Lin X, Cavanagh HD (2015) Ocular manifestations of graft-versus-host disease: 10 years' experience. Clin Ophthalmol 9:1209–1213. https://doi.org/10.2147/OPTH.S84704

Lipa M, Banu TJ (2016) A cross-sectional study to assess the morbidity pattern of ocular diseases in out-patient department of ophthalmology at a tertiary care hospital. Int J Res Med Sci 4:3797. https://doi.org/10.18203/2320-6012.ijrms20162823

Liu J, Li J, Huo J, Xie H (2011) Identification and quantitation of conjunctival aerobic bacterial flora from healthy residents at different ages in Southwest China. Afr J Microbiol Res 5:192

Mahendradas P, Kawali A, Luthra S et al (2020) Post-fever retinitis - newer concepts. Indian J Ophthalmol 68:1775–1786. https://doi.org/10.4103/ijo.IJO_1352_20

Maitray A, Rishi E, Rishi P, Gopal L, Bhende P, Ray R, Therese KL (2019) Endogenous endophthalmitis in children and adolescents: case series and literature review. Indian J Ophthalmol 67:795–800. https://doi.org/10.4103/ijo.IJO_710_18

Malik NN, Goh D, McLean C, Huchzermeyer P (2004) Orbital cellulitis caused by Peptostreptococcus. Eye (Lond) 18:643–644. https://doi.org/10.1038/sj.eye.6700657

Mandelbaum S, Forster RK, Gelender H, Culbertson W (1985) Late onset endophthalmitis associated with filtering blebs. Ophthalmology 92:964–972. https://doi.org/10.1016/s0161-6420(85)33947-7

Mantelli F, Moretti C, Macchi I, Massaro-Giordano G, Cozzupoli GM, Lambiase A, Bonini S (2016) Effects of sex hormones on ocular surface epithelia: lessons learned from polycystic ovary syndrome. J Cell Physiol 231:971–975. https://doi.org/10.1002/jcp.25221

Martins EN, Alvarenga LS, Hofling-Lima AL, Freitas D, Zorat-Yu MC, Farah ME, Mannis MJ (2004) Aerobic bacterial conjunctival flora in diabetic patients. Cornea 23:136–142. https://doi.org/10.1097/00003226-200403000-00006

McDermott AM (2013) Antimicrobial compounds in tears. Exp Eye Res 117:53–61. https://doi.org/10.1016/j.exer.2013.07.014

McKinley SH, Yen MT, Miller AM, Yen KG (2007) Microbiology of pediatric orbital cellulitis. Am J Ophthalmol 144:497–501. https://doi.org/10.1016/j.ajo.2007.04.049

McLaughlin-Borlace L, Stapleton F, Matheson M, Dart JK (1998) Bacterial biofilm on contact lenses and lens storage cases in wearers with microbial keratitis. J Appl Microbiol 84:827–838. https://doi.org/10.1046/j.1365-2672.1998.00418.x

McNatt J, Allen SD, Wilson LA, Dowell VR Jr (1978) Anaerobic flora of the normal human conjunctival sac. Arch Ophthalmol 96:1448–1450. https://doi.org/10.1001/archopht.1978.03910060196020

Miah MRA, Anwar S, Nahar N (2014) Conjunctival bacterial flora in diabetic patients. Ibrahim Medical College J 7:5–8. https://doi.org/10.3329/imcj.v7i1.17698

Mills DM, Bodman MG, Meyer DR, Morton AD III, Group ADS (2007) The microbiologic spectrum of dacryocystitis: a national study of acute versus chronic infection. Ophthalmic Plast Reconstr Surg 23:302–306. https://doi.org/10.1097/IOP.0b013e318070d237

Mirza N, Zierhut M, Korn A et al (2016) Graft versus self (GvS) against T-cell autoantigens is a mechanism of graft-host interaction. Proc Natl Acad Sci U S A 113:13827–13832. https://doi.org/10.1073/pnas.1609118113

Moeller CT, Branco BC, Yu MC, Farah ME, Santos MA, Hofling-Lima AL (2005) Evaluation of normal ocular bacterial flora with two different culture media. Can J Ophthalmol 40:448–453. https://doi.org/10.1139/i05-014

Moradi A, Salek S, Daniel E et al (2015) Clinical features and incidence rates of ocular complications in patients with ocular syphilis. Am J Ophthalmol 159:334–343.e331. https://doi.org/10.1016/j.ajo.2014.10.030

Moreno NP, Moreno RD, Sousa LB (2014) Aerobic bacterial microbiota of the conjunctiva in diabetic patients with normal and altered glycated hemoglobin levels in two regions in Brazil. Arq Bras Oftalmol 77:351–354. https://doi.org/10.5935/0004-2749.20140088

Nageswaran S, Woods CR, Benjamin DK Jr, Givner LB, Shetty AK (2006) Orbital cellulitis in children. Pediatr Infect Dis J 25:695–699. https://doi.org/10.1097/01.inf.0000227820.36036.f1

Nattis A, Perry HD, Rosenberg ED, Donnenfeld ED (2019) Influence of bacterial burden on meibomian gland dysfunction and ocular surface disease. Clin Ophthalmol 13:1225–1234. https://doi.org/10.2147/OPTH.S215071

Neumann M, Sjostrand J (1993) Central microbial keratitis in a Swedish city population. A three-year prospective study in Gothenburg. Acta Ophthalmol 71:160–164. https://doi.org/10.1111/j.1755-3768.1993.tb04982.x

Ogawa Y, Okamoto S, Wakui M et al (1999) Dry eye after haematopoietic stem cell transplantation. Br J Ophthalmol 83:1125–1130. https://doi.org/10.1136/bjo.83.10.1125

Ogawa Y, Kim SK, Dana R et al (2013) International chronic ocular graft-vs-host-disease (GVHD) consensus group: proposed diagnostic criteria for chronic GVHD (Part I). Sci Rep 3:3419. https://doi.org/10.1038/srep03419

Oguz H, Oguz E, Karadede S (2000) Effect of taurolidine on the normal eyelid and conjunctival flora. Curr Eye Res 21:851–855. https://doi.org/10.1076/ceyr.21.5.851.5540

Olson R, Donnenfeld E, Bucci FA Jr et al (2010) Methicillin resistance of Staphylococcus species among health care and nonhealth care workers undergoing cataract surgery. Clin Ophthalmol 4: 1505–1514. https://doi.org/10.2147/OPTH.S14333

Ormerod LD, Hertzmark E, Gomez DS, Stabiner RG, Schanzlin DJ, Smith RE (1987) Epidemiology of microbial keratitis in southern California. A multivariate analysis. Ophthalmology 94: 1322–1333. https://doi.org/10.1016/s0161-6420(87)80019-2

Ozkan J, Willcox MD (2019) The ocular microbiome: molecular characterisation of a unique and low microbial environment. Curr Eye Res 44:685–694. https://doi.org/10.1080/02713683.2019.1570526

Ozkan J, Zhu H, Gabriel M, Holden BA, Willcox MD (2012) Effect of prophylactic antibiotic drops on ocular microbiota and physiology during silicone hydrogel lens wear. Optom Vis Sci 89: 326–335. https://doi.org/10.1097/OPX.0b013e318243280e

de Paiva CS, Jones DB, Stern ME et al (2016) Altered mucosal microbiome diversity and disease severity in Sjogren syndrome. Sci Rep 6:23561. https://doi.org/10.1038/srep23561

Panchal B (2018) Commentary: post fever retinitis and vasculitis: a morphological conundrum. Indian J Ophthalmol 66:1844. https://doi.org/10.4103/ijo.IJO_1644_18

Parment PA (1997) The role of *Serratia marcescens* in soft contact lens associated ocular infections. A review. Acta Ophthalmol Scand 75:67–71. https://doi.org/10.1111/j.1600-0420.1997.tb00253.x

Partamian LG, Jay WM, Fritz KJ (1983) Anaerobic orbital cellulitis. Ann Ophthalmol 15:123–126

Paul G, Debadatta D, Ankita N, Mandal A, Bhutia TW (2021) Comparison of conjunctival microflora in patients undergoing cataract surgery. Indian J Clin Exp Ophthalmol 3:450–453

Peppard PE, Young T, Barnet JH, Palta M, Hagen EW, Hla KM (2013) Increased prevalence of sleep-disordered breathing in adults. Am J Epidemiol 177:1006–1014. https://doi.org/10.1093/aje/kws342

Perkins RE, Kundsin RB, Pratt MV, Abrahamsen I, Leibowitz HM (1975) Bacteriology of normal and infected conjunctiva. J Clin Microbiol 1:147–149. https://doi.org/10.1128/jcm.1.2.147-149.1975

Petrillo F, Pignataro D, Lavano MA et al (2020) Current evidence on the ocular surface microbiota and related diseases. Microorganisms 8:1033. https://doi.org/10.3390/microorganisms8071033

Phillips WB II, Tasman WS (1994) Postoperative endophthalmitis in association with diabetes mellitus. Ophthalmology 101:508–518. https://doi.org/10.1016/s0161-6420(13)31268-8

Prabhushanker M, Topiwalla TT, Ganesan G, Appandaraj S (2017) Bilateral retinitis following typhoid fever. Int J Retin Vitr 3:11. https://doi.org/10.1186/s40942-017-0065-z

Pradhan ZS, Jacob P (2013) Infectious scleritis: clinical spectrum and management outcomes in India. Indian J Ophthalmol 61:590–593. https://doi.org/10.4103/0301-4738.121085

Puliafito CA, Baker AS, Haaf J, Foster CS (1982) Infectious endophthalmitis. Review of 36 cases. Ophthalmology 89:921–929. https://doi.org/10.1016/S0161-6420(82)34704-1

Putnam CM (2016) Diagnosis and management of blepharitis: an optometrist's perspective. Clin Optom (Auckl) 8:71–78. https://doi.org/10.2147/OPTO.S84795

Quinet B, Mitanchez D, Salauze B et al (2010) Description et investigation d'une épidémie nosocomiale de colonisations et d'infections à Escherichia coli producteur d'une bêta-lactamase à spectre étendu dans un service de néonatologie. Arch Pediatr 17:S145–S149. https://doi.org/10.1016/s0929-693x(10)70916-7

Rajvanshi VS (1968) Bacterial flora of the conjunctiva. J All India Ophthalmol Soc 16:24–28

Ramachandran L, Sharma S, Sankaridurg PR et al (1995) Examination of the conjunctival microbiota after 8 hours of eye closure. CLAO J 21:195–199

Ramakrishnan R, Bharathi MJ, Shivkumar C, Mittal S, Meenakshi R, Khadeer MA, Avasthi A (2009) Microbiological profile of culture-proven cases of exogenous and endogenous endophthalmitis: a 10-year retrospective study. Eye (Lond) 23:945–956. https://doi.org/10.1038/eye.2008.197

Ranjith K, Arunasri K, Reddy GS, Adicherla H, Sharma S, Shivaji S (2017) Global gene expression in Escherichia coli, isolated from the diseased ocular surface of the human eye with a potential to form biofilm. Gut Pathog 9:15. https://doi.org/10.1186/s13099-017-0164-2

Rao PN, Rao KN (1972) Study of the normal conjunctival flora (bacterial and fungal) and its relations to external ocular infections. Indian J Ophthalmol 20:164–170

Rau G, Seedor JA, Shah MK, Ritterband DC, Koplin RS (2008) Incidence and clinical characteristics of enterococcus keratitis. Cornea 27:895–899. https://doi.org/10.1097/ICO.0b013e31816f633b

Reddy JC, Murthy SI, Reddy AK, Garg P (2015) Risk factors and clinical outcomes of bacterial and fungal scleritis at a tertiary eye care hospital. Mid E Afr J Ophthalmol 22:203–211. https://doi.org/10.4103/0974-9233.150634

Relhan N, Pathengay A, Albini T, Priya K, Jalali S, Flynn HW (2014) A case of vasculitis, retinitis and macular neurosensory detachment presenting post typhoid fever. J Ophthalmic Inflamm Infect 4:23. https://doi.org/10.1186/s12348-014-0023-y

Rietveld RP, ter Riet G, Bindels PJ, Sloos JH, van Weert HC (2004) Predicting bacterial cause in infectious conjunctivitis: cohort study on informativeness of combinations of signs and symptoms. BMJ 329:206–210. https://doi.org/10.1136/bmj.38128.631319.AE

Rishi E, Rishi P, Nandi K, Shroff D, Therese KL (2009) Endophthalmitis caused by Enterococcus faecalis: a case series. Retina 29:214–217. https://doi.org/10.1097/IAE.0b013e31818eccc7

Robert PY, Traccard I, Adenis JP, Denis F, Ranger-Rogez S (2002) Multiplex detection of herpes viruses in tear fluid using the "stair primers" PCR method: prospective study of 93 patients. J Med Virol 66:506–511. https://doi.org/10.1002/jmv.2173

Rowsey JJ, Newsom DL, Sexton DJ, Harms WK (1982) Endophthalmitis: current approaches. Ophthalmology 89:1055–1066. https://doi.org/10.1016/s0161-6420(82)34691-6

Sacchetti M, Lambiase A, Moretti C, Mantelli F, Bonini S (2015) Sex hormones in allergic conjunctivitis: altered levels of circulating androgens and estrogens in children and adolescents with vernal keratoconjunctivitis. J Immunol Res 2015:945317. https://doi.org/10.1155/2015/945317

Sadun AA (1999) Neuro-ophthalmic manifestations of diabetes. Ophthalmology 106:1047–1048. https://doi.org/10.1016/S0161-6420(99)90242-7

Sangwan VS (2010) Treatment of uveitis: beyond steroids. Indian J Ophthalmol 58:1–2. https://doi.org/10.4103/0301-4738.58466

Sankaridurg PR, Markoulli M, de la Jara PL, Harmis N, Varghese T, Willcox MD, Holden BA (2009) Lid and conjunctival micro biota during contact lens wear in children. Optom Vis Sci 86: 312–317. https://doi.org/10.1097/opx.0b013e318199d20c

Sathiamoorthi S, Smith WM (2016) The eye and tick-borne disease in the United States. Curr Opin Ophthalmol 27:530–537. https://doi.org/10.1097/ICU.0000000000000308

Seal DV, McGill JI, Mackie IA, Liakos GM, Jacobs P, Goulding NJ (1986) Bacteriology and tear protein profiles of the dry eye. Br J Ophthalmol 70:122–125. https://doi.org/10.1136/bjo.70.2.122

Shah SM, Howard RS, Sarkies NJ, Graham EM (1988) Tuberculosis presenting as retinal vasculitis. J R Soc Med 81:232–233

Sharma S (2012) Diagnosis of infectious diseases of the eye. Eye (Lond) 26:177–184. https://doi.org/10.1038/eye.2011.275

Sharma PD, Sharma N, Gupta RK, Singh P (2013a) Aerobic bacterial flora of the normal conjunctiva at high altitude area of Shimla Hills in India: a hospital based study. Int J Ophthalmol 6: 723–726. https://doi.org/10.3980/j.issn.2222-3959.2013.05.32

Sharma V, Chhangte L, Joshi V, Gupta S, Kalpana (2013b) A case of odontogenic orbital cellulitis causing blindness: a case report. Delhi J Ophthalmol 24:102–105

Shikari H, Antin JH, Dana R (2013) Ocular graft-versus-host disease: a review. Surv Ophthalmol 58:233–251. https://doi.org/10.1016/j.survophthal.2012.08.004

Shimizu E, Ogawa Y, Saijo Y et al (2019) Commensal microflora in human conjunctiva; characteristics of microflora in the patients with chronic ocular graft-versus-host disease. Ocul Surf 17: 265–271. https://doi.org/10.1016/j.jtos.2019.02.001

Shovlin JP, Argueso P, Carnt N et al (2013) 3. Ocular surface health with contact lens wear. Cont Lens Anterior Eye 36(Suppl 1):S14–S21. https://doi.org/10.1016/S1367-0484(13)60005-3

Singer TR, Isenberg SJ, Apt L (1988) Conjunctival anaerobic and aerobic bacterial flora in paediatric versus adult subjects. Br J Ophthalmol 72:448–451. https://doi.org/10.1136/bjo.72.6.448

Singh AG, Singh S, Matteson EL (2016) Rate, risk factors and causes of mortality in patients with Sjogren's syndrome: a systematic review and meta-analysis of cohort studies. Rheumatology (Oxford) 55:450–460. https://doi.org/10.1093/rheumatology/kev354

Smith CH (1954) Bacteriology of the healthy conjunctiva. Br J Ophthalmol 38:719–726. https://doi.org/10.1136/bjo.38.12.719

Sotozono C, Inagaki K, Fujita A, Koizumi N, Sano Y, Inatomi T, Kinoshita S (2002) Methicillin-resistant Staphylococcus aureus and methicillin-resistant Staphylococcus epidermidis infections in the cornea. Cornea 21:S94–S101. https://doi.org/10.1097/01.ico.0000263127.84015.3f

Soudakoff PS (1954) Bacteriologic examination of the conjunctiva*. Am J Ophthalmol 38:374–376. https://doi.org/10.1016/0002-9394(54)90854-5

Speaker MG, Milch FA, Shah MK, Eisner W, Kreiswirth BN (1991) Role of external bacterial flora in the pathogenesis of acute postoperative endophthalmitis. Ophthalmology 98:639–649. https://doi.org/10.1016/s0161-6420(91)32239-5. discussion 650

St Leger AJ, Desai JV, Drummond RA et al (2017) An ocular commensal protects against corneal infection by driving an interleukin-17 response from mucosal gammadelta T cells. Immunity 47: 148–158.e145. https://doi.org/10.1016/j.immuni.2017.06.014

Stapleton F, Carnt N (2012) Contact lens-related microbial keratitis: how have epidemiology and genetics helped us with pathogenesis and prophylaxis. Eye (Lond) 26:185–193. https://doi.org/10.1038/eye.2011.288

Stapleton F, Willcox MD, Fleming CM, Hickson S, Sweeney DF, Holden BA (1995) Changes to the ocular biota with time in extended- and daily-wear disposable contact lens use. Infect Immun 63:4501–4505. https://doi.org/10.1128/iai.63.11.4501-4505.1995

Stapleton F, Willcox MD, Sansey N, Holden BA (1997) Ocular microbiota and polymorphonuclear leucocyte recruitment during overnight contact lens wear. Aust N Z J Ophthalmol 25(Suppl 1): S33–S35. https://doi.org/10.1111/j.1442-9071.1997.tb01751.x

Stapleton F, Keay LJ, Sanfilippo PG, Katiyar S, Edwards KP, Naduvilath T (2007) Relationship between climate, disease severity, and causative organism for contact lens-associated microbial keratitis in Australia. Am J Ophthalmol 144:690–698. https://doi.org/10.1016/j.ajo.2007.06.037

Stapleton F, Keay L, Edwards K, Naduvilath T, Dart JK, Brian G, Holden BA (2008) The incidence of contact lens-related microbial keratitis in Australia. Ophthalmology 115:1655–1662. https://doi.org/10.1016/j.ophtha.2008.04.002

Stapleton F, Edwards K, Keay L, Naduvilath T, Dart JK, Brian G, Holden B (2012) Risk factors for moderate and severe microbial keratitis in daily wear contact lens users. Ophthalmology 119: 1516–1521. https://doi.org/10.1016/j.ophtha.2012.01.052

Sthapit P (2014) Conjunctival flora of normal human eye. JSM Ophthalmol 2:1021

Sunaric-Megevand G, Pournaras CJ (1997) Current approach to postoperative endophthalmitis. Br J Ophthalmol 81:1006–1015. https://doi.org/10.1136/bjo.81.11.1006

Suto C, Morinaga M, Yagi T, Tsuji C, Toshida H (2012) Conjunctival sac bacterial flora isolated prior to cataract surgery. Infect Drug Resist 5:37–41. https://doi.org/10.2147/IDR.S27937

Sward M, Kirk C, Kumar S, Nasir N, Adams W, Bouchard C (2018) Lax eyelid syndrome (LES), obstructive sleep apnea (OSA), and ocular surface inflammation. Ocul Surf 16:331–336. https://doi.org/10.1016/j.jtos.2018.04.003

Szkaradkiewicz A, Chudzicka-Strugala I, Karpinski TM et al (2012) Bacillus oleronius and Demodex mite infestation in patients with chronic blepharitis. Clin Microbiol Infect 18:1020–1025. https://doi.org/10.1111/j.1469-0691.2011.03704.x

Taylor PB, Tabbara KF, Burd EM (1988) Effect of preoperative fusidic acid on the normal eyelid and conjunctival bacterial flora. Br J Ophthalmol 72:206–209. https://doi.org/10.1136/bjo.72.3.206

Thiel HJ, Schumacher U (1994) [Normal flora of the human conjunctiva: examination of 135 persons of various ages]. Klin Monatsbl Augenheilkd 205:348–357. https://doi.org/10.1055/s-2008-1045542

Truong S, Cole N, Stapleton F, Golebiowski B (2014) Sex hormones and the dry eye. Clin Exp Optom 97:324–336. https://doi.org/10.1111/cxo.12147

Tseng CY, Liu PY, Shi ZY et al (1996) Endogenous endophthalmitis due to *Escherichia coli*: case report and review. Clin Infect Dis 22:1107–1108. https://doi.org/10.1093/clinids/22.6.1107

Turner L, Stinson I (1965) Mycobacterium fortuitum*. Am J Ophthalmol 60:329–331. https://doi.org/10.1016/0002-9394(65)90934-7

Ueta M, Kinoshita S (2010) Innate immunity of the ocular surface. Brain Res Bull 81:219–228. https://doi.org/10.1016/j.brainresbull.2009.10.001

Vafidis GC, Marsh RJ, Stacey AR (1984) Bacterial contamination of intraocular lens surgery. Br J Ophthalmol 68:520–523. https://doi.org/10.1136/bjo.68.8.520

Van Tyne D, Ciolino JB, Wang J, Durand ML, Gilmore MS (2016) Novel phagocytosis-resistant extended-spectrum beta-lactamase-producing Escherichia coli from keratitis. JAMA Ophthalmol 134:1306–1309. https://doi.org/10.1001/jamaophthalmol.2016.3283

Varaprasathan G, Miller K, Lietman T et al (2004) Trends in the etiology of infectious corneal ulcers at the F. I. Proctor Foundation. Cornea 23:360–364. https://doi.org/10.1097/00003226-200405000-00009

Venugopal R, Satpathy G, Sangwan S et al (2016) Conjunctival microbial flora in ocular Stevens-Johnson syndrome sequelae patients at a tertiary eye care center. Cornea 35:1117–1121. https://doi.org/10.1097/ICO.0000000000000857

Versura P, Giannaccare G, Campos EC (2015) Sex-steroid imbalance in females and dry eye. Curr Eye Res 40:162–175. https://doi.org/10.3109/02713683.2014.966847

Vishwanath S, Badami K, Sriprakash KS, Sujatha BL, Shashidhar SD, Shilpa YD (2014) Post-fever retinitis: a single center experience from south India. Int Ophthalmol 34:851–857. https://doi.org/10.1007/s10792-013-9891-7

Walker CB, Claoue CM (1986) Incidence of conjunctival colonization by bacteria capable of causing postoperative endophthalmitis. J R Soc Med 79:520–521. https://doi.org/10.1177/014107688607900907

Watters GA, Turnbull PR, Swift S, Petty A, Craig JP (2017) Ocular surface microbiome in meibomian gland dysfunction. Clin Exp Ophthalmol 45:105–111. https://doi.org/10.1111/ceo.12810

Weijtens O, Schoemaker RC, Romijn FP, Cohen AF, Lentjes EG, van Meurs JC (2002) Intraocular penetration and systemic absorption after topical application of dexamethasone disodium phosphate. Ophthalmology 109:1887–1891. https://doi.org/10.1016/s0161-6420(02)01176-4

Weiss A, Friendly D, Eglin K, Chang M, Gold B (1983) Bacterial periorbital and orbital cellulitis in childhood. Ophthalmology 90:195–203. https://doi.org/10.1016/s0161-6420(83)34573-5

Weiss A, Brinser JH, Nazar-Stewart V (1993) Acute conjunctivitis in childhood. J Pediatr 122:10–14. https://doi.org/10.1016/s0022-3476(05)83479-1

Wen X, Miao L, Deng Y et al (2017) The influence of age and sex on ocular surface microbiota in healthy adults. Invest Ophthalmol Vis Sci 58:6030–6037. https://doi.org/10.1167/iovs.17-22957

Weng Y, Liu J, Jin S, Guo W, Liang X, Hu Z (2017) Nanotechnology-based strategies for treatment of ocular disease. Acta Pharm Sin B 7:281–291. https://doi.org/10.1016/j.apsb.2016.09.001

Willcox MDP (2007) *Pseudomonas aeruginosa* infection and inflammation during contact lens wear: a review. Optom Vis Sci 84:273

Willcox MD (2013) Characterization of the normal microbiota of the ocular surface. Exp Eye Res 117:99–105. https://doi.org/10.1016/j.exer.2013.06.003

Williamson J, Haynes J (2018) The bugs behind infectious retinitis rev optometry. Rev Optom

Wong J (2000) Endogenous bacterial endophthalmitis An East Asian experience and a reappraisal of a severe ocular affliction. Ophthalmology 107:1483–1491. https://doi.org/10.1016/s0161-6420(00)00216-5

Wood MJ (1996) The comparative efficacy and safety of teicoplanin and vancomycin. J Antimicrob Chemother 37:209–222. https://doi.org/10.1093/jac/37.2.209

Xia Z, Zhu M, Zhang Y (2014) Effects of the probiotic Arthrobacter sp. CW9 on the survival and immune status of white shrimp (Penaeus vannamei). Lett Appl Microbiol 58:60–64. https://doi.org/10.1111/lam.12156

Youssef OH, Stefanyszyn MA, Bilyk JR (2008) Odontogenic orbital cellulitis. Ophthalmic Plast Reconstr Surg 24:29–35. https://doi.org/10.1097/IOP.0b013e318160c950

Zaheer M, Wang C, Bian F et al (2018) Protective role of commensal bacteria in Sjogren syndrome. J Autoimmun 93:45–56. https://doi.org/10.1016/j.jaut.2018.06.004

Zarate-Blades CR, Horai R, Caspi RR (2016) Regulation of autoimmunity by the microbiome. DNA Cell Biol 35:455–458. https://doi.org/10.1089/dna.2016.3432

Zhai H, Bispo PJM, Kobashi H, Jacobs DS, Gilmore MS, Ciolino JB (2018) Resolution of fluoroquinolone-resistant *Escherichia coli* keratitis with a PROSE device for enhanced targeted antibiotic delivery. Am J Ophthalmol Case Rep 12:73–75. https://doi.org/10.1016/j.ajoc.2018.09.006

Zhang Y, Liu ZR, Chen H et al (2013) Comparison on conjunctival sac bacterial flora of the seniors with dry eye in Ganzi autonomous prefecture. Int J Ophthalmol 6:452–457. https://doi.org/10.3980/j.issn.2222-3959.2013.04.08

Zhao Z, Yang M, Azar SR et al (2017) Viral retinopathy in experimental models of zika infection. Invest Ophthalmol Vis Sci 58:4355–4365. https://doi.org/10.1167/iovs.17-22016

Zilliox MJ, Gange WS, Kuffel G, Mores CR, Joyce C, de Bustros P, Bouchard CS (2020) Assessing the ocular surface microbiome in severe ocular surface diseases. Ocul Surf 18:706–712. https://doi.org/10.1016/j.jtos.2020.07.007

Chapter 3
The Microbiome of the Human Eye

Abstract Microbiome refers to the community of bacteria, fungi and viruses that are associated with our body. The most studied is the microbiome of the human gut which has a number of beneficial organisms that are essential to good health. Dysbiosis in the gut microbiome is associated with several diseases including cancers, inflammatory diseases and autoimmune diseases. Compared to the conventional cultivable approach which detected several cultivable bacteria, the advent of NGS technology led to the establishment of the ocular surface microbiome which revealed a greater diversity and abundance of bacteria which varied with disease status of the eye. Thus, characterisation of the ocular surface microbiome from the healthy and diseased eye could open new strategies for the treatment and prevention of eye diseases.

Keywords Microbiome · Ocular surface · Conjunctiva · Cornea · Ocular disease

3.1 Introduction

In 1985 it was observed by Staley and Konopka (1985) that the number of viable bacteria in an environmental or biological sample when cultured on a media plate is several orders of magnitude less than the number observed under the microscope (Harwani 2012). This anomaly between the cultivable bacteria and the observed number was referred to as 'The great plate count anomaly' (Staley and Konopka 1985). At best only 1% of the total existing bacteria are cultured. This anomaly clearly illustrated that the vast majority of the bacteria are not amenable to culture, and it is also an indication that some may be nonviable whereas could be viable but nonculturable (Li et al. 2014). The bacteria may not grow due to fastidious growth requirements and may require specific nutrients in the media, and the conditions for growth (pH, incubation temperature or levels of oxygen) may not be optimum (Kopke et al. 2005; Deming and Baross 2000). The growth of the bacteria may also be influenced by other bacteria which could compete for nutrients and inhibit growth and survival by releasing antimicrobial compounds (Tamaki et al. 2005). Thus, the challenge is to identify the vast majority of unculturable bacteria, and the popular approach that has developed over the years is the microbiome approach.

S. Shivaji, *Human Ocular Microbiome*,
https://doi.org/10.1007/978-981-19-1754-7_3

3.2 What Is a Microbiome?

The word 'microbiome' (*micro* meaning 'small' and *bíos* meaning 'life') refers to all the microorganisms (bacteria, fungi, viruses, protozoans, etc.) found in a specific environment. Currently, the microbiome is defined as 'a community of commensal, symbiotic, and pathogenic microorganisms within a body space or environment' (Lederberg and McCray 2001). With time, the definition also encompassed the genomes, gene expression patterns and proteomes of all the microorganisms (commensals, pathogenic and symbiotic, etc.) in each environment (Marchesi and Ravel 2015).

3.3 The Microbiome Approach

The microbiome approach is a culture-independent method to identify microbes based on the extraction of DNA from a sample, amplification of the variable regions 3 and 4 (V3-V4) of the 16S rRNA gene, a molecular phylogenetic marker for prokaryotes, sequencing of the amplicon and finally assigning the sequence to a taxon. This approach has been extensively used to unravel the microbiomes of environmental samples (Woese 1987; Keller and Zengler 2004; Rappe and Giovannoni 2003; Achtman and Wagner 2008), gut microbiomes (Rinninella et al. 2019; Turnbaugh et al. 2007, 2009) and other microbiomes of the human body like the skin, vagina, dental plaque and oral cavity (Liu et al. 2021; Byrd et al. 2018; Peterson et al. 2013) including the ocular microbiomes (Dong et al. 2011; Doan et al. 2016; Lee et al. 2012; Shin et al. 2016; Graham et al. 2007). One of the major outcomes of microbiome approach has been the identification of new uncultured and unexplored bacteria which represent a vast gene pool (Hugenholtz et al. 1998; Harwani 2012).

Two different sequencing strategies can be used as part of the microbiome approach (Gallon et al. 2019). In amplicon sequencing only a specific phylogenetic marker (like the 16S rRNA gene for prokaryotes or ITS2 region of nuclear ribosomal internal transcribed spacer region) in case of eukaryotes is amplified, and therefore the output only gives a view of the diversity of the microorganisms being studied and is suitable when only the microbial diversity is the goal (Mitra et al. 2011). In contrast, in the metagenome approach, the entire metagenomic DNA is sequenced, and the data not only unveils the taxonomy of the microbes being analysed but could also provide inputs into the genomes and genes of particular interest such as signalling pathways operating, virulence genes, antibiotic resistance genes, etc. (Fig. 3.1) (Mitra et al. 2011; Thomas et al. 2012). The principle of these approaches is summarised in Fig. 3.1. The various steps involved in amplicon sequencing are briefly outlined below (Fig. 3.2).

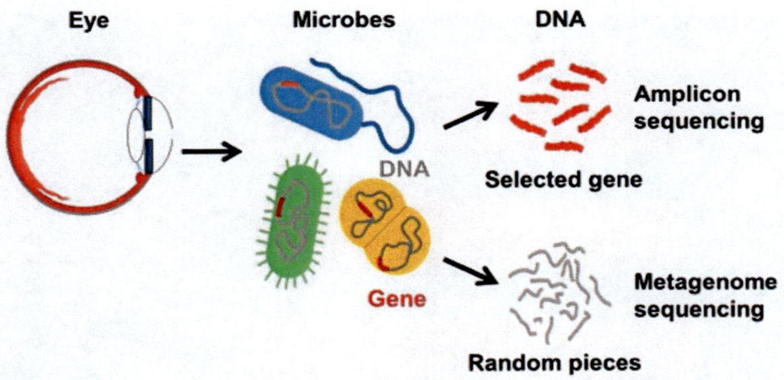

Fig. 3.1 Schematic representation of the two different sequencing strategies of the microbiome approach, namely, amplicon sequencing or targeted sequencing and whole metagenome sequencing

3.3.1 *Amplicon Sequencing*

1. **DNA Extraction:** Total genomic DNA is extracted under aseptic conditions from the biological sample, quality checked on 0.8% agarose gel, quantified and stored at −80 °C. Care is taken to avoid DNA contamination from the reagents used for extraction, from the PCR reagents and any other possible source including the lab surroundings.
2. **Amplification of the Target Gene:** Phylogenetic markers like V3-V4 region for bacteria and ITS2 for fungi are amplified using the genomic DNA as a template and specific primers. The 16S rRNA gene has nine variable regions which could be used as phylogenetic markers since the sequence of these regions is species/genera-specific (Fig. 3.3). However, there is a lack of consensus as to which region would be the most suitable and efficient to identify bacteria up to the species level. According to Bukin et al. (2019), bacteria could be best resolved at the species level using the sequence of the V2–V3, whereas others found V4-V6 region was better (Yang et al. 2016), whereas V2 and V8 regions were the least reliable as prokaryotic phylogenetic markers (Yang et al. 2016).
3. **Library Preparation:** Library preparation involved a two-step procedure. In the first step, specific target primers amplified the V3-V4 region, and the generated amplicon in the second PCR step was amplified using primers containing spacers, adaptors and barcodes.
4. **Sequencing:** Following purification and quantification, the amplicons were sequenced according to the guidelines of MiSeq™ Reagent Kit (Illumina, San Diego, CA, USA) (Caporaso et al. 2012).

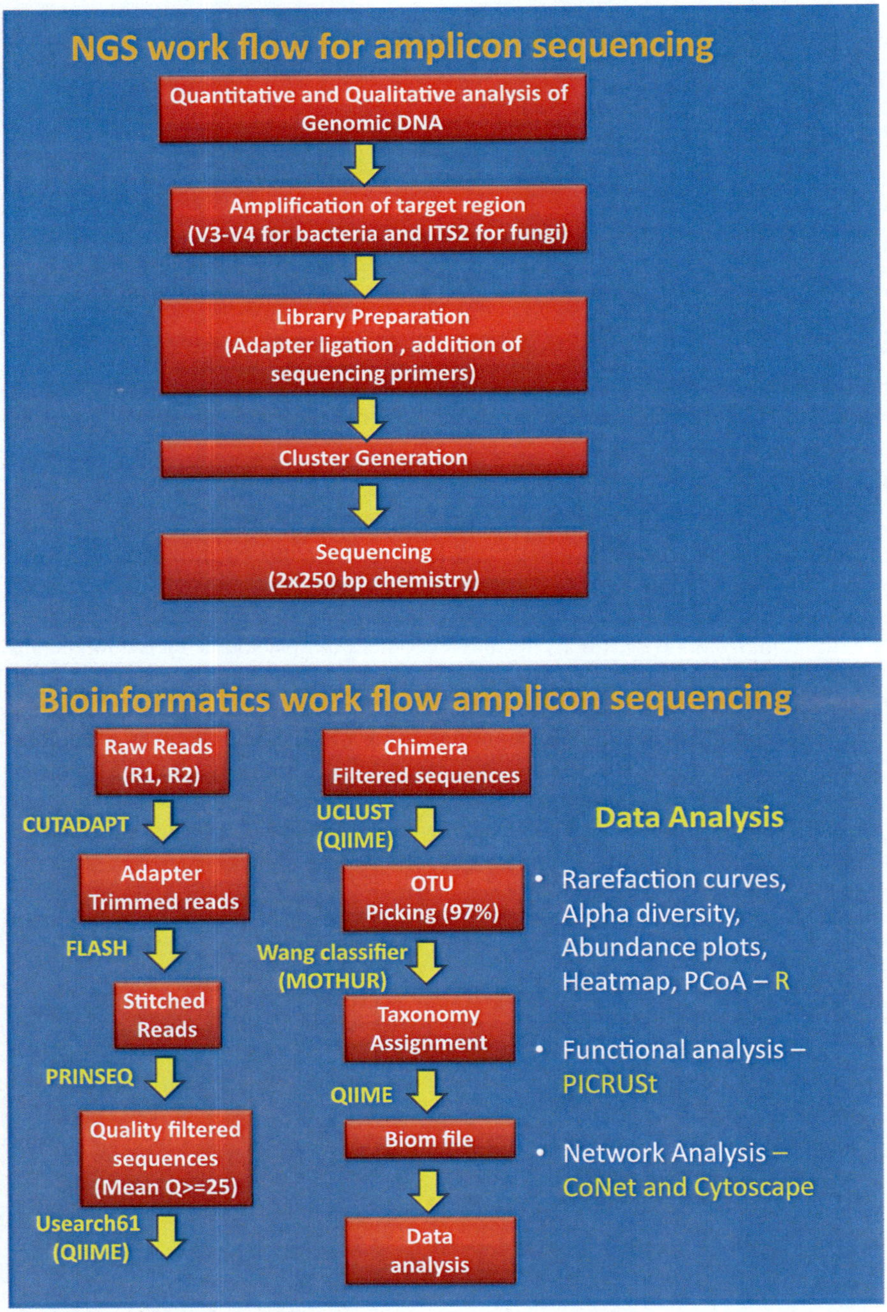

Fig. 3.2 Steps involved in amplicon sequencing and bioinformatics workflow. The various programmes used for the bioinformatics analyses are indicated in yellow adjacent to the arrow

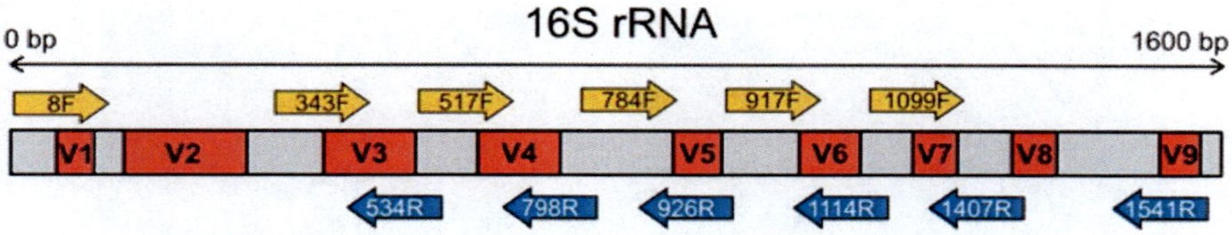

Region	Position	# b.p.
V1	69-99	30
V2	137-242	105
V3	338-533	195
V4	576-682	106
V5	822-879	57
V6	967-1046	79
V7	1117-1173	56
V8	1243-1294	51
V9	1435-1465	30

Fig. 3.3 Schematic representation of 16S rRNA gene indicating the position of the nine variable regions (V1 to V9), their length in bp and the forward and reverse primers that are used for PCR. The schematic figure is based on *E. coli* 16S rRNA gene (Nossa et al. 2010). The numbers within the yellow and blue arrows indicate the mapping position on *E. coli* 16S rRNA gene

3.3.2 Metagenome Sequencing

For metagenome sequencing DNA was prepared as above, fragmented, tagged with adapters and used for library preparation (Nextera reagent kit; Illumina Inc., USA). The quality of all the libraries were checked prior to sequencing using a 100 bp paired-end (PE) Illumina HiSeq platform (Fig. 3.4).

3.3.3 Data Analysis

3.3.3.1 Amplicon Sequencing Reads

This is the most complicated part of microbiome studies and needs a bioinformatics expert to finally translate the sequencing reads to a bacterium/fungi/virus/gene, etc. Different pipelines are used depending on whether one is using amplicon sequencing reads or reads obtained from metagenome sequencing. Figure 3.2 also captures briefly the various steps in bioinformatics involved in translating sequence reads

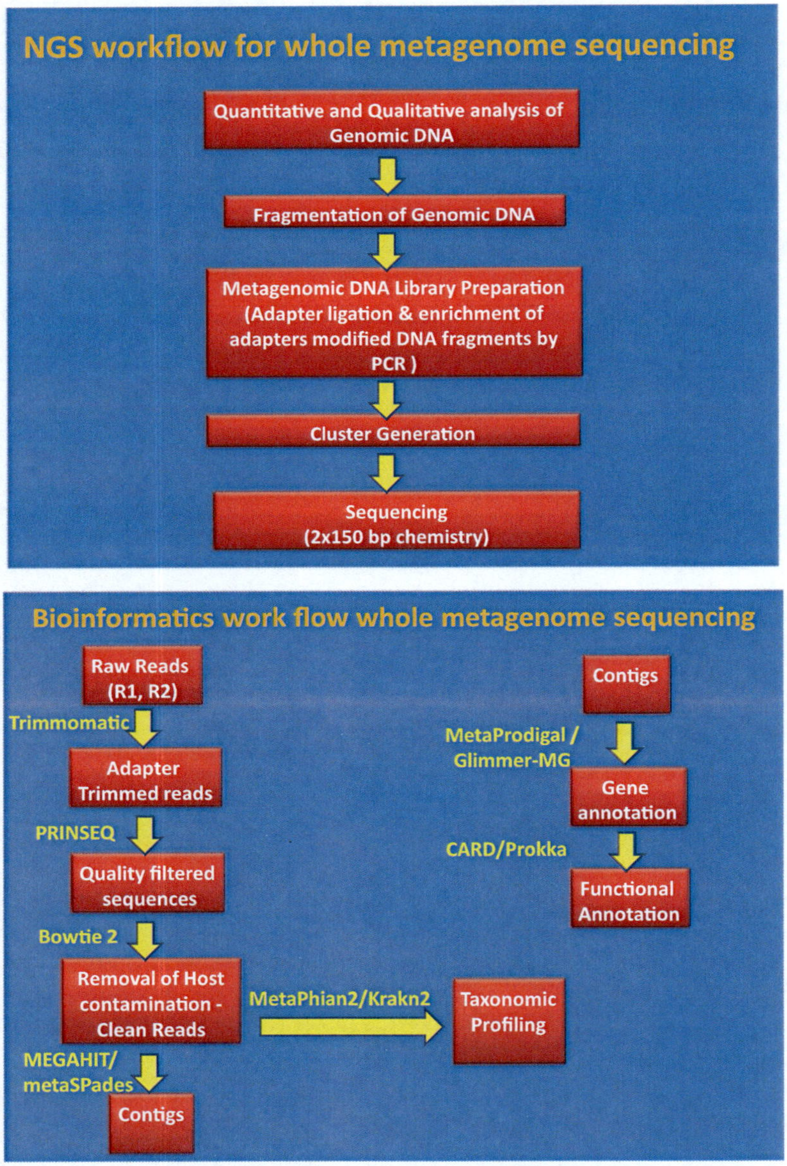

Fig. 3.4 Steps in metagenome sequencing and bioinformatics workflow. The various programmes used for the bioinformatics analyses are indicated in yellow adjacent to the arrow

up to the generation of the biome file which could then be used for data analyses such as α-diversity analysis, β-diversity analysis, rarefaction analysis, functional analysis, interaction networks, principal component analysis, heat-map analysis, etc.

3.3.3.2 Metagenome Sequencing Reads

The resulting Illumina raw reads are processed to remove adapter sequences and quality filtered to remove low-quality sequences and host DNA sequences, and the high-quality reads are assembled using Megahit v1.0.2 (Li et al. 2015) and classified as fungi, bacteria, viruses, protozoa and archaea using the Kraken database v2.0.9 (Wood et al. 2019) (Fig. 3.4). α-Diversity was normalised using cumulative sum scaling normalisation of the metagenome Seq package of Rv1.10.0 (https://www.rdocumentation.org/packages/metagenomeSeq/versions/1.10.0), and the normalised counts were analysed from the phylum to the species level. The Shannon and the Inverse Simpson indices of diversity were calculated with vegan package of R v2.5–6 (https://CRAN.R-project.org/package=vegan). For functional analysis contigs are aligned using specific databases like Resistance Gene Identifier tool v5.1.0 (https://toolshed.g2.bx.psu.edu/view/card/rgi/f96cbb663aa9) and MEGARes 2.0 database (Doster et al. 2020), which is a collation of multiple databases (CARD, ARG-ANNOT and ResFinder) to confirm the presence of resistance-conferring SNPs. A gene was designated as an antimicrobial resistance gene (AMR) only when a sequence exhibited ≥90% identity and a coverage of ≥25% to a reference AMR gene in the database. The Mycology Antifungal Resistance Database (MARDy) (http://www.mardy.net) is used to detect antifungal drug resistance genes. Counts of reads aligned to the ARGs are recorded and used for downstream comparative analyses. Thus, for a given class of genes, the contigs generated would be used to scan for the genes of interest using specific databases.

This chapter highlights the diversity of bacteria in the healthy human eye and under diseased conditions using the microbiome approach. Details related to the description of the diseases are given in Chap. 2.

3.4 Culture-independent Methods to Identify Ocular Bacteria

3.4.1 The Ocular Surface Microbiome Using Amplicon Sequencing

In Chap. 2 several aspects of the ocular surface culturable microbiome have been discussed, and features such as the bacterial load on the ocular surface and the viable community of bacteria on the ocular surface have been discussed. The microbiome approach confirms that the ocular surface has a microbiome of its own which is diverse and this approach detected not only the bacterial phylotypes that were earlier reported by the conventional cultivable-based method but also detected *Erwinia* spp., *Rhodococcus erythropolis*, *Rhodococcus* spp., *Pseudomonas* spp., *Klebsiella* spp. and *Propionibacterium* spp. (Graham et al. 2007; Dong et al. 2011) that were not identified earlier from the healthy eye. Further, in several microbiomes extremely high percentage of reads and OTUs could not be assigned to any known genus, and these represent possibly novel species-level phylotypes (Dong et al. 2011).

Abelson et al. (2015) based on studies on the human ocular microbiome commented that the 'ocular surface is actually teeming with life that might hold the secret to treating ophthalmic diseases'. Several studies that used NGS clearly demonstrated the presence of a unique resident commensal bacterial microbiome on the ocular surface (Dong et al. 2011a; Lee et al. 2012; Graham et al. 2007; Shin et al. 2016; Doan et al. 2016). All the bacteria could be assigned phylogenetically to three phyla, namely, *Firmicutes*, *Actinobacteria* and *Proteobacteria* (Huang et al. 2016; Dong et al. 2011; Ozkan et al. 2017; Zhou et al. 2014), and at the genera level, 59 genera (Dong et al. 2011) to a maximum of 600 genera (Zhou et al. 2014) were identified. Among these genera it was observed that several genera with >1% abundance were shared across all the microbiomes and are designated as the 'core genera' (Hamady and Knight 2009). Huang et al. (2016) reported that the human conjunctival microbiomes (100%) consistently revealed the presence of ten genera, namely, *Corynebacterium* spp., *Pseudomonas* spp., *Streptococcus* spp., *Anaerococcus* spp., *Staphylococcus* spp., *Millisia* spp., *Finegoldia* spp., *Acinetobacter* spp., *Simonsiella* spp. and *Veillonella* spp. with >1% mean abundance. These ten genera in the healthy human conjunctival microbiomes were designated as the core genera. Five of these ten core genera (*Pseudomonas* spp., *Acinetobacter* spp., *Corynebacteria* spp., *Streptococcus* spp. and *Staphylococcus* spp.) were also reported by Dong et al. (2011), who also reported seven other core genera that included *Propionibacterium* spp., *Brevundimonas* spp., *Bradyrhizobium* spp., *Aquabacterium* spp., *Simonsiella* spp., *Sphingomonas* spp. and *Methylobacterium* spp. in the conjunctiva. Interestingly it was observed that many of the core genera harbour species that are common ocular pathogens like those affiliated to the genera *Pseudomonas* spp., *Acinetobacter* spp., *Propionibacterium* spp., *Corynebacterium* spp., *Staphylococcus* spp., *Streptococcus* spp. and *Sphingomonas* spp. (Dong et al. 2011). In contrast, *Corynebacterium* was identified as the only core genus in the conjunctiva (Zhou et al. 2014). But Ozkan et al. (2017) demonstrated that when the ocular microbiota was analysed using swabs from the upper and then the lower bulbar conjunctiva, avoiding any possible contamination from the lid margin at three time points (baseline, 1 and 3 months), the majority of the OTUs (approximately 80%) could be assigned to 17 genera including *Corynebacterium* spp. (11.1%), *Acinetobacter* spp. (11.0%), *Pseudomonas* spp. (10.4%), *Sphingomonas* spp. (10.2%), *Streptococcus* spp. (4.8%), *Massilia* spp. (3.2%) and *Rothia* spp. (1.9%). This composition of genera of the conjunctiva remained relatively stable and unchanged even after 3 months. But Ozkan et al. (2017) detected the above genera with >1% abundance in approximately 80% of the conjunctival microbiomes and designated these as the 'putative core' genera in contrast to the core genera which are present with >1% abundance in all the microbiomes (Ozkan et al. 2017) (Table 3.1). Gomes et al. (2020) based on 11 publications including some publications discussed above on the eye microbiome found that *Corynebacterium* spp. was consistently present (11/11), followed by *Acinetobacter* spp. (9/11), *Pseudomonas* spp. (8/11; 19%), *Staphylococcus* spp. (7/11; 6%), *Propionibacterium*

Table 3.1 The core genera detected in the human conjunctival microbiome by next generation sequencing

Reference	Core genera
Huang et al. (2016)	*Corynebacterium* spp., *Acinetobacter* spp., *Pseudomonas* spp., *Staphylococcus* spp., *Streptococcus* spp., *Millisia* spp., *Anaerococcus* spp., *Finegoldia* spp., *Simonsiella* spp., and *Veillonella* spp.
Dong et al. (2011)	*Corynebacterium* spp., *Pseudomonas* spp., *Acinetobacter* spp., *Staphylococcus* spp., *Streptococcus* spp., *Bradyrhizobium* spp., *Propionibacterium* spp., *Aquabacterium* spp., *Brevundimonas* spp., *Simonsiella* spp., *Sphingomonas* spp., and *Methylobacterium* spp.
Zhou et al. (2014)	*Corynebacterium* spp.
Ozkan et al. (2017)	*Acinetobacter* spp., *Corynebacterium* spp., *Pseudomonas* spp., *Sphingomonas* spp., *Streptococcus* spp., *Massilia* spp. and *Rothia* spp.
Gomes et al. (2020)	*Acinetobacter* spp., *Corynebacterium* spp., *Pseudomonas* spp., *Propionibacterium* spp. *Staphylococcus* spp. and *Streptococcus* spp.

spp. (5/11; 7%) and *Streptococcus* spp. (5/11; 3%), and these were considered as representing the core microbiome. *Bacillus* spp. was not considered to be part of the core microbiome (4/11). It was anticipated that the core genera of the conjunctiva of healthy individuals would be unchanged across individuals. The observed differences in the core genera in the conjunctival microbiomes has been attributed to several factors including procedural methods like variations in the swab used, method of swabbing, the region of the conjunctiva swabbed and importantly to ethnic variations in the individuals sampled (Huang et al. 2016). The bacterial microbiomes could also vary due to exposure of the ocular surface to the environment and also to the prior use of tetracaine in sample collection in some studies (Ozkan et al. 2018; Dong et al. 2011; Shin et al. 2016) (Table 3.1). The ocular surface also has a remarkable diversity and abundance of potentially pathogenic bacteria whose pathogenicity is quelled by powerful mechanisms suppressing microbial pathogenicity on the ocular surface (Miller and Iovieno 2009; McClellan 1997; Papapanou 2002; Paster et al. 2001). But these ocular opportunistic pathogens may become the cause of infection after their virulence is enhanced by aging, trauma, stress and the depletion of resident commensal microbiota as reported for other human niches (van der Meulen et al. 2008; Cogen et al. 2008). NGS studies cannot confirm that the bacteria detected are viable residents of the ocular surface (Zegans and Van Gelder 2014) since the bacteria on the ocular surface could be acquired from the surrounding environment or from regions surrounding the eye and are thus considered 'colonisers'. St Leger and Caspi (2018) provided the first evidence that *Corynebacterium mastitidis*, from the conjunctiva of C57BL/6 mice, was a bona fide commensal on the ocular surface.

3.4.2 Impact of Age on Ocular Surface Microbiome

The ocular surface microbiome (OSM) was not influenced by age (22–64 years), and the alpha diversity indices of richness and Shannon diversity index (number of taxa and evenness of their distribution) remained unaltered (Ozkan et al. 2017). Other studies indicated differences in the ocular microbiome of adults over the age of 18 years (Wen et al. 2017; Zhou et al. 2014; Ozkan et al. 2017). Three phyla, namely, *Actinobacteria* (53%), *Proteobacteria* (39%) and *Firmicutes* (8%) were predominant and represented the core phyla in the paediatric group (younger than 18 years old) (Wen et al. 2017), whereas phylum *Bacteroidetes* was not part of the core phyla (Delbeke et al. 2021). Cavuoto et al. (2018) confirmed that *Proteobacteria* (39%) and *Firmicutes* (8%) were increased in paediatric eyes along with *Fusobacteria* and *Bacteroidetes* (Cavuoto et al. 2018). On the other hand, a twofold increase in *Actinobacteria* in adults was also observed, whereas *Cyanobacteria* did not exhibit any change between the two groups. At the genus level, *Streptococcus* spp., *Staphylococcus* spp. and *Brachybacterium* spp. were observed to be decreased in adults compared to paediatric samples, whereas *Corynebacterium* spp., *Paracoccus* spp. and *Propionibacterium* spp. were relatively more abundant in adults (Cavuoto et al. 2018; Zhou et al. 2014). β-Diversity analysis indicated that the paediatric and adult microbiomes were different and separated into distinct clusters. Further, the microbiomes did not vary between the right and left eyes, between the genders, across age groups, following surgery or when the patient was maintained on topical medications (Cavuoto et al. 2018). Several antibiotic resistance genes were enriched in the conjunctival microbiome of adults which could be attributed to changes in the immune homeostasis of the ocular surface microbiome (OSM) (Wen et al. 2017). The increased diversity in the microbiome in young individuals, compared to older age group, was attributed to decreased immunity, close contact between children and differences in hygiene (Zhou et al. 2014). It is known that sex-steroid imbalance in menopause initiates allergic ocular diseases including dry eye syndrome, and in addition profound changes in the ocular surface micro-environment predispose women to autoimmune and inflammatory diseases of the eye (Wen et al. 2017; Versura et al. 2015). Whether other factors like ethnicity, the geographical origin of the individuals and the environment alter the ocular surface microbiota is not clear (Ozkan et al. 2017; Zhou et al. 2014; Wen et al. 2017). Future studies should unravel the underlying mechanisms for these differences and their impact on OSM.

3.4.3 Impact of Sex on Ocular Surface Microbiome

Sex as a variable did not appear to impact the OSM. A few studies found no difference in the richness in diversity as judged by the absolute numbers of OTUs between males and females (Zhou et al. 2014; Ozkan et al. 2017), while others found the bacterial communities to be different between males and females (Shin et al.

2016; Wen et al. 2017). Wen et al. (2017) observed no significant difference in α-diversity, but the β-diversity was changed. *Staphylococcus epidermidis* and *Pseudomonas acnes* decreased significantly from male to female, while *E. coli* increased significantly in the female group. Shin et al. (2016) had earlier indicated that the conjunctival microbiota was influenced by gender with *Acinetobacter* and members of *Enterobacteriaceae* increased and those of *Anaerococcus* depleted in the ocular microbiota of females compared to males.

3.4.4 Impact of Geographical Location on Ocular Surface Cultivable Microbiome

Bacteria associated with ocular surface cultivable microbiome are known to vary depending on the geographical location of the individuals (Stapleton et al. 2007; Petrillo et al. 2020). But the ocular microbiome analysis of healthy individuals indicated a similarity between healthy individuals originating from distinct and different geographical regions and climatic conditions including the United States (Dong et al. 2011; Shin et al. 2016; Doan et al. 2016), China (Zhang et al. 2017; Huang et al. 2016; Wen et al. 2017), South Korea (Ham et al. 2018), Australia (Ozkan et al. 2017), the Gambia (Zhou et al. 2014) and Northern Ireland (Graham et al. 2007). The ocular microbiome were very similar at the phylum level, and the bacteria were affiliated to the phyla *Firmicutes*, *Proteobacteria* and *Actinobacteria* (Graham et al. 2007). Further, at the genus level, *Pseudomonas* spp., *Staphylococcus* spp., *Streptococcus* spp., *Corynebacterium* spp., *Propionibacterium* spp. and *Acinetobacter* spp. were detected consistently across geographic regions (Ozkan and Willcox 2019). Thus, these studies indicated that geographical location may not be a confounding factor in influencing the ocular microbiota in contrast to culture-based studies which found differences in the normal microbiota in different climate and geographical locations.

3.5 Microbiome of Various Microhabitats of the Eye

Various microhabitats of the eye including the skin of the eyelid, lid margin, fornix, conjunctival tissue and limbal tissue were analysed for their microbiome (Ozkan et al. 2018, 2019), and the studies indicated that these regions possess distinct bacterial communities. Quantitative PCR of the ocular surface samples for bacteria revealed that the ocular surface has a paucibacterial microbiome (Doan et al. 2016) with two orders of magnitude decreased compared to the bacteria from the facial skin or buccal mucosa. Comparison of the microbiomes indicated that conjunctival surface microbiome was significantly different when compared to the community in the fornix and limbus tissue. But the bacterial communities between the limbus and fornix were similar (Ozkan et al. 2018). *Pseudomonas* spp. (relative abundance 79.9%) was dominant in fornix and limbal samples compared to the conjunctival surfaces (6.3%). This is not surprising considering that *Pseudomonas* spp. is a known ocular pathogen, with *P. aeruginosa* being frequently isolated from contact lens-related keratitis patients (Stapleton et al. 2012; Hong et al. 2012). It was also found that the microbiome of the ocular surface was distinct from the cheek skin, the buccal mucosa and the skin below the eye (Doan et al. 2016; Shin et al. 2016). Further, the skin had the highest α-diversity and the conjunctiva the lowest diversity with the lid margin and the ocular surface showing moderate diversity. A few reports indicated that the microbiome of the ocular surface had a bacterial diversity higher than the surrounding skin (Zhou et al. 2013; Shin et al. 2016). These discrepancies were attributed to environmental and reagent contaminants. In the eyelid margin, microbiome *Propionibacterium* spp., *Corynebacterium* spp., *Enhydrobacter* spp. and *Staphylococcus* spp. (Ozkan et al. 2019; Lee et al. 2012) were consistently detected. NMDS plots also highlighted the differences in the bacterial community structure of the above regions of the eye. Ozkan et al. (2019) in a comprehensive study on the microbiome analysis of the eyelid margin tissue from patients with lid abnormalities, fornix and limbus conjunctival tissue from patients with pterygia, conjunctival swabs and facial skin swabs identified three broad groups of bacteria: Group 1 had a predominance of *Corynebacterium* spp. and *Staphylococcus* spp. that are residents of the skin (Grice et al. 2009) and lid margin, and they get transferred from the lid margin to the ocular surface (Doan et al. 2016; Ozkan et al. 2017; Lee et al. 2012; Shin et al. 2016). This transfer of the microbiota from the lid margin to the ocular surface is supported by the observation that the relative abundance of both these skin residents decreased progressively from the skin (12.2%) into the lid margin (7.6%) and onto the ocular surface (4.0%) (Ozkan et al. 2017, 2018; Shin et al. 2016; Doan et al. 2016; Lee et al. 2012). Group 2 consisted of *Acinetobacter* spp. and *Aeribacillus* spp. (Ozkan et al. 2019), and the relative abundance decreased from 12.3% on the ocular surface to 1.0% on the skin and further decreased to 0.5% in the lid margin (Groden et al. 1991) and conjunctiva. Both these bacteria probably originate from the air or water. The resilience of some species of *Acinetobacter* spp. to antimicrobials (Bonomo and Szabo 2006) may help them to associate with the ocular surface. Group 3 consisted of *Pseudomonas* spp., normally detected in the conjunctiva and within the lid margin and found rarely on the skin (Dekio et al. 2005) and ocular surface (Ozkan et al. 2017; Dong et al. 2011; Ham et al. 2018). The relative abundance of *Pseudomonas* spp. followed the trend conjunctiva (80%) > eyelid tissue (65%) > skin (6%) > ocular surface (2%) (Ozkan and Willcox 2019; Delbeke et al. 2021). Some OTUs of *Pseudomonas* spp. were specific to the lid margin (zOTU30) and conjunctiva (zOTU69, zOTU77 and zOTU191) (Ozkan et al. 2019). The presence of *Pseudomonas* spp., on the conjunctiva, which is a harsh and nutrient poor environment, is an indication of the resilience of *Pseudomonas* spp., in the conjunctival tissue (Mulcahy et al. 2014). *Pseudomonas* spp. was not affiliated to *Pseudomonas aeruginosa* (Zhang et al. 2017) and was localised to conjunctival tissue by in situ hybridisation implying that the *Pseudomonas* spp. is not a contaminant. Thus, diverse bacteria are localised to different niches of the eye, and this distinct biogeography, where some bacteria are confined to a specific niche whereas others are shared between multiple niches, is indicative that these niches provide a variety of habitats to support communities of varying diversity. Nevertheless, the ocular surface remains a low diverse environment compared to the vagina, gut, skin or oral cavity.

3.6 The Effect of Sampling Depth on the Ocular Surface Microbiome

Apart from sampling the microbiome of various microhabitats of the eye, Dong et al. (2011) studied the ocular surface microbiome after deep and soft sampling. In deep swabbing conjunctival swabs were collected using a dry cotton swab which was applied with pressure to obtain microbes from the depth of the conjunctiva. In soft swabbing a moist cotton swab was applied with minimal pressure to obtain the microbiome from the conjunctival surface (Dong et al. 2011). Higher abundance of *Proteobacteria* (*Bradyrhizobium* spp., *Delftia* spp. and *Sphingomonas* spp.) was detected in the deep sample, whereas by the soft swab, samples which represent the conjunctival/mucosal surface showed preponderance of *Actinobacteria* (*Corynebacterium* spp.) and *Firmicutes* (*Staphylococci* spp.) and a reduction in *Proteobacteria*. Opportunistic pathogens like *Herbaspirillum* spp. and *Rothia* spp. as well as environmental *Rhizobium* spp. and *Leptotrichia* spp. which were either minor or not detected in deep swab samples were also detected using the soft swabbing method. These results reflect that the conjunctival microbiome is stratified with *Staphylococci* spp. and *Corynebacterium* spp. confined to the mucosal layer and *Proteobacteria* spp. to the conjunctival epithelium. Dong et al. (2011) recommended deep swabbing to recover the full diversity of the conjunctival microbiota.

3.7 Ocular Microbiome Changes in the Diseased Eye

Several ocular diseases and changes in their cultivable microbiota have been discussed in the previous chapter. The emphasis was to detect culturable bacteria from an ocular sample retrieved from a patient's eye affected with one of the many ocular diseases so as to be able to diagnose the disease and also recommend therapy. As mentioned earlier success of culturing samples is low, and most of the times only one bacterium gets cultured and identified. But the possibility is that there may be others which may or may not contribute to the disease, and this could be resolved only if they could be identified. It is for this reason that attempts have been made more recently to identify the total microbiome of the diseased eye using the culture-independent microbiome approach (Baim et al. 2019; Kugadas and Gadjeva 2016;

Watane et al. 2019). This has been attempted for several ocular diseases, and the diversity changes are immense.

3.7.1 Ocular Microbiome of Individuals with Fungal Keratitis

Bacterial microbiome of the ocular surface of the eyes of healthy individuals lacking keratitis was dominated by *Thermus* spp., Corynebacterium spp., *Staphylococcus* spp., *Pseudomonas* spp. and *Achromobacter* spp. In contrast, in analysis of individuals affected with FK, the dominant genera included *Psychrobacter* spp., *Thermus* spp., *Pseudomonas* spp., *Achromobacter* spp., *Acinetobacter* spp. and *Caulobacter* spp. These alterations in the conjunctival microbiome were implicated in the pathogenesis of FK (Ge et al. 2019).

3.7.2 Ocular Microbiome of Individuals with Bacterial Keratitis

Ocular bacterial microbiomes were generated from the conjunctival swabs of healthy controls (HC-SW) and conjunctival swabs (BK-SW) and corneal scrapings (BK-CR) of bacterial keratitis (BK) patients using V3–V4 amplicon sequencing and compared for changes in diversity and abundance (Shivaji et al. 2021). In the microbiomes of conjunctival swabs of healthy controls, 149 distinct genera were identified, and consistently *Staphylococcus epidermidis* was shared across all the microbiomes analysed in the HC-SW (Ozkan et al. 2012, 2017; Huang et al. 2016; Willcox 2013; Sharma 2012; Shivaji et al. 2021). *Staphylococcus epidermidis* was also detected in conjunctival swabs of BK individuals but also revealed the presence of other pathogens (*S. aureus*, *Streptococcus pneumoniae* and *P. aeruginosa*). In comparison, to HC-SW and BK-SW, corneal scrapings of BK patients had a greater diversity (Shivaji et al. 2021) and included *Pseudomonas aeruginosa*, *S. pneumoniae*, *S. epidermidis*, *S. aureus*, and other taxa like *S. hominis*, *Streptococcus pyogenes*, *Nocardia* spp., *Corynebacterium amycolatum*, *C. pseudodiphtheriticum*, *C. striatum*, *Pseudomonas fluorescens* and *Klebsiella oxytoca* which have been reported earlier from individuals with keratitis (Egrilmez and Yildirim-Theveny 2020; Bourcier et al. 2003; Ormerod et al. 1987; Turner and Stinson 1965; Van Tyne et al. 2016; Neumann and Sjostrand 1993; Zhai et al. 2018) (Tables 3.2, 3.3, and 3.4). These differences in the microbiomes of the conjunctiva and corneal scrapings between the cohorts could discriminate BK-SW and BK-CR from HC-SW (Tables 3.2 and 3.3). Several genera like *Staphylococcus*, *Streptococcus*, *Paracoccus*, *Gordonia*, *Bosea*, *Capnocytophaga* and *Gemella* were increased in both conjunctival swabs and corneal scrapings of BK patients compared to the

Table 3.2 Median abundance of genera in the ocular microbiomes of conjunctival swabs of healthy controls (HC-SW, $n = 36$) and conjunctival swabs of bacterial keratitis patients (BK-SW, $n = 11$). Only genera exhibiting significant increase or decrease in abundance (BH corrected $p < 0.05$) are included

S. No.	Genus	Median abundance (%)[a] HC-SW	BK-SW	Wilcoxon test-*p* value (BH corrected *p* value <0.05)	Characteristics
Genera decreased in BK-SW					
1	*Enteractinococcus*	13.24	1.52	0.001	Not known
2	*Escherichia-Shigella*	8.7	1.5	0.001	Pathogen
3	*Pseudomonas*	5.71	0.56	0.001	Pathogen
4	*Acinetobacter*	1.15	0.25	0.001	Pathogen
5	*Ruminococcaceae_UCG-005*	0.45	0.09	0.002	Anti-inflammatory
6	*Haemophilus*	0.4	0.02	0.001	Pathogen
7	*Granulicatella*	0.28	0.03	0.001	Pathogen
8	*Rhodanobacter*	0.23	0.01	0.001	Antifungal
9	*Anaerococcus*	0.16	0.09	0.028	Pathogen
10	*Roseobacter* strain HIMB11	0.11	0.02	0.013	Not known
11	*Kocuria*	0.11	0.02	0.002	Pathogen
12	*Thermomonas*	0.11	0	0.001	Not known
13	*Propioniciclava*	0.1	0.02	0.001	Not known
Genera increased in BK-SW					
1	*Streptococcus*	7.33	27.7	0.001	Pathogen
2	*Corynebacterium*	3.91	10.95	0.001	Pathogen
3	*Staphylococcus*	2.89	16.05	0.001	Pathogen
4	*Methylobacterium*	0.15	0.51	0.001	Opportunistic pathogen
5	*Sphingomonas*	0.14	0.95	0.001	Pathogen
6	*Paracoccus*	0.1	1.05	0.001	Pathogen
7	*Bosea*	0.01	5.07	0.002	Pathogen
8	*Capnocytophaga*	0.02	0.55	0.001	Pathogen
9	*Abiotrophia*	0.02	0.48	0.002	Pathogen
10	*Gordonia*	0	0.35	0.025	Pathogen
11	*Roseomonas*	0.02	0.26	0.001	Opportunistic pathogen
12	*Enterobacter*	0.09	0.22	0.002	Pathogen
13	*Klebsiella*	0.07	0.19	0.001	Pathogen
14	*Gemella*	0.01	0.18	0.002	Pathogen
15	*Cutibacterium*	0.05	0.17	0.003	Pathogen
16	*Jeotgalicoccus*	0	0.12	0.002	Not known

Data from Shivaji et al. (2021)

[a] Only genera with a median abundance of >0.1% in at least one group of samples are listed

Table 3.3 Median abundance of genera in the ocular microbiomes of conjunctival swabs of healthy controls (HC-SW, $n = 36$) and corneal scrapings of bacterial keratitis patients (BK-CR, $n = 18$). Only genera exhibiting significant increase or decrease in abundance (BH corrected $p < 0.05$) are included

S. No.	Genus	Median abundance (%)[a]		Wilcoxon test-*p* value (BH corrected *p* value <0.05)	Characteristics
		HC-SW	BK-CR		
Genera decreased in BK-CR					
1	*Enteractinococcus*	13.24	0.62	0.001	Not known
2	*Escherichia-Shigella*	8.7	1.95	0.001	Pathogen
3	*Pseudomonas*	5.71	2.01	0.001	Pathogen
4	*Acinetobacter*	1.15	0.15	0.001	Pathogen
5	*Ornithinimicrobium*	0.67	0	0.001	Pathogen
6	*Ruminococcaceae_UCG-005*	0.45	0	0.001	Anti-inflammatory
7	*Haemophilus*	0.4	0.17	0.001	Pathogen
8	*Peptoniphilus*	0.39	0.01	0.001	Pathogen
9	*Granulicatella*	0.28	0	0.001	Pathogen
10	*Oceanobacillus*	0.22	0.14	0.011	Not known
11	*Anaerococcus*	0.16	0.06	0.001	Pathogen
12	*Sphingomonas*	0.14	0.05	0.001	Pathogen
13	*Brevundimonas*	0.11	0.01	0.001	Opportunistic pathogen
14	Roseobacter strainHIMB11	0.11	0	0.001	Not known
15	*Thermomonas*	0.11	0.04	0.001	Not known
16	*Marmoricola*	0.1	0	0.001	Not known
17	*Propioniciclava*	0.1	0.01	0.001	Not known
Genera increased in BK-CR					
1	*Streptococcus*	7.33	35.39	0.001	Pathogen
2	*Staphylococcus*	2.89	9.84	0.001	Pathogen
3	*Kocuria*	0.11	0.55	0.011	Pathogen
4	*Paracoccus*	0.1	1.32	0.001	Pathogen
5	*Bosea*	0.01	4.95	0.008	Pathogen
6	*Bifidobacterium*	0	0.6	0.001	Probiotic
7	*Gordonia*	0	0.37	0.001	Pathogen
8	*Gemella*	0.01	0.32	0.001	Pathogen
9	*Rosenbergiella*	0	0.28	0.001	Not known
10	*Capnocytophaga*	0.02	0.26	0.001	Pathogen
11	*Rothia*	0.05	0.18	0.014	Pathogen
12	*Virgibacillus*	0	0.15	0.001	Not known
13	*Meiothermus*	0.01	0.13	0.001	Not known
14	*Rhizobacteria*	0	0.12	0.001	Not known
15	*Aeromonas*	0	0.11	0.001	Pathogen

Data from Shivaji et al. (2021)

[a] Only genera with a median abundance >0.1% in at least one group of samples are listed

Table 3.4 Median abundance of genera in the ocular microbiomes of conjunctival swabs of bacterial keratitis patients (BK-SW, $n = 11$) and corneal scrapings of bacterial keratitis patients (BK-CR, $n = 18$). Only genera exhibiting significant increase or decrease in abundance (BH corrected $p < 0.05$) are included

S. No.	Genus	Median abundance (%)[a]		Wilcoxon test—p value (BH corrected p-value <0.05)	Characteristics
		BK-SW	BK-CR		
Genera decreased in BK-CR					
1	*Corynebacterium*	10.95	5.39	0.001	Pathogen
2	*Enteractinococcus*	1.52	0.62	0.025	Not known
3	*Peptoniphilus*	1.33	0.01	0.001	Pathogen
4	*Sphingomonas*	0.95	0.05	0.001	Pathogen
5	*Ornithinimicrobium*	0.71	0	0.001	Pathogen
6	*Methylobacterium*	0.51	0.18	0.003	Opportunistic pathogen
7	*Abiotrophia*	0.48	0	0.01	Pathogen
8	*Oceanobacillus*	0.27	0.14	0.044	Not known
9	*Roseomonas*	0.26	0.01	0.001	Opportunistic pathogen
10	*Acinetobacter*	0.25	0.15	0.04	Pathogen
11	*Enterobacter*	0.22	0.08	0.016	Pathogen
12	*Klebsiella*	0.19	0.06	0.003	Pathogen
13	*Cutibacterium*	0.17	0.04	0.003	Pathogen
14	*Marmoricola*	0.11	0	0.001	Not known
Genera increased in BK-CR					
1	*Paracoccus*	1.05	1.32	0.044	Pathogen
2	*Pseudomonas*	0.56	2.01	0.008	Pathogen
3	*Haemophilus*	0.02	0.17	0.001	Pathogen
4	*Kocuria*	0.02	0.55	0.012	Pathogen
5	*Rhodanobacter*	0.01	0.65	0.001	Antifungal
6	*Virgibacillus*	0.01	0.15	0.019	Not known
7	*Rosenbergiella*	0	0.28	0.016	Not known
8	*Rhizobacteria*	0	0.12	0.001	Not known
9	*Aeromonas*	0	0.11	0.001	Pathogen

Data from Shivaji et al. (2021)

[a] Differentially abundant genera having a median abundance >0.1% in at least one group of samples are listed

microbiomes of conjunctival swabs of HC (Tables 3.2 and 3.3), and concomitantly abundance of 11 genera decreased in both BK-SW and BK-CR compared to HC-SW microbiomes (compare Tables 3.2 and 3.3). It was suggested that the diversity and abundance in the conjunctiva and cornea of these specific genera in BK-SW and BK-CR compared to HC-SW are indicative that these genera are associated with bacterial keratitis. Several of these genera implicated in BK have earlier been identified in the lids or conjunctivae of patients with dry eye diseases (Graham et al. 2007), blepharitis (Ta et al. 2003; Groden et al. 1991), endophthalmitis, contact lens wearers (Shin et al. 2016), fungal Keratitis (Ge et al. 2019) and conjunctival scarring (Zhou et al. 2014). The bacterial microbiomes in the conjunctival swabs and corneal scrapings of the BK individuals were also different (Table 3.4). Beta diversity analysis of the bacterial microbiomes of HC-SW, BK-SW and BK-CR by heat map analysis (Fig. 3.5a) and NMDS plots (Fig. 3.5b) clearly separated the three cohorts into distinct microbiomes. Interaction networks further discriminated the HC-SW, BK-SW and BK-CR microbiomes based on the hub genera which showed that the negative interactions were increased in BK-CR and BK-SW compared to HC-SW. KEGG pathway analyses also clearly demonstrated dysbiotic changes in the bacterial microbiomes of BK-SW and BK-CR compared to HC-SW (Shivaji et al. 2021). Overall, the abundance of pathogenic bacteria was increased in the conjunctiva of BK patients compared to HC, and this may be related to causing or exacerbating ocular surface inflammation (Shivaji et al. 2021). In an interesting study, Cavuoto et al. (2021) studied observed that the α-diversity indices and principal coordinate analysis plots between the two eyes of individuals with unilateral keratitis were not different but differed with respect to the control individuals. Further, *Proteobacteria*, *Actinobacteria* and *Firmicutes* were tenfold greater in the eyes with keratitis and non-keratitis eye compared with controls. Several genera contributed to this difference and included *Micrococcus* spp. (100-fold greater), *Pseudomonas* spp. (700-fold higher abundance), *Ralstonia* spp. (5.7-fold lower) and *Cutibacterium* spp. (tenfold greater) in eyes with keratitis compared with controls. The genera *Methylobacterium* spp. and Delftia spp. also showed significant decrease in abundance, whereas *Staphylococcus* spp. showed decrease in keratitis compared to the controls. When the non-keratitis eye was compared to the controls, *Cutibacterium* spp. and *Micrococcus* spp. were 13- and 100-fold higher in abundance, whereas the abundance of *Ralstonia* spp. was 15-fold lower (Cavuoto et al. 2021). Overall, in individuals with unilateral keratitis, potential pathogens were increased in abundance, whereas commensal organisms decreased in abundance in both eyes compared with controls. Animal studies confirmed that ocular bacteria like *P. aeruginosa* could induce keratitis in mice (Kugadas et al. 2016; St Leger et al. 2017). Cavuoto et al. (2021) suggested that the ocular surface microbiome could be used as a diagnostic tool to identify individuals who develop keratitis and suffer worse visual morbidity. Further, increase in *Pseudomonas* spp. and a decrease in *Ralstonia* spp. abundance could serve as markers of keratitis.

3.7.3 Ocular Surface Microbiome of Individuals with Endophthalmitis

In endophthalmitis patients the vitreous was used for microbiome analysis, and the results indicated predominantly *Enterococcus faecalis*, *Staphylococcus epidermidis*,

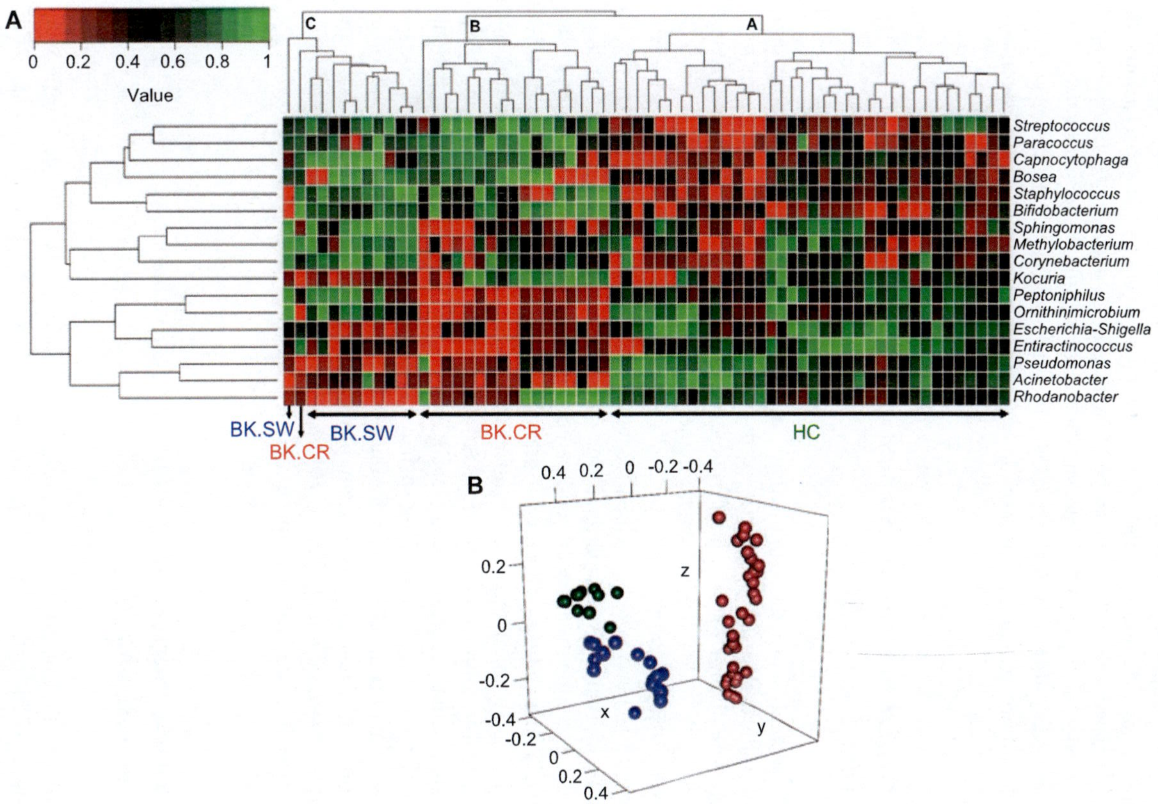

Fig. 3.5 Beta diversity analysis discriminating the ocular microbiomes by heat map analysis of 17 differentially abundant bacterial genera (**a**) and by NMDS plots based on Bray-Curtis dissimilarity of bacterial OTU abundance (**b**) in the conjunctival swabs of healthy controls (HC-SW, $n = 36$), in the conjunctival swabs of bacterial keratitis patients (BK-SW, $n = 11$) and in the corneal scraping of bacterial keratitis patients (BK-CR, $n = 18$). In (**b**) the HC-SW (green), BK-SW (blue) and BK-CR (purple) microbiomes varied significantly (PERMANOVA, $p = 0.001$). (Data from Shivaji et al. 2021)

Staphylococcus hominis, *Micrococcus luteus*, *Serratia marcescens*, *E. coli*, *Paenibacillus* spp., *Caulobacter* spp., *Moraxella catarrhalis* and *Comamonas testosteroni* (Kirstahler et al. 2018). BRiSK analysis of the vitreous of endophthalmitis patients confirmed the presence of *Pseudomonas* spp., *Staphylococcus* spp., *Streptococcus* spp. and *Moraxella* spp. (Lee et al. 2015a, b). Deshmukh et al. (2019) also confirmed the presence of *Pseudomonas* spp., *Staphylococcus* spp. and *Streptococcus* spp. in the vitreous but also identified *Acinetobacter* spp., *Gemella* spp. and *Haemophilus* spp. Thus, the microbiota analysis of the vitreous of endophthalmitis patients could provide leads for infectious disease management.

3.7.4 Ocular Surface Microbiome of Individuals with Conjunctivitis

Microbiome analysis of the conjunctival sacs of individuals with infectious conjunctivitis indicated that the bacteria were affiliated to the phyla *Proteobacteria* (26%), *Actinobacteria* (31%) and *Firmicutes* (35%) and included several genera (Mohamed et al. 2020). Seven genera, namely, *Staphylococcus* spp. (7%), *Streptococcus* spp. (10%), *Cutibacterium* spp. (10%), *Nocardioides* spp. (7%), *Corynebacterium* spp. (5%), *Anoxybacillus* spp. (5%) and *Acinetobacter* spp. (5%) were shared between the individuals with conjunctivitis (Mohamed et al. 2020). But, when cases with bilateral conjunctivitis were compared with individuals with unilateral conjunctivitis, statistically significant difference was observed at the genera level. Microbiome analysis of the ocular and nasopharyngeal swabs from controls and paediatric allergic rhinoconjunctivitis (ARC) patients indicated that the alpha diversity was significantly increased in the nasopharyngeal microbiomes of ARC patients than healthy controls ($p < 0.01$), but such an increase was not observed in the ocular microbiome (Yau et al. 2019). Further the ocular and nasopharyngeal microbiomes were significantly distinct and certain phyla (*Cyanobacteria*, *Bacteroidetes* and *Deinococcus-Thermus*) and genera (*Moraxella* and *Dolosigranulum*) and were detected in healthy controls but not in the ARC subjects (Yau et al. 2019).

3.7.5 Ocular Surface Microbiome of Individuals with Blepharitis

Microbiomes of tears samples and eyelashes of individuals with blepharitis were significantly different. The relative abundance of *Staphylococcus* spp., *Enhydrobacter* spp., *Streptophyta* spp. and *Corynebacterium* spp. was increased in blepharitis subjects indicative of the infestations of dust, soil and pollen grains (Lee et al. 2012). Decrease in abundance of *Propionibacterium* spp. in blepharitis subjects was also obvious (Lee et al. 2012).

3.7.6 Ocular Surface Microbiome of Individuals with Retinitis

Vitreous samples were analysed for microbiome analysis from post-fever retinitis (PFR) individuals and compared with control microbiomes. In the vitreous of PFR individuals, 18 genera were differentially abundant compared to control vitreous microbiomes, and the predominant genera were *Acinetobacter* spp., *Enterobacter* spp., *Clostridium* spp., *Lachnoclostridium* spp. and *Klebsiella* spp. (Arunasri et al. 2020a). Two genera, namely, *Tannerella* spp. and *Pimetobacter* spp. were enriched in PFR individuals, and both these discriminative genera were inferred to have antimicrobial, probiotic and anti-inflammatory functions (Arunasri et al. 2020a). The mycobiome of the vitreous of PFR patients also showed significant changes compared to the mycobiomes of healthy controls (Arunasri et al. 2020b).

3.7.7 Ocular Surface Microbiome of Individuals with Meibomian Gland Dysfunction

Microbiome analysis of the ocular surface microbiomes of gender- and aged-matched meibomian gland dysfunction (MGD) patients and corresponding controls indicated that the abundance of *Staphylococcus* spp. (20.71% vs. 7.88%) and *Sphingomonas* spp. (5.73% vs. 0.79%) was significantly higher in MGD patients, while the abundance of *Corynebacterium* (20.22% vs. 46.43%) was significantly lower (Dong et al. 2019). Watters et al. (2017) observed lowest abundance of *Staphylococcus* spp., and higher frequency of *Pseudomonas acnes* in individuals with moderate to severe MGD. Zhao et al. (2020) monitored the microbiome from the meibum, conjunctiva and eyelid skin in MGD patients and observed that the abundance of *Campylobacter jejuni*, *C. coli* and *Enterococcus faecium* which are pathogens was increased in the meibum of MGD patients; these genera were absent from the unaffected individuals. These pathogens *C. jejuni*, *C. coli* and *Enterococcus faecium* are known to cause eye infections (Janssen et al. 2008; Kuriyan et al. 2014). It was also observed that the microbiomes of the meibum of MGD patients displayed&spi2;stronger immune evasive virulence, compared to the controls (Zhao et al. 2020). It was also observed that *Corynebacterium* spp., *Pseudomonas acnes*, *Pseudomonas* spp. and bacteria affiliated to family *Neisseriaceae* were the most abundant species in the meibum of healthy controls. Further, the microbiomes in the MGD patients was age-dependent (Suzuki et al. 2020) but was not influenced by gender of the subjects. It was also observed that the microbiomes of meibum, conjunctival-sac and skin were distinct from each other (Suzuki et al. 2020).

3.7.8 Ocular Surface Microbiome of Individuals with Dry Eye Disease

Microbiome analysis indicated a distinct difference in the ocular surface microbiome in dry eye disease patients compared to controls. Graham et al. (2007) reported a greater abundance of *Rhodococcus erythropolis*, coagulase-negative *Staphylococci* spp., *Propionibacterium* spp., *Erwinia* spp., *Corynebacterium* spp. and K*lebsiella oxytoca* in dry eye patients. Several investigations also observed a nearly complete substitution of *Staphylococcus* spp. with *Pseudomonas* spp. in patients with dry eye (Willis et al. 2020; Doan et al. 2016; Graham et al. 2007). *Corynebacterium* spp. is one of the most frequently reported organisms in the eye microbiomes of DED patients and may reflect their role in DED (Zilliox et al. 2020). It was demonstrated that *Corynebacterium* has an inflammatory role in the host response to *Candida* spp. and *Pseudomonas* spp. (St Leger et al. 2017). In addition, *Acinetobacter* spp., *Pseudomonas* spp., *Pseudomonas acnes*, *Rhodococcus* spp. and *Bacillus* spp. were less common and inconsistently detected. Several other studies (Graham et al. 2007; Suto et al. 2012; Hori et al. 2008; Willis et al. 2020; Watters et al. 2017; Zhang et al. 2013) agree with Zilliox et al. (2020) with respect to the bacterial diversity in the eyes of DED patients.

3.7.9 Ocular Surface Microbiome of Individuals with Stevens-Johnson Syndrome

Pathogenic microorganisms including *Acinetobacter* spp., *Staphylococcus* spp., *Pseudomonas* spp. and *Streptococcus* spp. are abundant in the OSM of individuals with Stevens-Johnson syndrome (SJS) (Kittipibul et al. 2020; Moeller et al. 2005). In addition, it was observed that in 80% SJS patients using bandage contact lenses, *Staphylococcus* spp. was the dominant bacterium (Zilliox et al. 2020). Culture techniques have also indicated that *Corynebacterium* spp. and *Staphylococcus* spp. are dominant in SJS eyes (Venugopal et al. 2016; Frizon et al. 2014). Smaller amounts of *Prevotella* spp., *Lactobacillus* spp., *Fusobacterium* spp. and *Enterobacteriaceae* were also observed (Zilliox et al. 2020). This increase in abundance in pathogenic bacteria probably supports that chronic inflammation, and opportunistic infections are common in SJS group (Kittipibul et al. 2020). Altered microbiome in SJS individuals may also be influenced by the innate immunity status of the host (Ueta and Kinoshita 2012). Zilliox et al. (2020) observed significantly low α-diversity in SJS patients' eyes compared to the healthy subjects and suggested that it may indicate a role in disease pathogenesis.

3.7.10 Ocular Surface Microbiome of Individuals with Sjogren's Syndrome

A core microbiome represented by bacteria phylogenetically affiliated to the phyla *Firmicutes*, *Bacteroidetes*, *Actinobacteria* and *Proteobacteria* were identified in the inferior conjunctiva of rosacea patients and Sjogren's syndrome (SS) patients and compared with the microbiomes of control subjects (de Paiva et al. 2016). No significant differences were observed between the three cohorts when the composition or richness of the OSM was compared. But, when the tongue and stool microbiomes were compared, differences were obvious between the control and SS group individuals (de Paiva et al. 2016; Mandl et al. 2017). Subsequent studies also confirmed dysbiosis in the gut microbiome in Sjögren's syndrome patients compared to individuals with dry eye syndrome and controls (Mandl et al. 2017; Moon et al. 2020). Compositional changes in the gut microbiome of dry eye patients were somewhere between Sjögren's syndrome and controls (Moon et al. 2020).

3.7.11 Ocular Surface Microbiome of Individuals with Ocular Graft Versus Host Disease

Several bacteria (aerobic Gram-positive cocci and rods, *Streptococcus* spp., *Corynebacterium* spp., *Staphylococcus* spp., *Haemophilus influenzae*, *Propionibacterium* spp.) were observed in graft versus host disease (GVHD) patients using the cultivable approach (Shimizu et al. 2019). Microbiome analysis of the ocular surface of patients with ocular graft versus host disease (oGVHD) indicated that they had fewer bacterial sequences compared to the controls (Zilliox et al. 2020). Reads were detected only in 6/14 eyes analysed, and *Pasteurellaceae*, *Enterobacteriaceae* spp. and *Selenomonas* spp. were predominant in one eye and were absent in the other eyes. A good proportion of the oGVHD eyes (83%) also shared *Lactobacillus/Streptococcus* spp., with the healthy eyes, but the abundance in the oGVHD eyes was decreased (Shimizu et al. 2019). Zilliox et al. (2020) also observed decreased relative abundances of *Corynebacterium* spp. in oGVHD patients compared to DED patients, and the α-diversity for the healthy and diseased eyes was higher compared to the other diseases (chronic SJS, oGVHD, lax eyelid syndrome and DED).

3.7.12 Ocular Surface Microbiome of Individuals with Diabetes Mellitus

A good proportion of individuals with diabetes mellitus (DM) have ocular complications associated with DM such as cataract and diabetic retinopathy which cause visual impairment (Lee et al. 2015a, b; Sayin et al. 2015). Results on the ocular flora in DM patients are ambiguous. The bacterial detection rate in relation to standard diabetes parameters such as haemoglobin A1C levels, diabetic retinopathy or glycosuria was not significantly different (Suto et al. 2012; Moreno et al. 2014; Adam et al. 2015); but a few studies reported conflicting reports with higher rate of culturability of bacteria from diabetic patients (Nahar et al. 2014; Adam et al. 2015). Ham et al. (2018) observed that in DM patients, the overall bacterial diversity and phylogenetic diversity of the upper bulbar conjunctiva were more diverse than the lower bulbar conjunctiva and gender did not influence the microbiome (Ham et al. 2018). But, β-diversity analysis was distinctly different in DM patients compared to the controls, and this was not influenced by the gender of the individuals. At the phyla level *Firmicutes* were more abundant in the control group, whereas *Proteobacteria* were more abundant in the DM group (Ham et al. 2018). Li et al. (2019) confirmed that phyla *Proteobacteria* were increased in the diabetic group; in addition, bacteria affiliated to *Cyanobacteria* and *Firmicutes* were increased in the control group. Further, abundance differences were also recorded at the genus level with *Pseudomonas* spp., *Staphylococcus* spp., *Streptophyta* spp., *Acinetobacter* spp., *Corynebacterium* spp., *Novosphingobium* spp. and *Neisseriaceae* which were relatively more abundant in the controls compared to the DM group. Further, at the genus level, *Rheinheimera* spp., *Burkholderia* spp., *Acinetobacter* spp. and *Micrococcus* spp. were overrepresented in the diabetic group, and *Streptophyta* spp., *Staphylococcus* spp., *Novosphingobium* spp. and *Bradyrhizobiaceae* spp. were decreased in the healthy subjects. *Streptophyta* spp. and *Bradyrhizobiaceae* in OSM could be environmental contaminants. Among these opportunistic pathogens were also identified and included *Pseudomonas* spp., *Staphylococcus* spp. and *Corynebacterium* spp. (Dong et al. 2011). It is also of interest to note that *Acinetobacter* spp., *Sphingomonas* spp. and *Ralstonia* spp., which are predominant in diabetic patients, were previously detected in endophthalmitis and keratitis patients (Dong et al. 2011). A subsequent study also showed significantly increased abundance of *Pseudomonas* spp. in the DM group (Li et al. 2019). Thus, several studies confirmed that the OSM was significantly altered in diabetic patients compared to controls both at the phyla and genera level (Ham et al. 2018; Shin et al. 2016; Dong et al. 2011; Huang et al. 2016; Lee et al. 2012), and these changes in the OSM in DM patients could be attributed to the compromised immune function in DM patients. Immunity changes affect both the commensal and pathogenic bacteria which could have devastating effects and facilitate ocular infections such as bacterial keratitis or endophthalmitis. In a recent study, Das et al. (2021) demonstrated dysbiosis in the gut microbiomes in people with DM and diabetic retinopathy (DR) compared to control individuals.

3.7.13 Ocular Surface Microbiome of Individuals with Contact Lens Use

Results on the OSM of contact lens (CL) users versus non-lens users have not been very consistent. Shin et al. (2016) reported that the OSM in lens wearers showed higher abundances of *Pseudomonas* spp., *Methylobacterium* spp., *Lactobacillus* spp. and *Acinetobacter* spp. and lower abundances of *Corynebacterium* spp., *Staphylococcus* spp., *Streptococcus* spp. and *Haemophilus* spp., indicating that the diversity was similar to that of the skin under the eye of non-CL wearers. Culture-dependent methods (Erdogan et al. 2002; Larkin and Leeming 1991; Stapleton et al. 1997) also consistently showed the presence of *Propionibacterium* spp., *Bacillus* spp., *Corynebacterium* spp., *Micrococcus* spp., *Staphylococcus* spp., *Rothia* spp. and *Pseudomonas* spp. in cultures from contact lenses (Willcox 2013). It is possible that the CL transfers bacteria from the skin to the eye alternatively or the CL alters the ocular surface and facilitates the migration of bacteria from the skin to the eye (Shin et al. 2016; Boost et al. 2017; Gomes et al. 2020). The possibility that the lens may be the carrier of the microbes to the eye was also addressed by studies investigating the bacterial DNA load on the CL belonging to healthy wearers. The results indicated that when peroxide was used for disinfecting the CL, majority of the lenses (80%) were negative for bacterial DNA. But, when a multipurpose solution was used as the disinfectant, diversity and abundance were increased with *Streptococcus* spp., *Haemophilus* spp. and *Corynebacterium* spp. being more abundant (Retuerto et al. 2019; Delbeke et al. 2021). It was also observed that in CL wearers the distribution of Gram-positive bacteria showed a higher load on the lower lid margins compared to the upper bulbar conjunctiva (Cavuoto et al. 2019a, b; Sankaridurg et al. 2009). Further, *H. influenzae* which caused contact lens-associated red eye was more commonly isolated from patients who used extended-wear CL indicating that the organism isolated is probably dependent on the type of lens used (Sankaridurg et al. 1996). Zhang et al. (2017) did not detect differences in microbial diversity between contact lens wearers and non-wearers. But microbial diversity varied between soft contact lens (SCL) and orthokeratology lens (OKL) users. The abundance of *Bacillus* spp., *Lactobacillus* spp. and *Tatumella* spp. was less in OKL wearers than in non-wearers. In SCL wearers, *Delftia* abundance decreased, whereas in *Elizabethkingia* spp., levels increased. Cumulatively, the OSM is altered with type of CL wear and could be a risk factor for eye infection (Cavuoto et al. 2019a, b) and inflammation. Contact lenses are often associated with microbial contamination; therefore, the possibility is that the microbes that colonise the lenses like *Pseudomonas* spp. could lead to infection and inflammation (Willcox 2007). This is relevant from the point of view that the most common cause of microbial keratitis remains the Gram-negative bacterium *P. aeruginosa*.

3.7.14 Ocular Surface Microbiome of Individuals with Floppy Eyelid Syndrome

Severe cases of floppy eyelid syndrome (FES) can result in corneal inflammation, ulcerations and neovascularisation (Culbertson and Tseng 1994). In patients with lax eyelid syndrome (LES), which is a type of FES, the α-diversity was significantly reduced in the LES eyes compared to the healthy eye (Zilliox et al. 2020). Further the eyes of LES patients were dominated by *Corynebacterium* spp., and the other bacteria that were detected in some of the eyes include *Hydrotalea* spp., *Staphylococcus* spp., *Ralstonia* spp., *Rothia* spp. and *Clostridium XI* that were reasonably abundant (Zilliox et al. 2020). A comparison of the OSM of the eyes of LES and DED patients indicated that *Corynebacterium* spp. was the predominant bacterium (Zilliox et al. 2020).

3.7.15 Ocular Surface Microbiome of Individuals with Trachoma

Trachoma is an infectious disease of the eye, and the causative agent is *Chlamydia trachomatis* (Evans et al. 2019). The symptoms of trachoma include pain in the eyes, scarring of the eyelid, breakdown of the cornea, and eventual blindness. *Chlamydia trachomatis* spreads by either direct or indirect contact with the eye of an affected person or nose. Apart from personal hygiene which helps to prevent the spread, the treatment options include topical tetracycline and oral azithromycin. In 44 countries across Asia, Africa and South and Central America (Fenwick 2012), trachoma is a public health problem and affects about 80 million people. Analysis of the conjunctival microbiome of 220 residents of the Gambia found that in 80% of the healthy controls without trachoma, the bacterial diversity in the conjunctiva was predominantly represented by *Streptococcus* spp., *Staphylococcus* spp., *Propionibacterium* spp., *Ralstonia* spp., *Corynebacterium* spp. and *Bacillus* spp. In contrast, Zhou et al. (2014) reported a decreased diversity and an increased abundance of *Streptococcus* spp. and *Corynebacterium* spp. in the eyes of the trachoma patients.

3.8 Conclusions

The culture-dependent identification of ocular microbes from clinical samples is considered as a gold standard for the identification of the microbes and for subsequent diagnostics and therapy. This age old and time-tried method is based on phenotypic and a few biochemical characteristics to identify a bacterium up to the genus level. Further, this method allows only for the identification of the small minuscule fraction of bacteria (>1%) that are culturable. But, with the advent of amplification of phylogenetic markers and sequencing of the amplicons and in addition metagenomic sequencing approaches, the gross microbial diversity is now identified which includes not only the bacteria identified by culture-dependent methods but also several previously un-described bacteria indicating that the OSM is diverse. This also raises the question as to whether the observed effects on the eye are due to a single organism or group of organisms which act synergistically, a question that needs to be addressed. The NGS approach, apart from providing a comprehensive description of the OSM at the phylum and genera levels, also provides convincing evidence that the OSM is influenced by several confounding factors like sex, age, geographic distribution, microhabitat of the eye, etc. Now that it is accepted that the ocular surface has its own paucibacterial microbiome which differs with diseased state of the eye, it is not far-fetched to predict that with time, analysing the OSM in an ocular microbiology laboratory using NGS (Borroni et al. 2019) would become routine at least when the sample does not yield any growing culture.

References

Abelson MB, Lane K, Slocum C (2015) The secrets of ocular microbiomes. Rev Ophthalmol

Achtman M, Wagner M (2008) Microbial diversity and the genetic nature of microbial species. Nat Rev Microbiol 6:431–440. https://doi.org/10.1038/nrmicro1872

Adam M, Balci M, Bayhan HA et al (2015) Conjunctival flora in diabetic and nondiabetic individuals. Turk J Ophthalmol 45:193–196. https://doi.org/10.4274/tjo.33230

Arunasri K, Mahesh M, Sai Prashanthi G et al (2020a) Comparison of the vitreous fluid bacterial microbiomes between individuals with post fever retinitis and healthy controls. Microorganisms 8:751. https://doi.org/10.3390/microorganisms8050751

Arunasri K, Mahesh M, Sai Prashanthi G et al (2020b) Mycobiome changes in the vitreous of post fever retinitis patients. PLoS One 15:e0242138. https://doi.org/10.1371/journal.pone.0242138

Baim AD, Movahedan A, Farooq AV et al (2019) The microbiome and ophthalmic disease. Exp Biol Med (Maywood) 244:419–429. https://doi.org/10.1177/1535370218813616

Bonomo RA, Szabo D (2006) Mechanisms of multidrug resistance in *Acinetobacter* species and *Pseudomonas aeruginosa*. Clin Infect Dis 43(Suppl 2):S49–S56. https://doi.org/10.1086/504477

Boost M, Cho P, Wang Z (2017) Disturbing the balance: effect of contact lens use on the ocular proteome and microbiome. Clin Exp Optom 100:459–472. https://doi.org/10.1111/cxo.12582

Borroni D, Romano V, Kaye SB et al (2019) Metagenomics in ophthalmology: current findings and future prospectives. BMJ Open Ophthalmol 4:e000248. https://doi.org/10.1136/bmjophth-2018-000248

Bourcier T, Thomas F, Borderie V et al (2003) Bacterial keratitis: predisposing factors, clinical and microbiological review of 300 cases. Br J Ophthalmol 87:834–838. https://doi.org/10.1136/bjo.87.7.834

Bukin YS, Galachyants YP, Morozov IV et al (2019) The effect of 16S rRNA region choice on bacterial community metabarcoding results. Sci Data 6:190007. https://doi.org/10.1038/sdata.2019.7

Byrd AL, Belkaid Y, Segre JA (2018) The human skin microbiome. Nat Rev Microbiol 16:143–155. https://doi.org/10.1038/nrmicro.2017.157

Caporaso JG, Lauber CL, Walters WA et al (2012) Ultra-high-throughput microbial community analysis on the Illumina HiSeq and MiSeq platforms. ISME J 6:1621–1624. https://doi.org/10.1038/ismej.2012.8

Cavuoto KM, Banerjee S, Miller D et al (2018) Composition and comparison of the ocular surface microbiome in infants and older children. Transl Vis Sci Technol 7:16. https://doi.org/10.1167/tvst.7.6.16

Cavuoto KM, Banerjee S, Galor A (2019a) Relationship between the microbiome and ocular health. Ocul Surf 17:384–392. https://doi.org/10.1016/j.jtos.2019.05.006

Cavuoto KM, Stradiotto AC, Galor A (2019b) Role of the ocular surface microbiome in allergic disease. Curr Opin Allergy Clin Immunol 19:482–487. https://doi.org/10.1097/ACI.0000000000000559

Cavuoto KM, Galor A, Banerjee S (2021) Ocular surface microbiome alterations are found in both eyes of individuals with unilateral infectious keratitis. Transl Vis Sci Technol 10:19. https://doi.org/10.1167/tvst.10.2.19

Cogen AL, Nizet V, Gallo RL (2008) Skin microbiota: a source of disease or defence? Br J Dermatol 158:442–455. https://doi.org/10.1111/j.1365-2133.2008.08437.x

Culbertson WW, Tseng SC (1994) Corneal disorders in floppy eyelid syndrome. Cornea 13:33–42. https://doi.org/10.1097/00003226-199401000-00007

Das T, Jayasudha R, Chakravarthy S et al (2021) Alterations in the gut bacterial microbiome in people with type 2 diabetes mellitus and diabetic retinopathy. Sci Rep 11:2738. https://doi.org/10.1038/s41598-021-82538-0

Dekio I, Hayashi H, Sakamoto M et al (2005) Detection of potentially novel bacterial components of the human skin microbiota using culture-independent molecular profiling. J Med Microbiol 54:1231–1238. https://doi.org/10.1099/jmm.0.46075-0

Delbeke H, Younas S, Casteels I et al (2021) Current knowledge on the human eye microbiome: a systematic review of available amplicon and metagenomic sequencing data. Acta Ophthalmol 99:16–25. https://doi.org/10.1111/aos.14508

Deming JW, Baross JA (2000) Survival, dormancy, and nonculturable cells in extreme deep-sea environments. In: Colwell RR, Grimes DJ (eds) Nonculturable microorganisms in the environment. Springer, Boston, MA, pp 147–197. https://doi.org/10.1007/978-1-4757-0271-2_10

Deshmukh D, Joseph J, Chakrabarti M et al (2019) New insights into culture negative endophthalmitis by unbiased next generation sequencing. Sci Rep 9:844. https://doi.org/10.1038/s41598-018-37502-w

Doan T, Akileswaran L, Andersen D et al (2016) Paucibacterial microbiome and resident DNA virome of the healthy conjunctiva. Invest Ophthalmol Vis Sci 57:5116–5126. https://doi.org/10.1167/iovs.16-19803

Dong Q, Brulc JM, Iovieno A et al (2011) Diversity of bacteria at healthy human conjunctiva. Invest Ophthalmol Vis Sci 52:5408–5413. https://doi.org/10.1167/iovs.10-6939

Dong X, Wang Y, Wang W et al (2019) Composition and diversity of bacterial community on the ocular surface of patients with meibomian gland dysfunction. Invest Ophthalmol Vis Sci 60: 4774–4783. https://doi.org/10.1167/iovs.19-27719

Doster E, Lakin SM, Dean CJ et al (2020) MEGARes 2.0: a database for classification of antimicrobial drug, biocide and metal resistance determinants in metagenomic sequence data. Nucleic Acids Res 48:D561–D569. https://doi.org/10.1093/nar/gkz1010

Egrilmez S, Yildirim-Theveny S (2020) Treatment-resistant bacterial keratitis: challenges and solutions. Clin Ophthalmol 14:287–297. https://doi.org/10.2147/OPTH.S181997

Erdogan H, Kemal M, Toker MI et al (2002) Effect of frequent-replacement contact lenses on normal conjunctival flora. CLAO J 28:94–95

Evans JR, Solomon AW, Kumar R et al (2019) Antibiotics for trachoma. Cochrane Database Syst Rev 9:CD001860. https://doi.org/10.1002/14651858.CD001860.pub4

Fenwick A (2012) The global burden of neglected tropical diseases. Public Health 126:233–236. https://doi.org/10.1016/j.puhe.2011.11.015

Frizon L, Araujo MC, Andrade L et al (2014) Evaluation of conjunctival bacterial flora in patients with Stevens-Johnson Syndrome. Clinics (Sao Paulo) 69:168–172. https://doi.org/10.6061/clinics/2014(03)04

Gallon P, Parekh M, Ferrari S et al (2019) Metagenomics in ophthalmology: hypothesis or real prospective? Biotechnol Rep (Amst) 23:e00355. https://doi.org/10.1016/j.btre.2019.e00355

Ge C, Wei C, Yang BX et al (2019) Conjunctival microbiome changes associated with fungal keratitis: metagenomic analysis. Int J Ophthalmol 12:194–200. https://doi.org/10.18240/ijo.2019.02.02

Gomes JAP, Frizon L, Demeda VF (2020) Ocular surface microbiome in health and disease. Asia Pac J Ophthalmol (Phila) 9:505–511. https://doi.org/10.1097/APO.0000000000000330

Graham JE, Moore JE, Jiru X et al (2007) Ocular pathogen or commensal: a PCR-based study of surface bacterial flora in normal and dry eyes. Invest Ophthalmol Vis Sci 48:5616–5623. https://doi.org/10.1167/iovs.07-0588

Grice EA, Kong HH, Conlan S et al (2009) Topographical and temporal diversity of the human skin microbiome. Science 324:1190–1192. https://doi.org/10.1126/science.1171700

Groden LR, Murphy B, Rodnite J et al (1991) Lid flora in blepharitis. Cornea 10:50–53

Ham B, Hwang HB, Jung SH et al (2018) Distribution and diversity of ocular microbial communities in diabetic patients compared with healthy subjects. Curr Eye Res 43:314–324. https://doi.org/10.1080/02713683.2017.1406528

Hamady M, Knight R (2009) Microbial community profiling for human microbiome projects: tools, techniques, and challenges. Genome Res 19:1141–1152. https://doi.org/10.1101/gr.085464.108

Harwani D (2012) The great plate count anomaly and the unculturable bacteria. Int J Sci Res 2:350–351. https://doi.org/10.15373/22778179/sep2013/122

Hong J, Chen J, Sun X et al (2012) Paediatric bacterial keratitis cases in Shanghai: microbiological profile, antibiotic susceptibility and visual outcomes. Eye (Lond) 26:1571–1578. https://doi.org/10.1038/eye.2012.210

Hori Y, Maeda N, Sakamoto M et al (2008) Bacteriologic profile of the conjunctiva in the patients with dry eye. Am J Ophthalmol 146:729–734. https://doi.org/10.1016/j.ajo.2008.06.003

Huang Y, Yang B, Li W (2016) Defining the normal core microbiome of conjunctival microbial communities. Clin Microbiol Infect 22:643.e647–643.e612. https://doi.org/10.1016/j.cmi.2016.04.008

Hugenholtz P, Pitulle C, Hershberger KL et al (1998) Novel division level bacterial diversity in a Yellowstone hot spring. J Bacteriol 180:366–376. https://doi.org/10.1128/JB.180.2.366-376.1998

Janssen R, Krogfelt KA, Cawthraw SA et al (2008) Host-pathogen interactions in Campylobacter infections: the host perspective. Clin Microbiol Rev 21:505–518. https://doi.org/10.1128/CMR.00055-07

Keller M, Zengler K (2004) Tapping into microbial diversity. Nat Rev Microbiol 2:141–150. https://doi.org/10.1038/nrmicro819

Kirstahler P, Bjerrum SS, Friis-Moller A et al (2018) Genomics-based identification of microorganisms in human ocular body fluid. Sci Rep 8:4126. https://doi.org/10.1038/s41598-018-22416-4

Kittipibul T, Puangsricharern V, Chatsuwan T (2020) Comparison of the ocular microbiome between chronic Stevens-Johnson syndrome patients and healthy subjects. Sci Rep 10:4353. https://doi.org/10.1038/s41598-020-60794-w

Kopke B, Wilms R, Engelen B et al (2005) Microbial diversity in coastal subsurface sediments: a cultivation approach using various electron acceptors and substrate gradients. Appl Environ Microbiol 71:7819–7830. https://doi.org/10.1128/AEM.71.12.7819-7830.2005

Kugadas A, Gadjeva M (2016) Impact of microbiome on ocular health. Ocul Surf 14:342–349. https://doi.org/10.1016/j.jtos.2016.04.004

Kugadas A, Christiansen SH, Sankaranarayanan S et al (2016) Impact of microbiota on resistance to ocular *Pseudomonas aeruginosa*-induced keratitis. PLoS Pathog 12:e1005855. https://doi.org/10.1371/journal.ppat.1005855

Kuriyan AE, Sridhar J, Flynn HW Jr et al (2014) Endophthalmitis caused by Enterococcus faecalis: clinical features, antibiotic sensitivities, and outcomes. Am J Ophthalmol 158:1018–1023. https://doi.org/10.1016/j.ajo.2014.07.038

Larkin DF, Leeming JP (1991) Quantitative alterations of the commensal eye bacteria in contact lens wear. Eye (Lond) 5(Pt 1):70–74. https://doi.org/10.1038/eye.1991.14

Lederberg J, McCray A (2001) 'Ome sweet 'omics - a genealogical treasury of words. Genealogical treasury of words. Scientist 15:8

Lee SH, Oh DH, Jung JY et al (2012) Comparative ocular microbial communities in humans with and without blepharitis. Invest Ophthalmol Vis Sci 53:5585–5593. https://doi.org/10.1167/iovs.12-9922

Lee AY, Akileswaran L, Tibbetts MD et al (2015a) Identification of torque teno virus in culture-negative endophthalmitis by representational deep DNA sequencing. Ophthalmology 122:524–530. https://doi.org/10.1016/j.ophtha.2014.09.001

Lee R, Wong TY, Sabanayagam C (2015b) Epidemiology of diabetic retinopathy, diabetic macular edema and related vision loss. Eye Vis (Lond) 2:17. https://doi.org/10.1186/s40662-015-0026-2

Li L, Mendis N, Trigui H et al (2014) The importance of the viable but non-culturable state in human bacterial pathogens. Front Microbiol 5:258. https://doi.org/10.3389/fmicb.2014.00258

Li D, Liu CM, Luo R et al (2015) MEGAHIT: an ultra-fast single-node solution for large and complex metagenomics assembly via succinct de Bruijn graph. Bioinformatics 31:1674–1676. https://doi.org/10.1093/bioinformatics/btv033

Li S, Yi G, Peng H et al (2019) How ocular surface microbiota debuts in type 2 diabetes mellitus. Front Cell Infect Microbiol 9:202. https://doi.org/10.3389/fcimb.2019.00202

Liu F, Zhou Y, Zhu L et al (2021) Comparative metagenomic analysis of the vaginal microbiome in healthy women. Synth Syst Biotechnol 6:77–84. https://doi.org/10.1016/j.synbio.2021.04.002

Mandl T, Marsal J, Olsson P et al (2017) Severe intestinal dysbiosis is prevalent in primary Sjogren's syndrome and is associated with systemic disease activity. Arthritis Res Ther 19:237. https://doi.org/10.1186/s13075-017-1446-2

Marchesi JR, Ravel J (2015) The vocabulary of microbiome research: a proposal. Microbiome 3:31. https://doi.org/10.1186/s40168-015-0094-5

McClellan KA (1997) Mucosal defense of the outer eye. Surv Ophthalmol 42:233–246. https://doi.org/10.1016/s0039-6257(97)00090-8

van der Meulen IJ, van Rooij J, Nieuwendaal CP et al (2008) Age-related risk factors, culture outcomes, and prognosis in patients admitted with infectious keratitis to two Dutch tertiary referral centers. Cornea 27:539–544. https://doi.org/10.1097/ICO.0b013e318165b200

Miller D, Iovieno A (2009) The role of microbial flora on the ocular surface. Curr Opin Allergy Clin Immunol 9:466–470. https://doi.org/10.1097/ACI.0b013e3283303e1b

Mitra S, Stark M, Huson DH (2011) Analysis of 16S rRNA environmental sequences using MEGAN. BMC Genomics 12(Suppl 3):S17. https://doi.org/10.1186/1471-2164-12-S3-S17

Moeller CT, Branco BC, Yu MC et al (2005) Evaluation of normal ocular bacterial flora with two different culture media. Can J Ophthalmol 40:448–453. https://doi.org/10.1139/i05-014

Mohamed YH, Uematsu M, Morinaga Y et al (2020) Conjunctival sac microbiome in bacterial conjunctivitis. Invest Ophthalmol Vis Sci 61:4908–4908

Moon J, Choi SH, Yoon CH et al (2020) Gut dysbiosis is prevailing in Sjogren's syndrome and is related to dry eye severity. PLoS One 15:e0229029. https://doi.org/10.1371/journal.pone.0229029

Moreno NP, Moreno RD, Sousa LB (2014) Aerobic bacterial microbiota of the conjunctiva in diabetic patients with normal and altered glycated hemoglobin levels in two regions in Brazil. Arq Bras Oftalmol 77:351–354. https://doi.org/10.5935/0004-2749.20140088

Mulcahy LR, Isabella VM, Lewis K (2014) *Pseudomonas aeruginosa* biofilms in disease. Microb Ecol 68:1–12. https://doi.org/10.1007/s00248-013-0297-x

Nahar N, Anwar S, Miah MRA (2014) Conjunctival bacterial flora in diabetic patients. Ibrahim Medical College J 7:5–8. https://doi.org/10.3329/imcj.v7i1.17698

Neumann M, Sjostrand J (1993) Central microbial keratitis in a Swedish city population. A three-year prospective study in Gothenburg. Acta Ophthalmol 71:160–164. https://doi.org/10.1111/j.1755-3768.1993.tb04982.x

Nossa CW, Oberdorf WE, Yang L et al (2010) Design of 16S rRNA gene primers for 454 pyrosequencing of the human foregut microbiome. World J Gastroenterol 16:4135–4144. https://doi.org/10.3748/wjg.v16.i33.4135

Ormerod LD, Hertzmark E, Gomez DS et al (1987) Epidemiology of microbial keratitis in southern California. A multivariate analysis. Ophthalmology 94:1322–1333. https://doi.org/10.1016/s0161-6420(87)80019-2

Ozkan J, Willcox MD (2019) The ocular microbiome: molecular characterisation of a unique and low microbial environment. Curr Eye Res 44:685–694. https://doi.org/10.1080/02713683.2019.1570526

Ozkan J, Zhu H, Gabriel M et al (2012) Effect of prophylactic antibiotic drops on ocular microbiota and physiology during silicone hydrogel lens wear. Optom Vis Sci 89:326–335. https://doi.org/10.1097/OPX.0b013e318243280e

Ozkan J, Nielsen S, Diez-Vives C et al (2017) Temporal stability and composition of the ocular surface microbiome. Sci Rep 7:9880. https://doi.org/10.1038/s41598-017-10494-9

Ozkan J, Coroneo M, Willcox M et al (2018) Identification and visualization of a distinct microbiome in ocular surface conjunctival tissue. Invest Ophthalmol Vis Sci 59:4268–4276. https://doi.org/10.1167/iovs.18-24651

Ozkan J, Willcox M, Wemheuer B et al (2019) Biogeography of the human ocular microbiota. Ocul Surf 17:111–118. https://doi.org/10.1016/j.jtos.2018.11.005

de Paiva CS, Jones DB, Stern ME et al (2016) Altered mucosal microbiome diversity and disease severity in Sjogren syndrome. Sci Rep 6:23561. https://doi.org/10.1038/srep23561

Papapanou PN (2002) Population studies of microbial ecology in periodontal health and disease. Ann Periodontol 7:54–61. https://doi.org/10.1902/annals.2002.7.1.54

Paster BJ, Boches SK, Galvin JL et al (2001) Bacterial diversity in human subgingival plaque. J Bacteriol 183:3770–3783. https://doi.org/10.1128/JB.183.12.3770-3783.2001

Peterson SN, Snesrud E, Liu J et al (2013) The dental plaque microbiome in health and disease. PLoS One 8:e58487. https://doi.org/10.1371/journal.pone.0058487

Petrillo F, Pignataro D, Lavano MA et al (2020) Current evidence on the ocular surface microbiota and related diseases. Microorganisms 8:1033. https://doi.org/10.3390/microorganisms8071033

Rappe MS, Giovannoni SJ (2003) The uncultured microbial majority. Annu Rev Microbiol 57:369–394. https://doi.org/10.1146/annurev.micro.57.030502.090759

Retuerto MA, Szczotka-Flynn L, Mukherjee PK et al (2019) Diversity of ocular surface bacterial microbiome adherent to worn contact lenses and bacterial communities associated with care solution use. Eye Cont Lens 45:331–339. https://doi.org/10.1097/ICL.0000000000000578

Rinninella E, Raoul P, Cintoni M et al (2019) What is the healthy gut microbiota composition? A changing ecosystem across age, environment, diet, and diseases. Microorganisms 7:14. https://doi.org/10.3390/microorganisms7010014

Sankaridurg PR, Willcox MD, Sharma S et al (1996) Haemophilus influenzae adherent to contact lenses associated with production of acute ocular inflammation. J Clin Microbiol 34:2426–2431. https://doi.org/10.1128/jcm.34.10.2426-2431.1996

Sankaridurg PR, Markoulli M, de la Jara PL et al (2009) Lid and conjunctival micro biota during contact lens wear in children. Optom Vis Sci 86:312–317. https://doi.org/10.1097/opx.0b013e318199d20c

Sayin N, Kara N, Pekel G (2015) Ocular complications of diabetes mellitus. World J Diabetes 6:92–108. https://doi.org/10.4239/wjd.v6.i1.92

Sharma S (2012) Diagnosis of infectious diseases of the eye. Eye (Lond) 26:177–184. https://doi.org/10.1038/eye.2011.275

Shimizu E, Ogawa Y, Saijo Y et al (2019) Commensal microflora in human conjunctiva; characteristics of microflora in the patients with chronic ocular graft-versus-host disease. Ocul Surf 17:265–271. https://doi.org/10.1016/j.jtos.2019.02.001

Shin H, Price K, Albert L et al (2016) Changes in the eye microbiota associated with contact lens wearing. MBio 7:e00198. https://doi.org/10.1128/mBio.00198-16

Shivaji S, Jayasudha R, Chakravarthy SK et al (2021) Alterations in the conjunctival surface bacterial microbiome in bacterial keratitis patients. Exp Eye Res 203:108418. https://doi.org/10.1016/j.exer.2020.108418

St Leger AJ, Caspi RR (2018) Visions of eye commensals: the known and the unknown about how the microbiome affects eye disease. BioEssays 40:e1800046. https://doi.org/10.1002/bies.201800046

St Leger AJ, Desai JV, Drummond RA et al (2017) An ocular commensal protects against corneal infection by driving an interleukin-17 response from mucosal gammadelta T cells. Immunity 47:148–158.e145. https://doi.org/10.1016/j.immuni.2017.06.014

Staley JT, Konopka A (1985) Measurement of in situ activities of nonphotosynthetic microorganisms in aquatic and terrestrial habitats. Annu Rev Microbiol 39:321–346. https://doi.org/10.1146/annurev.mi.39.100185.001541

Stapleton F, Willcox MD, Sansey N et al (1997) Ocular microbiota and polymorphonuclear leucocyte recruitment during overnight contact lens wear. Aust N Z J Ophthalmol 25(Suppl 1):S33–S35. https://doi.org/10.1111/j.1442-9071.1997.tb01751.x

Stapleton F, Keay LJ, Sanfilippo PG et al (2007) Relationship between climate, disease severity, and causative organism for contact lens-associated microbial keratitis in Australia. Am J Ophthalmol 144:690–698. https://doi.org/10.1016/j.ajo.2007.06.037

Stapleton F, Edwards K, Keay L et al (2012) Risk factors for moderate and severe microbial keratitis in daily wear contact lens users. Ophthalmology 119:1516–1521. https://doi.org/10.1016/j.ophtha.2012.01.052

Suto C, Morinaga M, Yagi T et al (2012) Conjunctival sac bacterial flora isolated prior to cataract surgery. Infect Drug Resist 5:37–41. https://doi.org/10.2147/IDR.S27937

Suzuki T, Sutani T, Nakai H et al (2020) The microbiome of the meibum and ocular surface in healthy subjects. Invest Ophthalmol Vis Sci 61:18. https://doi.org/10.1167/iovs.61.2.18

Ta CN, Chang RT, Singh K et al (2003) Antibiotic resistance patterns of ocular bacterial flora: a prospective study of patients undergoing anterior segment surgery. Ophthalmology 110:1946–1951. https://doi.org/10.1016/s0161-6420(03)00735-8

Tamaki H, Sekiguchi Y, Hanada S et al (2005) Comparative analysis of bacterial diversity in freshwater sediment of a shallow eutrophic lake by molecular and improved cultivation-based techniques. Appl Environ Microbiol 71:2162–2169. https://doi.org/10.1128/AEM.71.4.2162-2169.2005

Thomas T, Gilbert J, Meyer F (2012) Metagenomics - a guide from sampling to data analysis. Microb Inform Exp 2:3. https://doi.org/10.1186/2042-5783-2-3

Turnbaugh PJ, Ley RE, Hamady M et al (2007) The human microbiome project. Nature 449:804–810. https://doi.org/10.1038/nature06244

Turnbaugh PJ, Hamady M, Yatsunenko T et al (2009) A core gut microbiome in obese and lean twins. Nature 457:480–484. https://doi.org/10.1038/nature07540

Turner L, Stinson I (1965) Mycobacterium fortuitum. Am J Ophthalmol 60:329–331. https://doi.org/10.1016/0002-9394(65)90934-7

Ueta M, Kinoshita S (2012) Ocular surface inflammation is regulated by innate immunity. Prog Retin Eye Res 31:551–575. https://doi.org/10.1016/j.preteyeres.2012.05.003

Van Tyne D, Ciolino JB, Wang J et al (2016) Novel phagocytosis-resistant extended-spectrum beta-lactamase-producing *Escherichia coli* from keratitis. JAMA Ophthalmol 134:1306–1309. https://doi.org/10.1001/jamaophthalmol.2016.3283

Venugopal R, Satpathy G, Sangwan S et al (2016) Conjunctival microbial flora in ocular Stevens-Johnson syndrome sequelae patients at a tertiary eye care center. Cornea 35:1117–1121. https://doi.org/10.1097/ICO.0000000000000857

Versura P, Giannaccare G, Campos EC (2015) Sex-steroid imbalance in females and dry eye. Curr Eye Res 40:162–175. https://doi.org/10.3109/02713683.2014.966847

Watane A, Cavuoto KM, Banerjee S et al (2019) The microbiome and ocular surface disease. Curr Ophthalmol Rep 7:196–203. https://doi.org/10.1007/s40135-019-00217-w

Watters GA, Turnbull PR, Swift S et al (2017) Ocular surface microbiome in meibomian gland dysfunction. Clin Exp Ophthalmol 45:105–111. https://doi.org/10.1111/ceo.12810

Wen X, Miao L, Deng Y et al (2017) The influence of age and sex on ocular surface microbiota in healthy adults. Invest Ophthalmol Vis Sci 58:6030–6037. https://doi.org/10.1167/iovs.17-22957

Willcox MD (2007) *Pseudomonas aeruginosa* infection and inflammation during contact lens wear: a review. Optom Vis Sci 84:273–278. https://doi.org/10.1097/OPX.0b013e3180439c3e

Willcox MD (2013) Characterization of the normal microbiota of the ocular surface. Exp Eye Res 117:99–105. https://doi.org/10.1016/j.exer.2013.06.003

Willis KA, Postnikoff CK, Freeman A et al (2020) The closed eye harbours a unique microbiome in dry eye disease. Sci Rep 10:12035. https://doi.org/10.1038/s41598-020-68952-w

Woese CR (1987) Bacterial evolution. Microbiol Rev 51:221–271. https://doi.org/10.1128/mr.51.2.221-271.1987

Wood DE, Lu J, Langmead B (2019) Improved metagenomic analysis with Kraken 2. Genome Biol 20:257. https://doi.org/10.1186/s13059-019-1891-0

Yang B, Wang Y, Qian PY (2016) Sensitivity and correlation of hypervariable regions in 16S rRNA genes in phylogenetic analysis. BMC Bioinformatics 17:135. https://doi.org/10.1186/s12859-016-0992-y

Yau JW, Hou J, Tsui SKW et al (2019) Characterization of ocular and nasopharyngeal microbiome in allergic rhinoconjunctivitis. Pediatr Allergy Immunol 30:624–631. https://doi.org/10.1111/pai.13088

Zegans ME, Van Gelder RN (2014) Considerations in understanding the ocular surface microbiome. Am J Ophthalmol 158:420–422. https://doi.org/10.1016/j.ajo.2014.06.014

Zhai H, Bispo PJM, Kobashi H et al (2018) Resolution of fluoroquinolone-resistant Escherichia coli keratitis with a PROSE device for enhanced targeted antibiotic delivery. Am J Ophthalmol Case Rep 12:73–75. https://doi.org/10.1016/j.ajoc.2018.09.006

Zhang Y, Liu ZR, Chen H et al (2013) Comparison on conjunctival sac bacterial flora of the seniors with dry eye in Ganzi autonomous prefecture. Int J Ophthalmol 6:452–457. https://doi.org/10.3980/j.issn.2222-3959.2013.04.08

Zhang H, Zhao F, Hutchinson DS et al (2017) Conjunctival microbiome changes associated with soft contact lens and orthokeratology lens wearing. Invest Ophthalmol Vis Sci 58:128–136. https://doi.org/10.1167/iovs.16-20231

Zhao F, Zhang D, Ge C et al (2020) Metagenomic profiling of ocular surface microbiome changes in meibomian gland dysfunction. Invest Ophthalmol Vis Sci 61:22. https://doi.org/10.1167/iovs.61.8.22

Zhou Y, Gao H, Mihindukulasuriya KA et al (2013) Biogeography of the ecosystems of the healthy human body. Genome Biol 14:R1. https://doi.org/10.1186/gb-2013-14-1-r1

Zhou Y, Holland MJ, Makalo P et al (2014) The conjunctival microbiome in health and trachomatous disease: a case control study. Genome Med 6:99. https://doi.org/10.1186/s13073-014-0099-x

Zilliox MJ, Gange WS, Kuffel G et al (2020) Assessing the ocular surface microbiome in severe ocular surface diseases. Ocul Surf 18:706–712. https://doi.org/10.1016/j.jtos.2020.07.007

Chapter 4
Antimicrobial Resistance in Ocular Bacteria

Abstract Every year a significant number of people die due to drug-resistant infectious diseases. Several countries have put together efforts to overcome antimicrobial resistance so as to overcome the loss of human life and the accompanying economic burden. The human ocular surface has a paucibacterial microbiome, and several ocular bacteria have acquired resistance to different classes of antibiotics. This chapter is to understand antimicrobial resistance in ocular bacteria with reference to their identity, the drugs to which they are resistant to and strategies to overcome resistance.

Keywords Antimicrobial resistance · Ocular bacteria · Biofilm · Antibiotics · AMR strategies

4.1 Introduction

One of the major challenges that the clinical world faces is to overcome antimicrobial resistance (AMR). AMR is a process by which microbes (bacteria, fungi, viruses, etc.) become resistant to the lethal effects of an antimicrobial agent. Globally drug-resistant infectious diseases are fatal approximately to 0.7 million people (Davies et al. 2013). In the United Kingdom it is estimated that approximately ten million people would die annually from complications due to AMR (de Kraker et al. 2016). The annual costs of fighting resistant bacterial infections are estimated to be between \$21 and \$34 billion. Thus, to overcome the loss of human life and the economic burden, various agencies such as World Health Organization's Global Action Plan on AMR, the Trans-Atlantic Task Force on Antimicrobial Resistance, Pan American Health Organization on Antimicrobial Resistance, etc. have been active with respect to surveillance and for educating the public on AMR. To emphasise on the gravity of the situation, the World Health Organisation (WHO) published a list of 12 deadliest drug-resistant bacteria 'the dirty dozen' which were also ranked based on their severity (Table 4.1).

S. Shivaji, *Human Ocular Microbiome*,
https://doi.org/10.1007/978-981-19-1754-7_4

Table 4.1 The dirty dozen of bacteria exhibiting antimicrobial resistance

Bacterium	Ranking	Resistant to
Pseudomonas aeruginosa	Critical	Carbapenem
Acinetobacter baumannii	Critical	Carbapenem
Enterobacteriaceae	Critical	Carbapenem, ESBL-producing
Enterococcus faecium	High	Vancomycin
Methicillin-resistant *Staphylococcus aureus* (MRSA)	High	Vancomycin
Helicobacter pylori	High	Clarithromycin
Salmonellae spp.	High	Fluoroquinolone
Campylobacter spp.	High	Fluoroquinolone
Neisseria gonorrhoeae	High	Cephalosporin, fluoroquinolone
Streptococcus pneumoniae	Medium	Penicillin
Haemophilus influenzae	Medium	Ampicillin
Shigella spp.	Medium	Fluoroquinolone

Modified from www.sciencealert.com/the-12-deadliest-drug-resistant-bacteria-have-officially-been-ranked

4.2 AMR in Ocular Bacteria

Ocular microbes that are normally eliminated by using a particular drug are now reported to have become resistant to topical antibiotics like macrolides, drugs containing polymyxin B as a combination, fluoroquinolone and aminoglycosides (Bharathi et al. 2010; Kim and Toma 2011; Lee et al. 2019; Shimizu et al. 2013). For instance, Kunimoto et al. (1999), screened 1558 isolates from the cornea of bacterial keratitis patients and 4417 isolates from various infections of the eye like corneal infections, endophthalmitis, conjunctivitis and blepharitis were found to be resistant to some or all of the following drugs, namely, ciprofloxacin, gentamicin, tobramycin, polymyxin, gatifloxacin, amikacin, azithromycin, moxifloxacin, levofloxacin, norfloxacin, cephalothin, ampicillin, neomycin and tetracyclines (Bharathi et al. 2010; Block et al. 2000). The resistant bacteria included *Staphylococcus epidermidis*, spp., *Staphylococcus* spp., *S. aureus*, coagulase-negative *Staphylococcus* spp., alpha-haemolytic *Streptococcus* spp., *Streptococcus pneumoniae*, *Pseudomonas aeruginosa*, *Corynebacterium macginleyi*, *Corynebacterium* spp., *Enterococcus faecalis* and *Haemophilus influenza* (Bharathi et al. 2010; Block et al. 2000) isolated from individuals with either keratitis, endophthalmitis and corneal ulcers (Mayo et al. 1986; Morrissey et al. 2004; Chaudhry et al. 1999; Harper et al. 2007; Eguchi et al. 2008; Cavuoto et al. 2008; Rau et al. 2008; Das et al. 2019; Kim and Toma 2011). That both Gram-negative and Gram-positive ocular organisms showed increase in antimicrobial resistance was also confirmed by Joseph et al. (2019). All the above studies confirmed that ocular bacteria were emerging to become antimicrobial resistant (Chalita et al. 2004).

4.3 Factors Driving Antibiotic Resistance in Ocular Bacteria

Several factors either extrinsic or intrinsic to the microbe are responsible for the observed AMR in microbes, and humans play a major role in contributing to the acquisition of AMR in microbes. The various reasons that contribute to AMR include (a) improper dosing of antibiotics, (b) use of broad-spectrum antibiotics, (c) indiscriminate use of antibiotics in other organisms as in veterinary and agriculture activities (Buckner et al. 2018), (d) improper disposal of residual antibiotics, (e) use of substandard antibiotics and (f) inability to prevent spread of AMR bacteria in healthcare facilities. Some or all of the above issues have been reviewed (Kumarasamy et al. 2010; Muloi et al. 2019; Allcock et al. 2017; Castro-Sanchez et al. 2016; Ranjith et al. 2021a).

Antibiotics generally kill bacteria by disrupting the membrane structure, by inhibiting cell wall biosynthesis and by inhibiting replication, transcription or translation, folate biosynthesis, nucleic acid metabolism and repair (Fig. 4.1) (Li and Webster 2018). Table 4.2 catalogues the different types of antibiotics, their corresponding targets and the resistance strategies of the bacteria. Thus, it is obvious, that if bacteria could develop strategies to protect the above targets (Fig. 4.2), they could become resistant to the killing effects of the drug. Several strategies that bacteria have developed to overcome the killing effect of the drug (Ranjith et al. 2021a) include the following (Table 4.2):

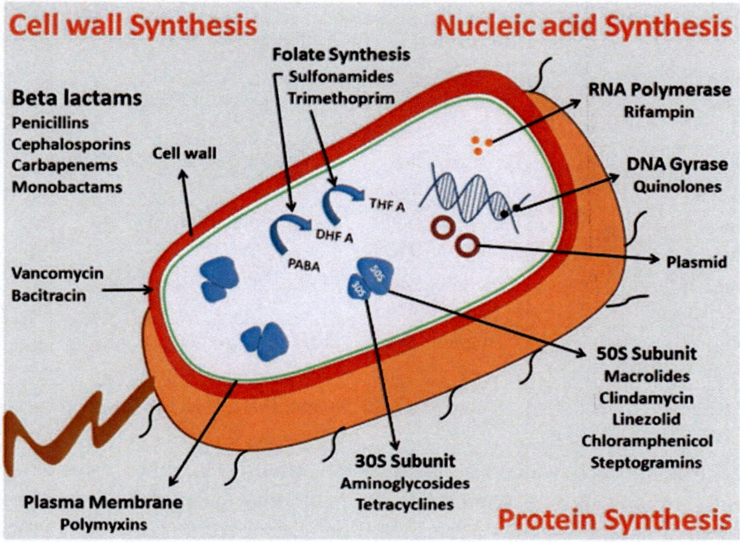

Fig. 4.1 Antimicrobial agents normally target cell wall synthesis, protein and nucleic acid synthesis or disrupt the membrane structure to render the cells inviable. (Modified from Li and Webster 2018 and from Ranjith et al. 2021a)

Table 4.2 Targets of different antibiotics and corresponding resistance strategies of bacteria

Antibiotic class/antibiotic	Target	Resistance strategy
β-Lactams:	Cell wall, peptidoglycan biosynthesis	Hydrolysis, efflux, altered target
Ampicillin, cephamycin, meropenem, aztreonam		
Glycopeptides:	Cell wall, peptidoglycan biosynthesis	Reprogramming peptidoglycan biosynthesis
Vancomycin, teicoplanin		
Lipopeptides:	Cell membrane	Altered target
Daptomycin		
Cationic peptides:	Cell membrane	Efflux, altered target
Colistin		
Aminoglycosides: Gentamicin, streptomycin, spectinomycin	Translation	Efflux, altered target, phosphorylation, acetylation, nucleotidylation
Tetracyclines:	Translation	Efflux, altered target, monooxygenation
Minocycline, Tigecycline		
Macrolides:	Translation	Efflux, altered target, hydrolysis, glycosylation, phosphorylation
Erythromycin, azithromycin		
Lincosamides:	Translation	Efflux, altered target, nucleotidylation
Clindamycin		
Streptogramins:	Translation	Efflux, altered target, C-O lyase (type B streptogramins), acetylation (type A streptogramins)
Synercid		
Oxazolidinones:	Translation	Efflux, altered target
Linezolid		
Phenicols:	Translation	Efflux, altered target, acetylation
Chloramphenicol		
Rifamycins:	Transcription	Efflux, altered target, ADP-ribosylation
Rifampin		
Quinolones:	DNA replication	Efflux, altered target, acetylation
Ciprofloxacin		
Pyrimidines:	Dihydrofolate reductase (DHFR)	Efflux, altered DHFR
Trimethoprim		
Sulfonamides: Sulfamethoxazole	Dihydropteroate synthase (DHPS)	Efflux, altered DHPS
Lipopeptides:	Cell membrane	Efflux, altered target
Daptomycin, colistin		

Modified based on Silver (2016), Kapoor et al. (2017) and Kumar and Singh (2013)

1. Rendering the cell wall impermeable to antibiotics like the β-lactams which include cephalosporins (cephamycin), penicillins (ampicillin), monobactams (aztreonam), penems (meropenem) and the glycopeptide antibiotics (vancomycin and teicoplanin) (Džidić et al. 2008).

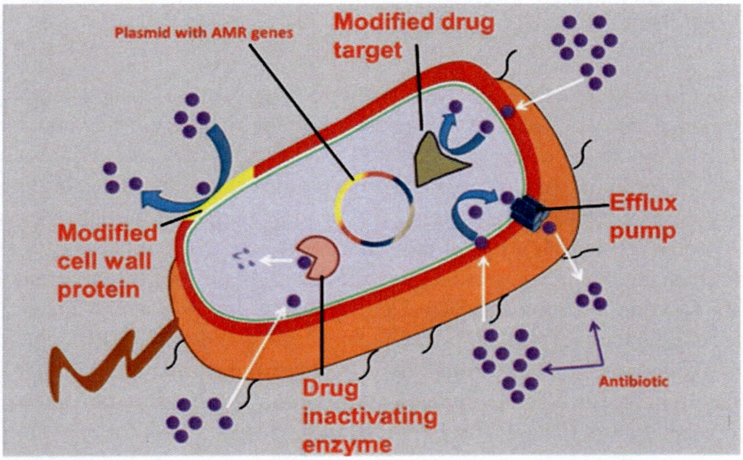

Fig. 4.2 Multiple mechanisms which make bacteria resistant to antimicrobial agents. (Modified from Ranjith et al. 2021a and Dantas and Sommer 2014)

2. By altering the cell membrane, it could become resistant to daptomycin which disrupts bacterial cell membrane (Pogliano et al. 2012) and colistin which binds to LPSs and phospholipids in the cell membrane (Andrade et al. 2020).
3. By modifying the translational machinery (Witzky et al. 2019) (Table 4.2) by phosphorylation, acetylation and nucleotidylation so as to prevent the effects of the antibiotics like minocycline, gentamicin, azithromycin, streptomycin, tigecycline, spectinomycin, clindamycin, linezolid erythromycin, synercid and chloramphenicol which act on the translational machinery (Witzky et al. 2019) (Table 4.2).
4. By altering the transcriptional machinery so as to prevent antibiotics like rifampin which target transcription (Wehrli 1983).
5. By altering the replication machinery so as to prevent antibiotics like ciprofloxacin that inhibit replication or activation of efflux pumps (Hooper et al. 1987).
6. By modifying target pathways or activation of efflux pumps (Brolund et al. 2010) so as to become resistant to trimethoprim and sulfamethoxazole.
7. By inactivating the antibiotic agent by chemical modification (Reygaert 2018).
8. By inducing active efflux of the antibiotic (Soto 2013).
9. By acquiring genes and plasmids which confer resistance to the antimicrobials (Bennett 2008).
10. The development of the potential to form a biofilm by the bacteria.

The mechanism by which ocular bacteria have become resistant has been mostly derived from studies on other clinical isolates, and very rarely attempts have been made to understand the strategy in ocular isolates. One aspect that has been studied is the ability of ocular bacteria to form biofilms so as to overcome the lethal effects of the antimicrobial agent.

4.4 What Is a Biofilm?

Anthony van Leeuwenhoek in 1684 was the first to report small living "animalcules" in dental plaque and suggested that embedded in the plaque are bacteria that are protected from the action of vinegar that was used as a mouthwash (Van Leeuwenhoek 1684). The dental plaque was the first biofilm described. The biofilm is a community of sessile microbes (bacterial and/or fungal cells) of the same species or multiple species. These microbes in the biofilm get sequestered in the EPS (extra-polymeric substance) which is secreted by the bacteria in the biofilm stage and which protects the bacterium from the lethal effects of an antimicrobial agent (Fig. 4.3) and also protects the bacteria from the host factors (Davies et al. 1998). EPS constitutes the abiotic component of the biofilm, and it maintains the structural integrity of the biofilm, facilitates cell-to-cell communication within the biofilm and anchors the biofilm to a substratum (Characklis 1990). Biofilm is also crucial for the viability of cells since it modulates oxygen diffusion, substrate absorption and also transport nutrients within the biofilm (Donlan and Costerton 2002; Flemming and Wingender 2010). The biofilm community is exposed to diverse conditions within the biofilm due to unequal availability of nutrients and metabolic by-products thus resulting in subpopulations within the biofilm which differ in their tolerance to antimicrobials, environmental stresses, the host immune system and predation by other microorganisms (Donlan and Costerton 2002; Stewart and Franklin 2008; Watnick and Kolter 2000). Within a biofilm the bacteria are sessile and adapt to a sedentary life

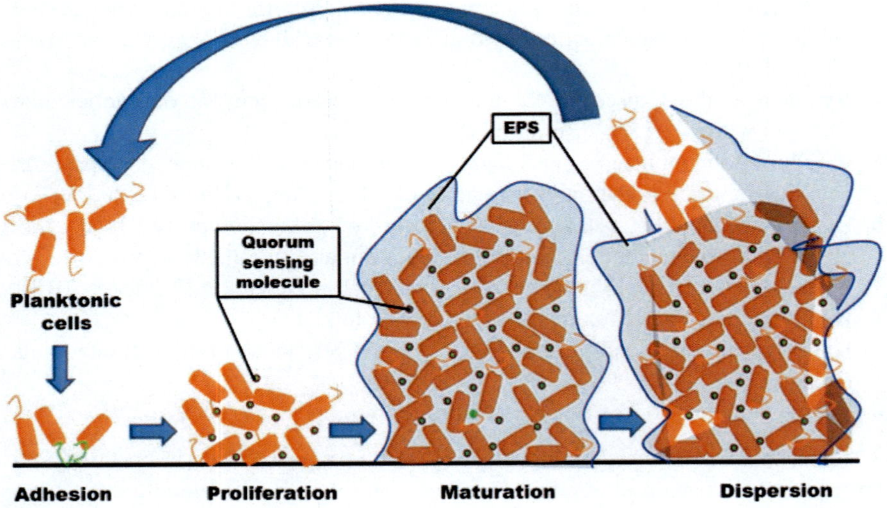

Fig. 4.3 Schematic diagram depicting the adhesion or attachment, proliferation or microcolony formation, maturation and detachment or dispersion in the development of a biofilm in bacteria clearly exhibiting the transition of the free-swimming planktonic phase to a sessile phase. EPS refers to extra-polymeric substance. (Modified from Joo and Otto 2012)

style on the substratum (O'Toole and Kolter 1998). Biofilm formation is a stagewise process and occurs in four discernible stages: adhesion, microcolony formation, maturation and dispersion (Behlau and Gilmore 2008; Ranjith et al. 2021a; Joo and Otto 2012) (Fig. 4.3).

4.4.1 Adhesion

Changes in environmental cues such as temperature, pH, iron concentration, osmolarity, oxygen levels, etc. act as triggers of biofilm formation in bacteria (O'Toole and Kolter 1998). These changes trigger the adhesion of cells to a substratum. Initially the planktonic cells adhere to the substratum in a reversible fashion aided by physicochemical forces such as hydrophobicity, hydrophobic interactions and van der Waals forces (O'Toole et al. 2000). Subsequently, irreversible adhesion occurs (Behlau and Gilmore 2008). In *S. epidermidis* and *S. aureus* surface proteins like autolysins (*atl* and *atlE*), motility organelles like fimbriae (Gross et al. 2001; Buttner et al. 2015; Ormerod et al. 1987; Blumer et al. 2005) and pili (Deziel et al. 2001), extracellular DNA (Pakkulnan et al. 2019), exo-polysaccharides (Ma et al. 2009) and MSCRAMMs, and the bacterial microbial surface components (Foster and Hook 1998; Clarke and Foster 2006) that interact with matrix proteins such as fibronectin (Maxe et al. 1986; McElroy et al. 2002) and fibrinogen (Pei et al. 1999) are used by the bacteria to adhere to the substratum.

4.4.2 Microcolony Formation

Following adhesion, in stage 2, more planktonic cells are recruited, and the cells proliferated (Sauer et al. 2002; Pilchová et al. 2014; Donlan and Costerton 2002) and get organised into multilayered cellular microcolonies. Polysaccharides like intercellular adhesin as in *S. epidermidis* and *S. aureus* (Rohde et al. 2007; Mack et al. 1999), surface proteins like Aapas in *S. epidermidis* (Bowden et al. 2005; Schaeffer et al. 2015) and the *S. aureus* surface protein G (SasG) (Bateman et al. 2005) mediate this step. These biofilms when organised into three-dimensional structures create different microniches by gradients of pH, oxygen and nutrients (Davies 2003), and cells metabolically active are located at the surface, whereas the dormant cells are more internal. The surface cells are more susceptible to the effect of drugs compared to the internal cells (Davies 2003) which also depends on the diffusion of antimicrobial molecules into the biofilm. These factors are likely to be responsible for the resistance of biofilm cells to antimicrobials (Davies 2003).

4.4.3 Maturation

In stage 3, the biofilm appears as vertical column like structures or mushroom-like assemblies with distinct water channels and covered with EPS which is a complex matrix made up of polysaccharides, nucleic acids, proteins, lipids, dead bacterial cells and other polymeric substances hydrated to 85–95% with water (Flemming and Wingender 2010; Costerton et al. 1981). EPS protects the bacteria from the action of antimicrobial agents, from the host immune responses and from predation by phagocytic amoeba and bacteriophages (Donlan and Costerton 2002). The accumulation-associated protein (Aap); the polysaccharide intercellular adhesin (PIA, encoded by the *icaADBC* locus); and the biofilm-associated homologue protein (Bhp, termed Bap in *S. aureus*) contribute to this matrix (Lasa and Penades 2006). The water channels in the biofilm provide nutrients to the deeper biofilm layers and also facilitate waste removal (O'Toole et al. 2000). The mature biofilms prepare for the dispersal phase.

4.4.4 Dispersal

In this phase the sessile biofilm cells transform into planktonic cells a process that involves sensing environmental cues, cell-to-cell signalling and their transduction through complex regulatory networks and facilitates the release of individual cells and/or multicellular aggregates from the biofilm (Guilhen et al. 2017). In this phase, bacterial cells revert to the free-swimming planktonic phase, a process dependent on several enzymes like beta-mannanase which degrade polysaccharide (Dow et al. 2003), exopolysaccharide lyase (Allison et al. 1998), chitinases (Baty et al. 2000), nucleases (Tang et al. 2008; Tetz et al. 2009), alginate lyase (Boyd and Chakrabarty 1994) and several proteases (Cui et al. 2016; Lister and Horswill 2014). In *S. aureus* biofilms, a thermo-nuclease degrades extracellular DNA that holds *S. aureus* biofilms together, while the alginate matrix which is important for *P. aeruginosa* biofilm is degraded by alginate lyase (Boyd and Chakrabarty 1994: Mann et al. 2009). Small signalling molecules have also been implicated in the dispersion phase, and these appear to be species-specific and include *N*-butanoyl-L-homoserine lactone (C4HSL, as in *P. aeruginosa* biofilms) (Dong et al. 2008), *Pseudomonas* quinolone signal (PQS, as in *Pseudomonas* spp.), the furanosyl borate autoinducer (AI-2, as in *Vibrio cholera*) and the staphylococcal peptides δ-toxin and autoinducer peptide (AIP-I, as in *Staphylococcus* spp.) (Kaplan 2010). Dispersal phase is also positively influenced by release of surfactants (like rhamnolipids) and phenol-soluble modulins (Ma et al. 2009; Aleksic et al. 2017; Angelini et al. 2009), by chemical factors and environmental changes (Kaplan 2010). Quorum sensing may also be involved in this critical stage.

4.5 Biofilm Formation in Ocular Bacteria

The human ocular surface has a paucibacterial microbiome dominated by propionibacteria, corynebacteria and coagulase-negative staphylococci compared to the buccal mucosa or facial skin which have 100-fold more bacterial abundance (Doan et al. 2016). The Gram-positive bacteria, especially coagulase-negative *Staphylococcus* (CoNS), *S. epidermidis*, *Staphylococcus aureus* and *Streptococcus viridians* cause several ocular diseases like acute postoperative endophthalmitis, keratitis, lacrimal system infections, periorbital infections, and infection of the anophthalmic cavity of ocular prosthesis users and by infecting ocular devices like lacrimal intubation devices scleral buckles, conjunctival plugs, intraocular lenses, etc. (Table 4.3). It was also observed that biofilm formation by ocular isolates was dependent on the nature of the lens. For instance, García-Sáenz et al. (2000) demonstrated that in a biofilm-positive *Staphylococcus epidermidis*, the adhesion to lenses occurred depending on the chemistry of the lenses. Adhesion occurred in a decreasing order on the following types of lenses: silicone, acrylic, heparin surface-modified poly(methyl methacrylate) (HSM PMMA), single-piece PMMA with polypropylene loops and fluorine surface-modified PMMA and hydrogel (García-Sáenz et al. 2000). *Enterococcus faecalis* also forms robust biofilms on PMMA and hydrophobic acrylic IOLs, but the biofilm biomass was less on silicone IOLs (Kobayakawa et al. 2005; Miller et al. 2011). Gram-positive pathogens prefer acrylic IOLs for biofilm formation, but the Gram-negative bacterium *Pseudomonas aeruginosa* prefers PMMA and silicone lenses for biofilm formation (Gabriel et al. 1998). The surface of staphylococci is highly hydrophilic, while that of *P. aeruginosa* is highly hydrophobic (Bruinsma et al. 2001) thus implying that hydrophobicity is an important determinant of biofilm formation (Kodjikian et al. 2003; Schauersberger et al. 2003). However, it may be noted that results on biofilm formation were not consistent with respect to the chemistry of IOLs used and the strain involved (García-Sáenz et al. 2000; Huang et al. 2007; Raskin et al. 1993; Portolés et al. 1993). The virulence of the above *Staphylococcus* spp. has been attributed to its ability to form a biofilm which confers antimicrobial resistance (Fridkin et al. 2005; Weigel et al. 2003; Behlau and Gilmore 2008; Paharik and Horswill 2016). Most studies on biofilm formation in *S. aureus* and *S. epidermidis* used non-clinical strains (Kirchhoff et al. 2020). In a recent study, clinical strains of both *S. aureus* and *S. epidermidis* were characterised from patients with infectious endophthalmitis, and it was confirmed that both ocular *S. epidermidis* and *S. aureus* exhibited biofilm-forming capacity by using different methods such as the qualitative Congo red agar method and the quantitative crystal violet and were also imaged by scanning electron microscopy (SEM) (Hou et al. 2012; Okajima et al. 2006; Xu et al. 2016; Saraswathi and Beuerman 2015). The temporal dynamics of biofilm formation was monitored in *S. aureus* and *S. epidermidis* up to 120 h, and this revealed that the adhesion to the substratum occurred irrespective of the substratum (covers slip or cornea) by 4 h (Figs. 4.4 and 4.5). Further, in both the dispersal phase occurred between 96 and 120 h depending on the substratum (covers slip or cornea)

Table 4.3 Ocular disease-causing microorganisms and biofilm formation

Disease	Infectious agents in the biofilms	Biofilm localization
Endophthalmitis	Coagulase-negative staphylococci spp., *Staphylococcus epidermidis*, *Propionibacterium acnes*, *Pseudomonas aeruginosa* and *Enterococcus faecalis*	Intraocular lens, posterior chamber, posterior capsular bag (Han et al. 1996; Benz et al. 2004; Shirodkar et al. 2012; Doyle et al. 1995; Adan et al. 2008; García-Sáenz et al. 2000; Griffiths et al. 1989; Okajima et al. 2006; Shimizu et al. 2006; Teichmann 2000; Pathengay et al. 2012; Ramappa et al. 2012; Kresloff et al. 1998; Miller et al. 2007; Bispo et al. 2008)
Keratitis	Staphylococcal species, *Staphylococcus aureus*, *Pseudomonas aeruginosa*, Serratia spp., *Fusarium* spp. and *Acanthamoeba*	Contact lens wearers (Cheng et al. 1999; Passos et al. 2010; Lichtinger et al. 2012; Keay et al. 2011; Hammersmith 2006; Paranhos et al. 2007; Guiotti et al. 2018; Stapleton and Dart 1995; Dutta et al. 2012; Klotz et al. 1989)
Infectious crystalline keratopathy	*Streptococcus* spp., *Streptococcus mitis*, *S. gordonii*, *S. sanguis*, *Enterobacter aerogenes*, *Acinetobacter lwoffi*, *Citrobacter koseri*, *Stenotrophomonas maltophilia*, *Pseudomonas aeruginosa*, *Candida albicans*, *C. guilliermondii* and *C. parapsilosis* and *Acanthamoeba*	Corneal stroma (crystalline keratopathy) (Osakabe et al. 2006; Ainbinder et al. 1998; Khater et al. 1997; Rhem et al. 1996; Hunts et al. 1993)
Scleral buckle infection	*Staphylococcus epidermidis*, *S. aureus*, coagulase-negative *Staphylococci* spp., *Staphylococcus saprophyticus*, *Streptococcus pneumoniae*, *Corynebacterium* spp., *Pseudomonas aeruginosa*, *Bacillus cereus*, *Acinetobacter calcoaceticus*, *Aspergillus flavus*, *Citrobacter freundii*, *Escherichia coli*, *Alcaligenes faecalis*, *Bacillus megaterium*, *Aspergillus terreus*, nontuberculous *Mycobacterium* spp., *Mycobacterium chelonae* and *Proteus mirabilis*	Smiddy et al. (1993), Holland et al. (1991), Pathengay et al. (2004), Paranhos et al. (2007), Guiotti et al. (2018), Duggirala et al. (2007), Catalanotti et al. (2005), Behlau and Gilmore (2008), Bispo et al. (2015), Darouiche (2004), Moreno et al. (2020), Paharik and Horswill (2016), Otto (2008)
Disease	Infectious agents in the biofilms	Biofilm localization
Conjunctival plug infection	*S. haemolyticus* and *Candida tropicalis*	Sugita et al. 2001; Yokoi et al. 2000
Lacrimal intubation devices	*Staphylococcus* spp., *S. epidermidis*, *P. aeruginosa and M. chelona*	Balikoglu-Yilmaz et al. (2014), Parsa et al. (2010), Samimi et al. (2013), Paranhos et al. (2007), Guiotti et al. (2018), Duggirala et al. (2007), Catalanotti et al. (2005), Behlau and

(continued)

Table 4.3 (continued)

Disease	Infectious agents in the biofilms	Biofilm localization
		Gilmore (2008), Bispo et al. (2015), Darouiche (2004), Moreno et al. (2020), Paharik and Horswill (2016), Otto (2008)
Orbital implants	*Staphylococcus* spp., *S. epidermidis*, *S. aureus*, *M. chelonae*, *Achromobacter xylosoxidans*, *P. aeruginosa* and yeasts *Candida* spp., and *Trichosporon* spp.	Silicone stents, glass Jones tubes, orbital plate implants made of nylon, porous polyethylene, silicone, and metal and anophthalmic socket sphere implants (Samimi et al. 2013; Balikoglu-Yilmaz et al. 2014; Parsa et al. 2010)

(Figs. 4.4 and 4.5). These studies confirm biofilm formation in ocular *S. aureus* and *S. epidermidis* which was earlier demonstrated using synthetic material like silicone, polymethylmethacrylate, hydrophilic acrylic, hydrophobic acrylic, etc. (Okajima et al. 2006; Baillif et al. 2013; Moreno et al. 2020) which are used in the manufacture of indwelling ocular devices including scleral buckles, contact lenses, valvular tubes, sutures and keratoprostheses (Darouiche 2004; Moreno et al. 2020; Paharik and Horswill 2016; Otto 2008; Guiotti et al. 2018; Paranhos et al. 2007; Jassim et al. 2015; Behlau and Gilmore 2008; Saraswathi and Beuerman 2015). Monitoring biofilm formation of both the clinical ocular isolates of *S. epidermidis* and *S. aureus* on ex vivo cornea is important because these bacteria cause hospital-acquired infections and are involved in corneal biofilms in experimental induced keratitis mice (Saraswathi and Beuerman 2015), in patients with infectious crystalline keratopathy (Elder et al. 1996; Fulcher et al. 2001; Georgiou et al. 2002) or pterygium scleritis (Mihara et al. 2004). Biofilms also occurred in the absence of prosthetic material and in the absence of active corneal inflammation or infection (Cordoba et al. 2019). Further, clinical *Staphylococcal* isolates have binding proteins on the surface to bind specifically to host molecules like fibrinogen, fibronectin and host extracellular matrix proteins (Herrmann et al. 1988: McGavin et al. 1993). Studies have also indicated that when biofilms formed on coverslips were compared with those formed on cadaveric cornea, EPS secretion was more copious on cornea than on cover slips (Figs. 4.4 and 4.5). Temporal changes in the thickness of the biofilm were monitored by confocal microscopy (Jefferson et al. 2005; Baidamshina et al. 2017; Ranjith et al. 2021b) and the biofilm thickness of *S. aureus* and *S. epidermidis* increased from 4 to 120 h. These observations were also confirmed by crystal violet and XTT methods (Ranjith et al. 2021b). It was consistently demonstrated that cells in the biofilm phase are several fold more resistant to antibiotics (Fridkin et al. 2005; Weigel et al. 2003; Behlau and Gilmore 2008; Paharik and Horswill 2016; Ranjith et al. 2017, 2019).

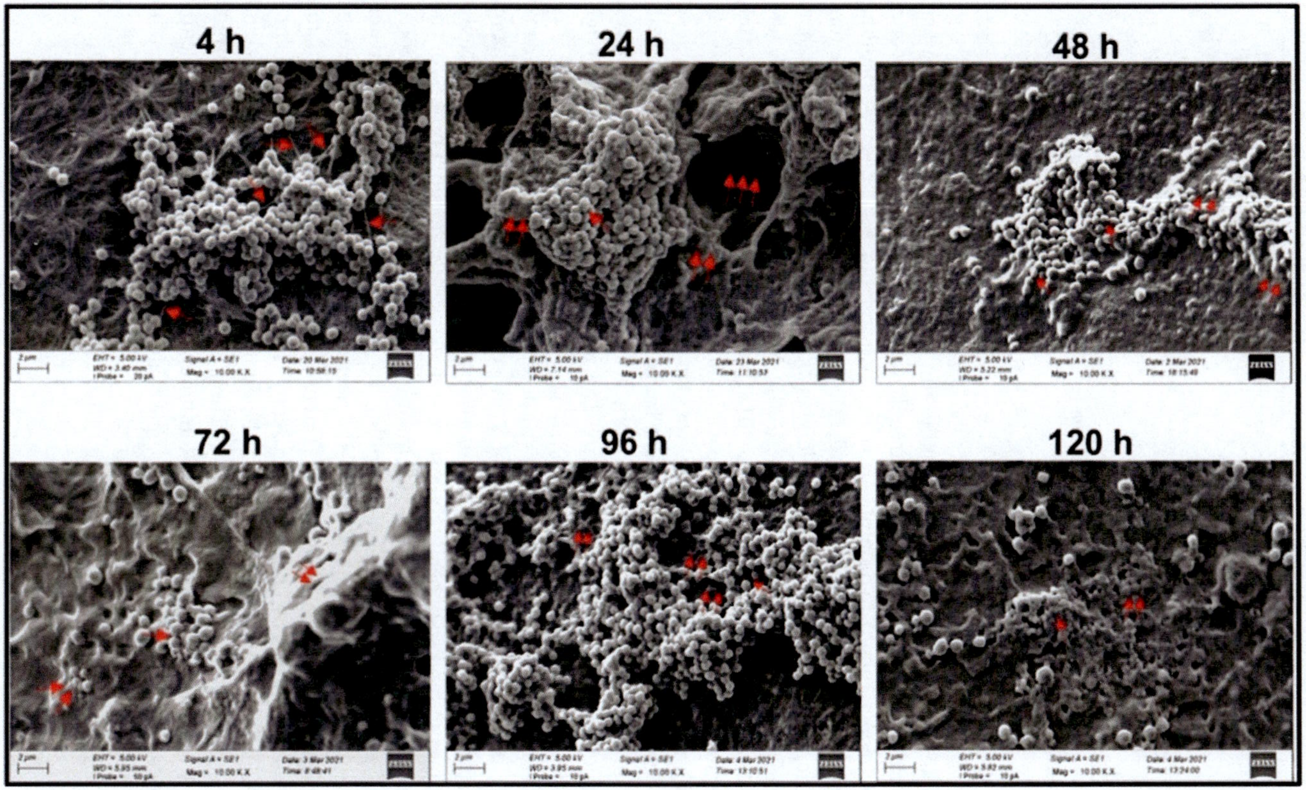

Fig. 4.4 Scanning electron microscope photomicrographs showing temporal changes in the biofilm formation in ocular *Staphylococcus aureus* L1054/2020 on cadaveric cornea between 4 and 120 h of biofilm formation. (From Ranjith et al. 2021b)

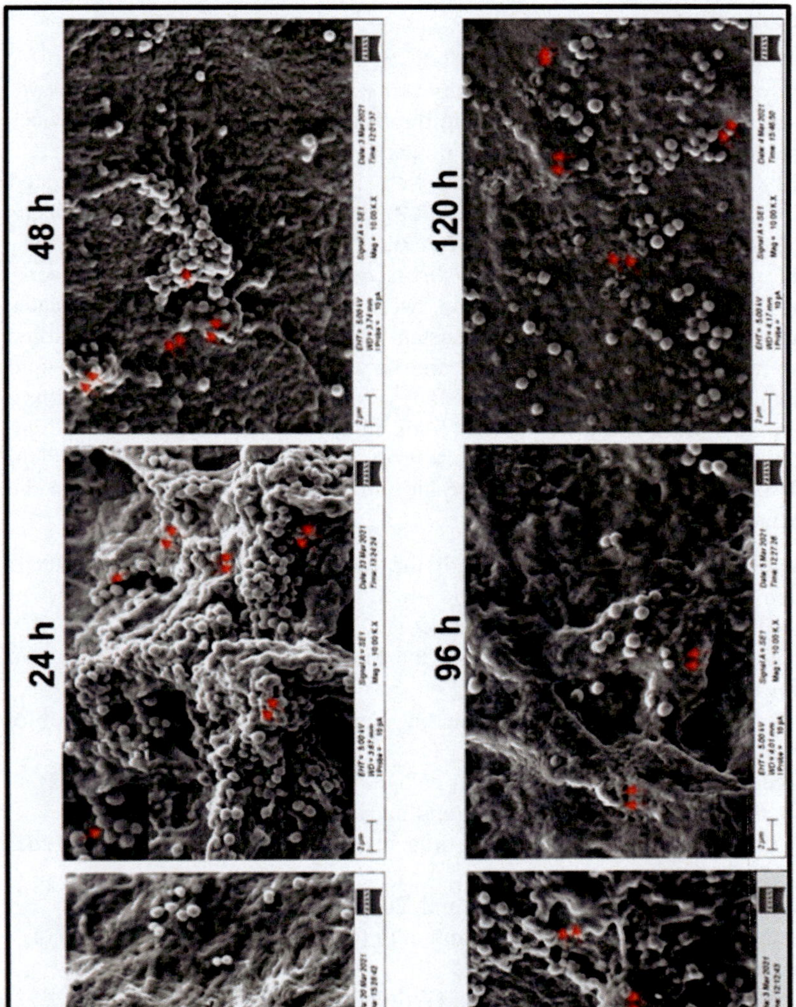

Fig. 4.5 Scanning electron microscope photomicrographs showing temporal changes in the biofilm formation in ocular *Staphylococcus epidermidis* L1058/2020 on cadaveric cornea between 4 and 120 h of biofilm formation. (From Ranjith et al. 2021b)

4.6 AMR in Ocular Bacteria Is Associated with Biofilm Formation

The ability of ocular bacteria *Pseudomonas aeruginosa, S. epidermidis, Streptococcus* spp. and *Enterobacter* spp. to form biofilm (Elder et al. 1995) on ocular devices like contact lenses, intraocular lenses, lid implants, suture material, etc. (Behlau and Gilmore 2008; Zegans et al. 2002; Bispo et al. 2015) has been demonstrated (Table 4.3). Katiyar and Vishwakarma (2012) demonstrated that *Staphylococcus aureus, S. epidermidis, Micrococcus luteus, P. aeruginosa, Moraxella* spp., *Neisseria* spp., *Proteus mirabilis S. marcescens, Bacillus* spp., *Enterobacter agglomerans, E. coli* and *Klebsiella* spp. were the predominant (85%) isolates from intraocular lenses. These were resistant to antibiotics, and the planktonic cells were more susceptible compared to the cells in the biofilm phase, and the minimum inhibitory concentration (MIC) of the antibiotic required for killing the cells in the biofilm phase was several folds greater than that required for killing the cells in the planktonic phase (Kobayakawa et al. 2005; Jassim et al. 2015; Pichi et al. 2014). This increase in MIC in the biofilm phase could be attributed to the following factors:

1. Compromised drug penetration in to the biofilm (Mah and O'Toole 2001; Gordon et al. 1988) due to EPS which effectively limits the entry of aminoglycosides into the biofilm (Fux et al. 2005; Khan et al. 2010)
2. Suboptimal effect of the drug within the biofilm (Xu et al. 2000; Tack and Sabath 1985)
3. Transformation of cells into viable but nonculturable state in the biofilm (Li et al. 2014)
4. Emergence of drug-resistant persister cells (Gerdes and Maisonneuve 2012)
5. Adaptation to survive under oxygen and nutrient limiting conditions
6. Upregulation of genes required for drug resistance (Hoyle et al. 1992; Elder et al. 1995)
7. Inactivation of antibiotics (Billings et al. 2015)
8. Gene transfer within the biofilm community (de Kievit 2009)
9. Acquiring resistance to phagocytosis
10. Induction of LPS modification genes (Wilton et al. 2016)

In ocular *E. coli*, the presence of several virulent genes (*fimB* to *fimI*, *papB* to *papX*, etc.) and prophages (Enterobacteria phage P1, Enterobacteria phage HK97, *Escherichia* phage *D108*, etc.) contributes to the high concordance observed between AMR genes and antibiotic resistance in a biofilm (Ranjith et al. 2017, 2020).

4.7 Biofilm in the Diseased Eye

4.7.1 Biofilms in Endophthalmitis

Inflammation in the posterior region of the eye due to a microbial pathogen is a characteristic feature of endophthalmitis. The pathogen generally gains access to the ocular globe following cataract surgery, intravitreal injections, penetrating open globe injury, keratitis, etc. (Sadaka et al. 2012; Han et al. 1996; Benz et al. 2004). For more details on different types of endophthalmitis, the common symptoms, the incidence following various ocular surgeries, the predominant bacterial species associated with endophthalmitis, etc., see Chap. 2.

4.7.1.1 Genes Associated with Biofilm Formation in Endophthalmitis

S. epidermidis strains carrying the *ica* locus form biofilms efficiently on different IOL surfaces (García-Sáenz et al. 2000; Shimizu et al. 2006). In Japan, approximately 40% of the biofilm-forming *S. epidermidis* isolates (Suzuki et al. 2005) possessed the *ica*A gene, whereas in Mexico, 26.7% of commensal conjunctival isolates which were biofilm-positive were also positive for *ica*A and/or *ica*D (Juarez-Verdayes et al. 2013). In India, 36% for CoNS species other than *S. epidermidis* had *ica* genes (Makki et al. 2011). The presence of *ica*AD genes in bacteria-forming biofilm in South Florida was high (86%), but the gene was present only in isolates that also possessed the accessory gene regulator locus (*agr*) type I (Bispo and Miller 2012; Bispo et al. 2015).

4.7.2 Biofilms in Keratitis

Keratitis is an infection of the cornea by bacteria and fungi, and the predisposing factors are trauma and improper care and cleaning of contact lenses leading to loss of vision (Bourcier et al. 2003; Keay et al. 2006). The preponderant bacteria associated with contact lens-associated infection include *S. aureus*, other staphylococcal species, *Serratia* spp. and *P. aeruginosa* (Cheng et al. 1999; Passos et al. 2010; Lichtinger et al. 2012). For other details on incidence of keratitis, see Chap. 2.

4.7.2.1 Factors Associated with Contact Lens Keratitis

Several factors are known to positively influence microbial keratitis in contact lens wearers such as:

1. The ability of the lens to modify the corneal epithelium
2. The ability of the lens to support growth of organisms on the ocular surface

3. By the lens interfering with the natural clearance mechanisms (Fleiszig and Evans 2010; Willcox et al. 2010)
4. By the induction of hypoxia and hypercapnia following lens use (Ren et al. 1999)
5. By the ability of the lens to limit tear fluid exchange and limiting the antimicrobial properties of tears
6. By providing a substratum for biofilm formation on a previously damaged corneal epithelium (Willcox et al. 2010)
7. By the ability of the organisms to form biofilms on contact lenses and lens storage cases

The first step in the development of bacterial keratitis is binding of the microbes to damaged corneal epithelium (Dart and Seal 1988; Ramphal et al. 1981). *P. aeruginosa* has been associated with biofilm formation on the posterior surface of the contact lens both in animal models (Tam et al. 2010) and humans (Stapleton and Dart 1995). The process of biofilm formation in bacteria is facilitated by a repertoire of genes that are required for adaptation, survival (Stover et al. 2000) and adhesion to different contact lens materials (Dutta et al. 2012; Klotz et al. 1989).

4.7.3 Biofilms in Infectious Crystalline Keratopathy

Infectious crystalline keratopathy disease is associated with deposits of crystalline metabolic products in the stroma and corneal epithelium and stroma caused by corneal infections and systemic diseases (Scheie and Grayson 1971). ICK is difficult to treat (Meisler et al. 1984: Reiss et al. 1986). The predisposing factors of ICK include topical use of steroids and corneal surgery especially penetrating keratoplasty (Georgiou et al. 2002; Meisler et al. 1984: Osakabe et al. 2006; Elder et al. 1996; Fulcher et al. 2001; Reiss et al. 1986). Gram-positive *Streptococcus* spp., *S. mitis* and *S. gordonii* and Gram-negative *Pseudomonas aeruginosa*, *Citrobacter koseri*, *Enterobacter aerogenes*, *Acinetobacter lwoffii and Stenotrophomonas maltophilia* and the yeasts *Candida guilliermondii* and *Candida parapsilosis* are the main pathogens associated with ICK. *Acanthamoeba* may also cause ICK (Rhem et al. 1996; Osakabe et al. 2006; Khater et al. 1997; Ainbinder et al. 1998). ICK is indolent, does not exhibit immune response and is recalcitrant to antimicrobial therapy probably because ICK is associated with biofilm-forming organisms on the corneal tissue. Transmission electron microscopy of corneal biopsy specimens in three cases of ICK confirmed the presence of *Candida albicans* and bacteria within a biofilm (Elder et al. 1996; Fulcher et al. 2001; Georgiou et al. 2002), whereas in five cases of chronic microbial keratitis without crystalline changes, these microorganisms were not detected. In rabbits *Streptococcus sanguis* type II-induced crystalline lesions also had high concentrations of extracellular polysaccharides on the cornea but were absent in the eyes that developed suppurative stromal lesion (Hunts et al. 1993). Use of lasers to disrupt biofilms (Daneshvar et al.

2000; Masselos et al. 2009) and treatment with antimicrobials were effective in the treatment of ICK.

4.8 Biofilms Associated with Ocular Implants

An implanted device or an object in the eye is an impetus to biofilm formation since the foreign body could serve as the substratum for the formation of biofilm (Nickel et al. 1989). Apart from contact lenses and IOLs, ocular infections are also associated with several other implants of the eye like orbital implants, conjunctival plugs and lacrimal intubation devices. Many of these infections are recalcitrant to treatment due to their ability to form biofilms (Table 4.3).

4.8.1 Biofilms Associated with Scleral Buckles

Scleral buckling is a surgical treatment for retinal detachment to restore vision. In this procedure retinal detachment is repaired by positioning the scleral buckles between the conjunctiva and sclera to indent the globe (Zegans et al. 2002). The scleral buckles are commonly made of silicone, rubber or semi-hard plastic and may have a sponge or a solid form. Gram-positive cocci, especially coagulase-negative staphylococci, and nontuberculous *Mycobacterium* spp., *Mycobacterium chelonae* and *Proteus mirabilis* (Holland et al. 1991), are the commonly reported scleral buckle-associated infections, but the infections exhibit a delayed onset and are usually refractory to antimicrobial therapy (Pathengay et al. 2004; Holland et al. 1991; Smiddy et al. 1993). But the diversity associated with the buckles is much greater as indicated in a study involving 102 eyes of 101 patients (Kazi et al. 2015) (Table 4.3). In the infected scleral buckles, scanning electron microscopy demonstrated biofilm positivity was low (18%) in patients with conjunctival erosion, infection or following repeat surgery (Asaria et al. 1999) and was maximum (65%) when buckles were removed due to infection and extrusion of the buckles (Holland et al. 1991). Bispo et al. (2015) suggested that bacteria may attach to the buckles during surgery, form biofilms and cause either infection lacking any overt signs or indolent infections (Bispo et al. 2015). Thus, scleral buckles harbour latent bacterial biofilms and function as reservoirs for pathogenic bacteria (Zegans et al. 2002).

4.8.2 Biofilms Associated with Conjunctival Plug

Punctal or conjunctival plugs are used primarily for the treatment of ocular surface dryness (Balaram et al. 2001) by occluding the lacrimal ducts to block or reduce the drainage down the nasolacrimal ducts and retain the tears in the eye. Punctal plugs are made of silicone, biodegradable polyesters, collagen and hydrogels. These plugs when implanted could cause complications like canaliculitis, dacryocystitis and acute conjunctivitis (Bourkiza and Lee 2012; Yokoi et al. 2000) and when recovered from patients were not susceptible to antimicrobial treatment, implying the possibility of biofilm-related infections (Joganathan et al. 2010). Biofilms were also detected in punctual plugs from patients who did not show signs of infection (Sugita et al. 2001). Yokoi et al. (2000) reported biofilm formation on a punctual plug in a conjunctivitis patient, and *Candida tropicalis* and *S. haemolyticus* were present on the plug.

4.8.3 Biofilms Associated with Lacrimal Intubation Devices and Orbital Implants

Lacrimal intubation devices (Jones tubes and lacrimal stents) are the method of choice to treat obstruction of the nasolacrimal duct which could cause excessive production of tears and epiphora. Both the lacrimal stents and Jones tubes removed after surgery harboured coccoid and/or both rod and coccoid bacteria sequestered in a biofilm (Balikoglu-Yilmaz et al. 2014; Ibanez et al. 2011; Samimi et al. 2013) with *S. epidermidis* and *P. aeruginosa* (Balikoglu-Yilmaz et al. 2014) associated with the stents, whereas recalcitrant culture-negative bacterial biofilms were associated with the lacrimal intubation devices (Parsa et al. 2010). SEM of both Jones tube and silicone stent revealed that the biofilms were polymicrobial consisting of spirochetes, fusiforms, short rods and cocci. Biofilms were also detected by confocal laser scanning microscopy along the internal surfaces of the silicone stent (Parsa et al. 2010). In a comprehensive study, a diverse array of microorganisms was identified as colonizing biomaterials implanted in the lacrimal drainage system (which included silicone stents, Jones tubes, orbital plate implants and anophthalmic socket sphere implants). Interestingly all the implants supported microbial growth, and *Mycobacterium chelonae*, *Staphylococcus aureus* and *Pseudomonas aeruginosa* were most common with the silicone stents (Samimi et al. 2013). Further, the orbital plates were culture positive for bacteria (*M. chelonae*, *Staphylococcus* spp., *P. aeruginosa* and *Achromobacter xylosoxidans*) and yeasts (*Trichosporon* spp. and *Candida* spp.). Scanning electron microscopy analysis indicated that 83% of the implants had organisms in a biofilm encased in a glycocalyx (Samimi et al. 2013).

4.9 Quorum Sensing and Biofilm Formation in Ocular Bacteria

Biofilm formation is preceded by a density-dependent control of gene expression in bacteria known as "quorum sensing" (QS) which upregulates the expression of several genes (>600) (Davies et al. 1998). QS regulates various physiological activities of cells which include virulence, motility, sporulation, symbiosis, conjugation and biofilm formation. In Gram-negative bacteria, the autoinducer signalling molecules are the Acyl-homoserine lactones (AHLs), which act as the triggers for QS and are synthesised within the cell. When the threshold concentration of AHLs is attained, AHLs diffuse out from the cell, enter nearby cells and bind to LuxR or LuxL the response regulators. The complex of AHL and LuxR or LuxL (AHL-LuxR/LUXL) drives the transcription of genes involved in QS (Fig. 4.6) (Papenfort and Bassler 2016; Ranjith et al. 2021a). In contrast, in Gram-positive bacteria the autoinducing cyclic peptides (AIP) which are either acyclic oligopeptides or cyclic lactone peptides act as the quorum sensing molecules (QSM). AIP is synthesised as a precursor which is then transformed into a mature AIP, and when the AIPs attain a threshold concentration in the extracellular environment, they bind to a sensory histidine kinase on the membrane. This histidine kinase gets activated following the binding of AIP and undergoes autophosphorylation and subsequently phosphorylates a response regulator in the

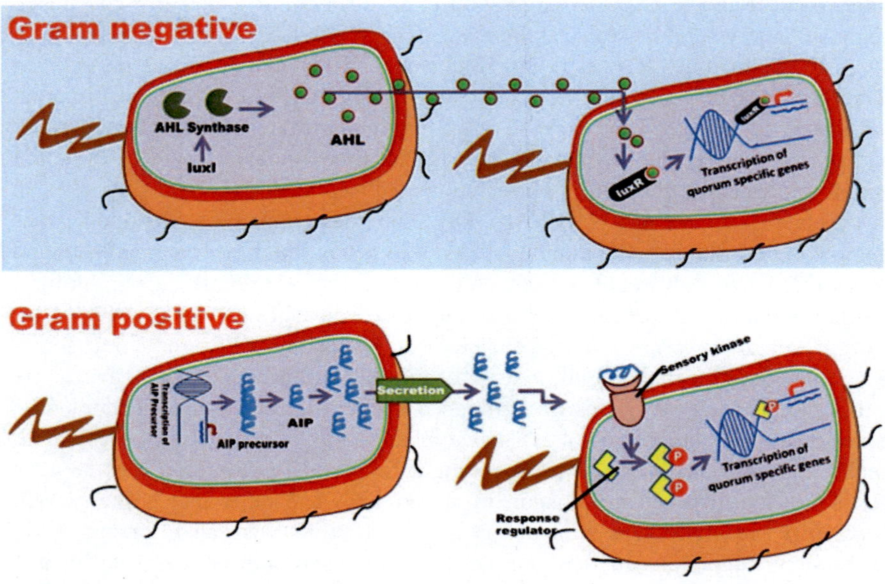

Fig. 4.6 Quorum sensing mechanism in Gram-positive and Gram-negative bacteria showing the involvement of LuxR (the response regulator), the sensory kinase, AHL (acyl-homoserine lactones) and AIP (autoinducing peptides). (From Ranjith et al. 2021a)

cytoplasm which then binds to DNA and drives the transcription of quorum-specific genes (Rutherford and Bassler 2012; Papenfort and Bassler 2016) thus triggering biofilm formation. In ocular bacteria QS is yet to be studied (Elder et al. 1995). In *Aeromonas hydrophila*, *P. aeruginosa*, *Serratia liquefaciens* and *S. marcescens*, N-acyl-homoserine lactone has been identified as the QSM, but such molecules are yet to be identified in *Acinetobacter* spp., *E. coli*, *Haemophilus influenza*, *Klebsiella oxytoca* and *Stenotrophomonas maltophilia* (Zhu et al. 2001, 2002). In *P. aeruginosa* isolated from microbial keratitis, patients' QS involving LASL and LASR or *RHL*R and *RHL*AB and the QSM AHL have been detected (Zhu et al. 2001, 2002).

4.10 Candidate Genes Involved in Biofilm Formation in Ocular Bacteria

Several approaches have led to the identification of genes involved in biofilm formation, and these methods include large-scale mutant screens (Oluyombo et al. 2019), proteome analyses (Sauer et al. 2002), transcription profiling (Schembri et al. 2003; Junker et al. 2007; Ren et al. 2004; Hancock and Klemm 2007) and promoter identification assays (Amores et al. 2017). These genes include *ica* and *icaB* in *Staphylococcus warneri*, *S. capitis*, *S. hominis*, *S. haemolyticus*, coagulase-negative staphylococci and *S. lugdunensis* (Makki et al. 2011); *atlE* and *ica* in coagulase-negative *Staphylococcus* spp. (Jassim et al. 2015); *icaA*, *ica*D, *opp*3AB and *arc*A in *S. epidermidis* (Flores-Paez et al. 2015; Jena et al. 2017); and *ladS* in *P. aeruginosa* (Dave et al. 2020). In ocular *E. coli*, several genes involved in biofilm formation have been identified (Table 4.4). It was also demonstrated that biofilm-forming ocular *E. coli* L-1216/2010 (Fig. 4.7) was 100 times more resistant to the eight antibiotics tested in the biofilm phase, compared to planktonic phase (Fig. 4.7). In this isolate of *E. coli* (L-1216/2010), DNA microarray analysis identified genes involved in biofilm formation and included 426 genes which were upregulated and 866 which were downregulated compared to *E. coli* L-1339/2013 a non-biofilm-forming strain. The upregulated genes included cell adhesion genes required for biofilm formation, genes required for biofilm architecture, LPS production genes, genes encoding for the integral proteins of the cell membrane, extracellular matrix remodelling genes and genes that influence antigen presentation. Concomitantly, genes that confer antimicrobial resistance including toxin production genes (*chp*S, *chp*B, *sat*, *yjg*K and *ygj*N), genes encoding antimicrobial efflux (*mdt*M and *cyc*A), genes that are involved in transport of amino acids and other metabolites (*cbr*B, *cbr*C, *his*I and *mgl*B) and virulence genes (*ins*Q, *yjg*K) were also upregulated in expression (Ranjith et al. 2017). Further, genes involved in intracellular trafficking, vesicular transport and secretion (U category of genes) and genes involved in defence mechanisms (V) were downregulated, whereas genes in DNA replication, recombination and repair (L) were upregulated in expression both ocular and

Table 4.4 Genes involved in biofilm formation in ocular *Escherichia coli*[a]

S. No.	Gene	Description of gene
1	Motility genes	
	fhiA	Encodes flagellar system
	ECs1554	Encodes a putative minor tail protein
2.	*Adhesion genes*	
	yadL, *yad*M, Z5029 and Z1651	Encoding for putative fimbrial-like adhesin proteins
3.	Lipopolysaccharide genes	
	waaB, *waaP*, *waaJ* and *waaR*	Encoding for lipid A of LPS
	waaU and *waaY*	Encoding for the phosphorylated core oligosaccharide of LPS
	waaL and *wzzB*	Encoding for the O antigen of LPS
4.	Biofilm architecture	
	yhjN	Encoding for cellulose synthase regulator
	wcaL and *wcaM*	Encoding for colanic acid synthesis
5.	Drug resistance, transport and active transport	
	*mdl*A and *mdt*M and *mdt*O	Encoding for multidrug drug efflux pumps
	*emr*B	Confers resistance to cyanide *m*-chlorophenyl-hydrazone, tetrachlorosalicylanilide, organo-mercurials, nalidixic acid and thiolactomycin
	sat	Encoding for the secreted auto-transporter toxin
	ompC	Encoding for secretion of extracellular proteins,
	htrE	Encoding putative chaperone-usher fimbrial protein
	mhpA, *mhpB*, *mhpF*, *yiaY*, *yajO*/*ydbK* and *yjjN*	Encoding dehydrogenases
6.	Pathogenicity and virulence genes	
	ECs3276	Encoding a virulence protein populating the cluster O80301
	sat, *yjgK*, *chpS*, *chpB* and *ygjN*	Encoding for toxin production and secretions that governs virulence in pathogenic bacteria
	*wzzB*and *waaL*	Encoding proteins that influence antigen presentation
	yiiX	Encoding extracellular matrix remodelling
	cbrB, *cbrC*, *hisI* and mgl*B*	Transport of amino acids and other metabolites
7.	DNA repair and stress genes	
	iraD, *uvrD* and *recC*	Encoding genes involved in DNA damage repair
	speB, *ybeW*, *hscC* and *djlC*	Encoding genes involved in stationary phase stress
	dsmB, *glpB* and *nrdD*	Encoding genes involved in anaerobic stress
	uspC	Encoding a universal stress protein (Beloin et al. 2004; Hancock and Klemm 2007; Ren et al. 2004; Schembri et al. 2003)
8.	*ryfA*	Encoding a small RNA (Bak et al. 2015)
9.	*tolA*	Encoding for the inner membrane protein-cell envelope integrity
10.	*bdcA* and *bdcR*	Encoding *bdc*R the regulatory gene of *bdcA*

(continued)

Table 4.4 (continued)

S. No.	Gene	Description of gene
11.	*atlE*	Encoding for autolysins
12.	*icaA* and *icaB*	Encoding for biofilm formation
13.	*icaA*, *icaD*, *arcA* and *opp*3AB	Encoding for biofilm formation
14.	*ladS*	Encoding for biofilm formation

[a] Data from Dave et al. (2020), Jassim et al. (2015), Makki et al. (2011), Flores-Paez et al. (2015), Jena et al. (2017), Ranjith et al. (2017, 2019)

pathogenic ABU *E. coli*. Additionally, ocular *E. coli* and *E. coli* K12 both biofilm-forming strains could be differentiated with respect to differences in the regulation of genes involved in mitosis and meiosis and cell cycle control (D), cell motility (N) and the above categories of U and V genes. Thus, expression of genes in ocular *E. coli* was similar to that of pathogenic *E. coli* rather than non-pathogenic *E. coli* K12 (Ranjith et al. 2017). In a subsequent study, homologous recombination was used to mutate seven genes (*bdc*R, *mhp*A, *mhp*B, *ryf*A, *tol*A, *ytf*R and *mdt*O) in ocular *E. coli* L-1216/2010 (Ranjith et al. 2017). Biofilm formation was inhibited in five of the mutants (Δ*bdc*R, Δ*mhp*A, Δ*mhp*B, Δ*ryf*A, and Δ*tol*A) and the thickness of the biofilm reduced from 17.2 mm in the wild type to 1.5 from 4.8 mm in the mutants. However, mutants Δ*ytf*R and Δ*mdt*O retained the potential to form biofilm. Further, complementation of the mutant genes with the wild-type gene restored biofilm formation in all mutants except in Δ*mhp*B. In the five mutants which lost their ability to form biofilm (Δ*bdc*R, Δ*mhp*A, Δ*mhp*B, Δ*ryf*A and Δ*tol*A), their susceptibility to antibiotics streptomycin, ceftazidime, cefuroxime, ciprofloxacin, erythromycin, gentamicin, cefotaxime, sulfamethoxazole and imipenem in the planktonic phase compared to wild-type ocular *E. coli* in the planktonic phase remained unchanged. ΔmdtO was the only mutant with altered MIC to sulfamethoxazole, erythromycin, imipenem and streptomycin both in the planktonic and biofilm phase. Involvement of *mhp*A, *mhp*B and *bdc*R in ocular *E. coli* biofilm formation is demonstrated for the first time. In addition, evidence was provided to indicate that *tol*A and *ryf*A are required for biofilm formation, while *ytf*R and *mdt*O are not required. It was also observed that inhibition of biofilm formation in the five *E. coli* mutants could also be attributed to inhibition of motility (as in Δ*tol*A) or swimming/swarming/EPS production as in Δ*bdc*R, Δ*mhp*A, Δ*mhp*B and Δ*ryf*A (Ranjith et al. 2019). Thus, targeted inactivation of genes could be a strategy to mitigate biofilm formation to overcome antibiotic resistance.

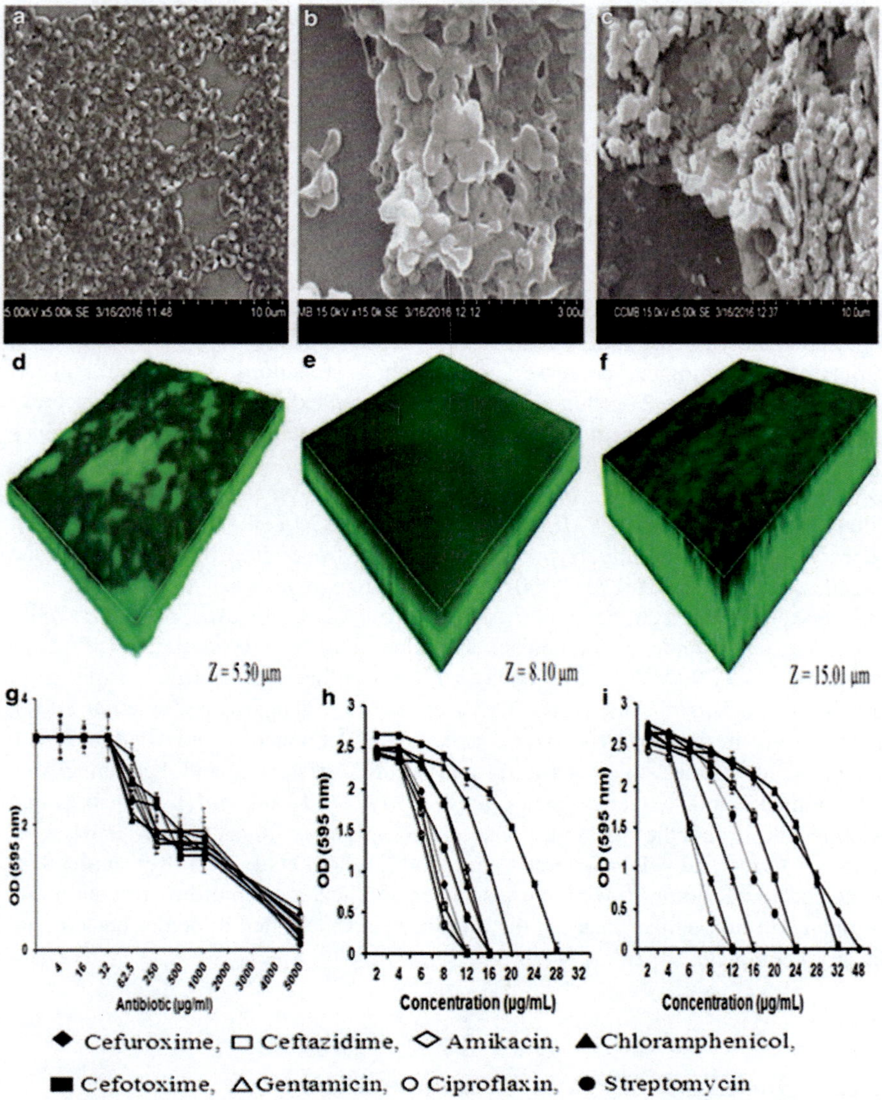

Fig. 4.7 Biofilm formation in ocular *E. coli* L-1216/2010. Scanning electron microscopy photomicrographs of biofilm formation after 24 h (**a**), 48 h (**b**) and 72 h (**c**). The thickness of the biofilm (*Z* axis, 5.30, 8.10 and 15.01 μm) by confocal scanning laser microscopy after 24 h (**d**), 48 h (**e**) and 72 h (**f**) of biofilm growth. *E. coli* was stained with Syto9 in (**d**)–(**f**). Antibiotic susceptibility of ocular *E. coli* L-1216/2010 in the biofilm phase (**g**), the planktonic phase (**h**) and non-biofilm-forming *E. coli* L-1339/2013 (**i**). (From Ranjith et al. 2017)

4.11 Comparison of Candidate Genes Involved in Biofilm Formation in Non-ocular Bacteria

Several genes such as *tolA* (Vianney et al. 2005; Whiteley et al. 2001), *csrA* (Jackson et al. 2002), *ymdB* (Kim et al. 2013), *ryfA* (Bak et al. 2015; Whiteley et al. 2001), *bdcA* (Ma et al. 2011), *ytfR* (Pinweha et al. 2018) and *ydgG* (Herzberg et al. 2006) and *spot* and *relA* (de la Fuente-Nunez et al. 2014) are required for biofilm formation in non-ocular bacteria (*P. aeruginosa*, *E. coli* or *Burkholderia pseudomallei*) (Vianney et al. 2005). Niba et al. (2007) identified 110 genes involved in biofilm formation in *E. coli*, and these included genes coding for motility (40 nos.), for type I fimbriae (7 nos.), Curli genes (6 nos.), LPS genes (8 nos.) and genes coding for a variety of physiological processes and metabolic functions. Alav et al. (2018) reported that 16 of the 21 efflux pump genes in *E. coli* exhibited increase or decrease in expression during biofilm formation. AMR is associated with efflux pump genes (Levy 2002), and several of these genes (*fsr*, *acrB*, *acrD*, *acrE*, *emrE*, *emrY*, *yde and tolC* and) were important for biofilm formation (De Kievit et al. 2001; Junker et al. 2006). The transport genes *MexAB-OprM* and *MexCD-OprJ* (Gillis et al. 2005), genes involved in motility (*crp* and *cya*A), genes encoding biosynthesis of the flagella (Chilcott and Hughes 2000), type I fimbriae genes (*fim* A to *fim*I), the *fim* like genes *fhi*A and *ECs1554* and Curli genes (*csg*A to *csg*G, *cya*, *yie*O, *ihf*B, *rpo*S, *lon*, *mlr*A, *omp*R and *nlp*D) (Hancock and Klemm 2007; Niba et al. 2007; Prigent-Combaret et al. 2000; Ren et al. 2005), LPS biosynthesis genes (*ipc*A, *rfa*D, *rfa*E, *rfa*F, *rfa*G, *rfa*H, *rfa*P and *gmh*B) (Niba et al. 2007; Hancock and Klemm 2007), cellulose synthesis genes (*bcs*ABZC and *bcs*EFG) (Hancock and Klemm 2007), colanic acid synthesis genes (*wca*L and *wca*M) (Hancock and Klemm 2007), pathogenicity and virulence genes (*c2418 to c2440* and *rfaH*) (Hancock and Klemm 2007) and the stress response encoding genes (*pph*A, *csp*G, *csp*H, *hha*, *sox*S, *ibp*A, *ibp* and *yfi*D) (Schembri et al. 2003; Beloin et al. 2004; Ren et al. 2004; Hancock and Klemm 2007) have also been implicated in biofilm formation. As compared to non-ocular bacteria, the biofilm genes identified in ocular bacteria are far from complete.

4.12 Hacking Biofilms to Overcome AMR

Eradication of biofilm-forming pathogens have essentially followed a two-way approach: in the first phase, efforts were made to prevent the formation of the biofilm either by inhibiting quorum sensing (QS) or by preventing the adhesion of cells to the substratum, and in the second approach, the strategy was to dismantle the already formed biofilm. In both these approaches, novel anti-biofilm agents have been screened to inhibit biofilm formation.

4.12.1 Using Inhibitors of Quorum Sensing

(Z)-5-(bromomethylene) furan-2(5H)-one, a synthetic furanone, has been used to inhibit biofilm formation in *Staphylococcus intermedius*, *S. epidermidis*, *S. mutans* and *Streptococcus anginosus*, by targeting the signalling pathway in biofilm formation (Lönn-Stensrud et al. 2009). Acyclic amines (Kaur et al. 2017) also inhibit biofilm formation in *S. aureus* by interacting with staphylococcal accessory regulator (Balamurugan et al. 2017; Arya and Princy 2012) of the virulence genes *hla*, *hld* and *fnbA*.

4.12.2 CRISPRi Technology to Mitigate Quorum Sensing

CRISPRi technology has been used to inactivate *lux*S or *fim*H so as to inhibit biofilm formation (Zuberi et al. 2017a). Both these genes following CRISPRi knockdown inhibited *E. coli* biofilm formation (Zuberi et al. 2017a, b).

4.12.3 Use of Biocides to Inhibit Adhesion

Biocides such as poly(4-vinyl-*N*-alkylpyridinium bromide) and *N*-hexylated poly (4-vinylpyridine) were used to coat glass slides which were used for biofilm formation, and this coating caused a 100-fold reduction in cell counts when *P. aeruginosa*, *S. epidermidis*, *S. aureus* and or *E. coli* were allowed to colonise on them (Tiller et al. 2001). Further, antibiotics such as rifampin, doxycycline and clarithromycin were also used as surface coatings to prevent in vitro biofilm formation (Rose et al. 2015). An IOL coated with norfloxacin has also inhibited post-bacterial infection after cataract surgery and was successfully tested also in a rabbit model (Garty et al. 2011).

4.12.4 Use of Antimicrobial Compounds Including Peptides to Inhibit Adhesion

Antimicrobial peptides including antimicrobial compounds like gallium and silver have also been used to inhibit biofilm formation (Kazemzadeh-Narbat et al. 2013). The advantage of this approach is that the antimicrobial agents could be released all through the adhesion phase and thus could interfere and inhibit biofilm formation, but the compounds should be compatible with the host (Desrousseaux et al. 2013). But long-term exposure to antimicrobial peptides and antimicrobial compounds may favour selection of spontaneous resistant mutants and have undesirable effects on the ocular surface microbiome and are not recommended.

4.12.5 Modifying the Surface Structure of Ocular Devices to Inhibit Adhesion

An alternate approach to using biocides to coat ocular devices would be to modify the surface of the ocular device so that bacteria do not adhere to the surface. But the challenge in this approach is to retain the optical properties of the devices. Polymers, such as polyethylene glycol, polyacrylamide and dextran, which are known to form nanostructures which interfere with ability of the microbes to adhere to the substratum could also be used (May et al. 2014; Salwiczek et al. 2014). Other materials that have been tried for this purpose include nanopores, nanotubes and nanopillars made of titanium dioxide, anodised aluminium or polymethylmethacrylate (Desrousseaux et al. 2013).

4.12.6 D-amino Acids to Disperse Biofilms

Biofilms of *E. coli*, *S. aureus*, *Bacillus subtilis* and *P. aeruginosa* could be dispersed using D-amino acids (Kolodkin-Gal et al. 2010; Romero et al. 2011; Hochbaum et al. 2011). D-amino acids prevent cell surface-associated proteins to connect neighbouring cells in to large aggregates and thus trigger biofilm disassembly by (Geoghegan et al. 2010).

4.12.7 Antibiotics to Disperse Biofilms

Antibiotics are the first choice for killing cells in the planktonic phase but are ineffective against dormant cells in a biofilm. But a combination of antibiotics (amikacin, ciprofloxacin and clarithromycin) with *N*-acetylcysteine and Tween 80 disperse bacteria biofilms (Munoz-Egea et al. 2016). Nanoparticles of NiO, ZnO, CuO and Sb_2O_3 were also evaluated for toxicity effects in *Streptococcus aureus*, *E. coli* and *Bacillus subtilis*, and CuO nanoparticles were the most toxic (Baek and An 2011). Nanoparticles overcome drug resistance by increasing efflux, decreasing uptake and inhibiting biofilm formation (Pelgrift and Friedman 2013). The acyldepsipeptide antibiotic, ADEP4, binds to ClpP a specific protease of *S. aureus* and transforms ClpP into a nonspecific protease that degrades over 400 proteins killing both growing and dormant cells (Conlon et al. 2013). It was also demonstrated that in a mouse biofilm infection model, ADEP4 in combination with rifampicin reduced the number of *S. aureus* below detectable limits, while neither vancomycin, rifampicin nor ADEP4 was able to do so individually (Conlon et al. 2013).

4.12.8 Enzymatic Degradation of the Biofilm Matrix to Disperse Biofilms

Disruption of the EPS matrix could disperse mature biofilms. The endogenous enzyme dispersin B in biofilm cells disperses the biofilm and facilitates subsequent formation of new biofilms by degrading EPS (Flemming et al. 2016; Chaignon et al. 2007). Dispersin B effectively disperses biofilms of *Achromobacter xylosoxidans*, *S. epidermidis* and *Burkholderia cenocepacia* (Dobrynina et al. 2015). DNase I is yet another enzyme which could individually and in combination with several antibiotics (Tetz et al. 2009) disrupt the biofilm matrix and thus increase the killing effect of antibiotics (Torelli et al. 2017; Tetz et al. 2009) in *Enterococcus faecalis* and *E. faecium*. Oils derived from specific plants (*Cinnamomum tamala* and *Pogostemon heyneanus*) along with DNases also eradicate already existing biofilms (Rubini et al. 2018) in *Staphylococcus aureus* which are methicillin-resistant (MRSA). Several other enzymes like proteases in combination ethylenediamine tetra-acetic acid (EDTA) and antiseptics (Lefebvre et al. 2016), glycoside hydrolases (Baker et al. 2016), proteinase K (Kumar Shukla and Rao 2013) and α-amylase (Watters et al. 2016) also disrupt biofilm matrix components. Studies in animal models are required for validation of the above results.

4.12.9 Use of Natural Products to Disperse Biofilms

Natural products like azithromycin, usnic acid, ginseng and garlic extract also inhibit bacterial and fungal biofilms (Bjarnsholt et al. 2005; Wu et al. 2015; Ding et al. 2018).

4.12.10 Use of Signalling Molecules to Disperse Biofilms

As mentioned earlier small signalling molecules have also been implicated in the dispersion phase, and these appear to be species-specific and include *N*-butanoyl-L-homoserine lactone (as in *P. aeruginosa* biofilms) and *Pseudomonas* quinolone signal (PQS) (as in *Pseudomonas*) (Dong et al. 2008), the staphylococcal peptides δ-toxin and AIP-I (as in *Staphylococci*) (Kaplan 2010) and the furanosyl borate autoinducer AI-2 (as in *Vibrio cholera*). Nitric oxide (NO) decreases in intracellular c-di-GMP levels (Barraud et al. 2009b) and thus is used to regulate dispersal of biofilms in *S. aureus* (Schlag et al. 2007) and *P. aeruginosa* (Barraud et al. 2006) and to disperse various single and multispecies biofilms (Barraud et al. 2009a, b).

4.12.11 Suppression of Enzymes and Inhibition of Biofilm Formation

Biofilm formation could also be inhibited when enzymes known to be important for biofilm formation like those associated with cell adhesion, acid tolerance and production, glucan synthesis and quorum sensing were inhibited (Kulshrestha et al. 2016). In *Streptococcus mutans* biofilm formation could be inhibited by calcium fluoride nanoparticles which inhibited EPS synthesis, acid production and acid tolerance. Further, the treatment with calcium fluoride caused significant downregulation of virulence genes (*spa*P, *ftf*, *com*DE, *gtf*C and *vicR*) associated with biofilm formation.

4.12.12 Use of Photodynamic Therapy to Eliminate Biofilms

Photodynamic therapy (PDT) is a mechanism which generates reactive oxygen species (ROS) in the presence of visible light, oxygen and a photosensitiser and thus kills Gram-negative and Gram-positive bacteria and inhibits biofilm formation as in *Streptococcus mutants* (Misba et al. 2016). Compared to the Gram-negative bacteria (*Klebsiella pneumonia*), the Gram-positive bacteria (*Enterococcus faecalis*) are more susceptible to PDT due to higher photosensitiser binding efficiency and increased ROS-mediated DNA degradation (Misba et al. 2017; Misba and Khan 2018). The expression of biofilm-related genes is reduced.

4.12.13 Phage Therapy to Inhibit Biofilm Formation

Lytic phages that destroy or lyse cells are more effective in inhibiting bacteria in mice compared to antibiotics (Smith and Huggins 1982; Alemayehu et al. 2012), and a combination of phage and antibiotics was more effective in killing bacteria like *P. aeruginosa* (Lu and Collins 2009; Chaudhry et al. 2017; Torres-Barcelo et al. 2014) and inhibiting biofilms (Danis-Wlodarczyk et al. 2016). Engineered phages expressing anti-biofilm enzymes like bacteriophage endolysins (Schmelcher et al. 2015) and dispersin B (DspB) (Lu and Collins 2007) were effective in killing both planktonic as well as biofilms of *S. aureus* and *E. coli*.

4.12.14 Other Strategies to Inhibit Biofilm Formation

Dispersal of biofilms could also be achieved by inducing autolysis of cells (Ma et al. 2009), by passive detachment of cells from the biofilm by either chemicals, by force or by environmental changes (Kaplan 2010) and by the release of rhamnolipids or phenol-soluble modulins (Aleksic et al. 2017; Bonnichsen et al. 2015; Angelini et al. 2009). Thus, understanding the molecular basis of dispersal could form the basis for the development of new agents that will efficiently hack biofilms.

4.12.15 Limitations in Hacking Strategies

Several strategies have been tried to overcome biofilm formation and associated AMR. Major limitations of these strategies are that they are restricted to in vitro experiments. Several strategies described above like the use of signalling molecules, nanoparticles, D-amino acids, halogenated furanones and natural products could make the transition from laboratory use to the clinic at least to treat skin disorders. But phage therapy without successful trials in humans would not be implemented in the clinic. But phage therapy along with antibiotic therapy could be tried (Gupta 2011).

4.13 The Gap Areas

In ocular bacteria biofilm formation, AMR and strategies to hack biofilms have not been documented in detail compared to the data available for non-ocular bacterial isolates. A few of the gap areas include:

1. Identification of ocular bacteria up to the species level.
2. Comprehensive studies involving screening of ocular isolates for biofilm formation potential, antimicrobial resistance and the relationship between biofilm formation and antimicrobial resistance are required for all ocular bacteria.
3. The molecular basis of AMR and biofilm formation in ocular isolates.
4. Understand the competition between the diverse genera and species in a biofilm.
5. Screen naturally occurring and synthetic compounds for dispersing biofilms.
6. Initiate studies on polymicrobial biofilms.

4.14 Conclusions

Biofilm formation and AMR need to be addressed in greater detail in specific ocular isolates rather than type strains. The concept of polymicrobial biofilms and their response to antibiotics is an area that needs to be investigated. It is needed to understand competition between different bacteria in a biofilm and the impact it has on AMR. Systematically screen naturally occurring and synthetic compounds as agents for dispersing biofilms so as to overcome AMR.

References

Adan A, Casaroli-Marano RP, Gris O et al (2008) Pathological findings in the lens capsules and intraocular lens in chronic pseudophakic endophthalmitis: an electron microscopy study. Eye (Lond) 22:113–119. https://doi.org/10.1038/sj.eye.6702615

Ainbinder DJ, Parmley VC, Mader TH et al (1998) Infectious crystalline keratopathy caused by Candida guilliermondii. Am J Ophthalmol 125:723–725. https://doi.org/10.1016/s0002-9394(98)00024-5

Alav I, Sutton JM, Rahman KM (2018) Role of bacterial efflux pumps in biofilm formation. J Antimicrob Chemother 73:2003–2020. https://doi.org/10.1093/jac/dky042

Aleksic I, Petkovic M, Jovanovic M et al (2017) Anti-biofilm properties of bacterial di-rhamnolipids and their semi-synthetic amide derivatives. Front Microbiol 8:2454. https://doi.org/10.3389/fmicb.2017.02454

Alemayehu D, Casey PG, McAuliffe O et al (2012) Bacteriophages phiMR299-2 and phiNH-4 can eliminate Pseudomonas aeruginosa in the murine lung and on cystic fibrosis lung airway cells. MBio 3:e00029-00012. https://doi.org/10.1128/mBio.00029-12

Allcock S, Young EH, Holmes M et al (2017) Antimicrobial resistance in human populations: challenges and opportunities. Glob Health Epidemiol Genom 2:e4. https://doi.org/10.1017/gheg.2017.4

Allison DG, Ruiz B, SanJose C et al (1998) Extracellular products as mediators of the formation and detachment of Pseudomonas fluorescens biofilms. FEMS Microbiol Lett 167:179–184. https://doi.org/10.1111/j.1574-6968.1998.tb13225.x

Amores GR, de Las HA, Sanches-Medeiros A et al (2017) Systematic identification of novel regulatory interactions controlling biofilm formation in the bacterium Escherichia coli. Sci Rep 7:16768. https://doi.org/10.1038/s41598-017-17114-6

Andrade FF, Silva D, Rodrigues A et al (2020) Colistin update on its mechanism of action and resistance, present and future challenges. Microorganisms 8:1716. https://doi.org/10.3390/microorganisms8111716

Angelini TE, Roper M, Kolter R et al (2009) Bacillus subtilis spreads by surfing on waves of surfactant. Proc Natl Acad Sci U S A 106:18109–18113. https://doi.org/10.1073/pnas.0905890106

Arya R, Princy SA (2012) Computational approach to design small molecule inhibitors and identify SarA as a potential therapeutic candidate. Med Chem Res 22:1856–1865. https://doi.org/10.1007/s00044-012-0185-9

Asaria RH, Downie JA, McLauglin-Borlace L et al (1999) Biofilm on scleral explants with and without clinical infection. Retina 19:447–450. https://doi.org/10.1097/00006982-199909000-00014

Baek YW, An YJ (2011) Microbial toxicity of metal oxide nanoparticles (CuO, NiO, ZnO, and Sb2O3) to Escherichia coli, Bacillus subtilis, and Streptococcus aureus. Sci Total Environ 409: 1603–1608. https://doi.org/10.1016/j.scitotenv.2011.01.014

Baidamshina DR, Trizna EY, Holyavka MG et al (2017) Targeting microbial biofilms using Ficin, a nonspecific plant protease. Sci Rep 7:46068. https://doi.org/10.1038/srep46068

Baillif S, Leduff F, Hartmann DJ et al (2013) Staphylococcus epidermidis biofilm formation and structural organization on different types of intraocular lenses under in vitro flow conditions. Ophthalmic Res 50:83–90. https://doi.org/10.1159/000350550

Bak G, Lee J, Suk S et al (2015) Identification of novel sRNAs involved in biofilm formation, motility, and fimbriae formation in Escherichia coli. Sci Rep 5:15287. https://doi.org/10.1038/srep15287

Baker P, Hill PJ, Snarr BD et al (2016) Exopolysaccharide biosynthetic glycoside hydrolases can be utilized to disrupt and prevent Pseudomonas aeruginosa biofilms. Sci Adv 2:e1501632. https://doi.org/10.1126/sciadv.1501632

Balamurugan P, Praveen Krishna V, Bharath D et al (2017) Staphylococcus aureus quorum regulator SarA targeted compound, 2-[(methylamino)methyl]phenol inhibits biofilm and down-regulates virulence genes. Front Microbiol 8:1290. https://doi.org/10.3389/fmicb.2017.01290

Balaram M, Schaumberg DA, Dana MR (2001) Efficacy and tolerability outcomes after punctal occlusion with silicone plugs in dry eye syndrome. Am J Ophthalmol 131:30–36. https://doi.org/10.1016/s0002-9394(00)00620-6

Balikoglu-Yilmaz M, Yilmaz T, Cetinel S et al (2014) Comparison of scanning electron microscopy findings regarding biofilm colonization with microbiological results in nasolacrimal stents for external, endoscopic and transcanalicular dacryocystorhinostomy. Int J Ophthalmol 7:534–540. https://doi.org/10.3980/j.issn.2222-3959.2014.03.27

Barraud N, Hassett DJ, Hwang SH et al (2006) Involvement of nitric oxide in biofilm dispersal of Pseudomonas aeruginosa. J Bacteriol 188:7344–7353. https://doi.org/10.1128/JB.00779-06

Barraud N, Schleheck D, Klebensberger J et al (2009a) Nitric oxide signaling in Pseudomonas aeruginosa biofilms mediates phosphodiesterase activity, decreased cyclic di-GMP levels, and enhanced dispersal. J Bacteriol 191:7333–7342. https://doi.org/10.1128/JB.00975-09

Barraud N, Storey MV, Moore ZP et al (2009b) Nitric oxide-mediated dispersal in single- and multi-species biofilms of clinically and industrially relevant microorganisms. Microb Biotechnol 2:370–378. https://doi.org/10.1111/j.1751-7915.2009.00098.x

Bateman A, Holden MT, Yeats C (2005) The G5 domain: a potential N-acetylglucosamine recognition domain involved in biofilm formation. Bioinformatics 21:1301–1303. https://doi.org/10.1093/bioinformatics/bti206

Baty AM III, Eastburn CC, Techkarnjanaruk S et al (2000) Spatial and temporal variations in chitinolytic gene expression and bacterial biomass production during chitin degradation. Appl Environ Microbiol 66:3574–3585. https://doi.org/10.1128/AEM.66.8.3574-3585.2000

Behlau I, Gilmore MS (2008) Microbial biofilms in ophthalmology and infectious disease. Arch Ophthalmol 126:1572–1581. https://doi.org/10.1001/archopht.126.11.1572

Beloin C, Valle J, Latour-Lambert P et al (2004) Global impact of mature biofilm lifestyle on Escherichia coli K-12 gene expression. Mol Microbiol 51:659–674. https://doi.org/10.1046/j.1365-2958.2003.03865.x

Bennett PM (2008) Plasmid encoded antibiotic resistance: acquisition and transfer of antibiotic resistance genes in bacteria. Br J Pharmacol 153(Suppl 1):S347–S357. https://doi.org/10.1038/sj.bjp.0707607

Benz MS, Scott IU, Flynn HW Jr et al (2004) Endophthalmitis isolates and antibiotic sensitivities: a 6-year review of culture-proven cases. Am J Ophthalmol 137:38–42. https://doi.org/10.1016/s0002-9394(03)00896-1

Bharathi MJ, Ramakrishnan R, Shivakumar C et al (2010) Etiology and antibacterial susceptibility pattern of community-acquired bacterial ocular infections in a tertiary eye care hospital in south India. Indian J Ophthalmol 58:497–507. https://doi.org/10.4103/0301-4738.71678

Billings N, Birjiniuk A, Samad TS et al (2015) Material properties of biofilms-a review of methods for understanding permeability and mechanics. Rep Prog Phys 78:036601. https://doi.org/10.1088/0034-4885/78/3/036601

Bispo PJ, Miller D (2012) Distinct frequency of biofilm (bf)-related genes among ciprofloxacin (cip) susceptible and resistant S. epidermidis (sepi) agr types i (ti) and ii (tii) isolates from endophthalmitis (end). In: Proceedings of the Interscience Conference on Antimicrobial Agents and Chemotherapy, San Francisco, CA, USA, p 912

Bispo PJ, Melo GB, d'Azevedo PA et al (2008) [Culture proven bacterial endophthalmitis: a 6-year review]. Arq Bras Oftalmol 71:617–622. https://doi.org/10.1590/s0004-27492008000500002

Bispo PJ, Haas W, Gilmore MS (2015) Biofilms in infections of the eye. Pathogens 4:111–136. https://doi.org/10.3390/pathogens4010111

Bjarnsholt T, Jensen PO, Rasmussen TB et al (2005) Garlic blocks quorum sensing and promotes rapid clearing of pulmonary Pseudomonas aeruginosa infections. Microbiology (Reading) 151: 3873–3880. https://doi.org/10.1099/mic.0.27955-0

Block SL, Hedrick J, Tyler R et al (2000) Increasing bacterial resistance in pediatric acute conjunctivitis (1997-1998). Antimicrob Agents Chemother 44:1650–1654. https://doi.org/10.1128/AAC.44.6.1650-1654.2000

Blumer C, Kleefeld A, Lehnen D et al (2005) Regulation of type 1 fimbriae synthesis and biofilm formation by the transcriptional regulator LrhA of Escherichia coli. Microbiology (Reading) 151:3287–3298. https://doi.org/10.1099/mic.0.28098-0

Bonnichsen L, Bygvraa Svenningsen N, Rybtke M et al (2015) Lipopeptide biosurfactant viscosin enhances dispersal of Pseudomonas fluorescens SBW25 biofilms. Microbiology (Reading) 161: 2289–2297. https://doi.org/10.1099/mic.0.000191

Bourcier T, Thomas F, Borderie V et al (2003) Bacterial keratitis: predisposing factors, clinical and microbiological review of 300 cases. Br J Ophthalmol 87:834–838. https://doi.org/10.1136/bjo.87.7.834

Bourkiza R, Lee V (2012) A review of the complications of lacrimal occlusion with punctal and canalicular plugs. Orbit 31:86–93. https://doi.org/10.3109/01676830.2011.648802

Bowden MG, Chen W, Singvall J et al (2005) Identification and preliminary characterization of cell-wall-anchored proteins of Staphylococcus epidermidis. Microbiology (Reading) 151:1453–1464. https://doi.org/10.1099/mic.0.27534-0

Boyd A, Chakrabarty AM (1994) Role of alginate lyase in cell detachment of Pseudomonas aeruginosa. Appl Environ Microbiol 60:2355–2359. https://doi.org/10.1128/aem.60.7.2355-2359.1994

Brolund A, Sundqvist M, Kahlmeter G et al (2010) Molecular characterisation of trimethoprim resistance in Escherichia coli and Klebsiella pneumoniae during a two year intervention on trimethoprim use. PLoS One 5:e9233. https://doi.org/10.1371/journal.pone.0009233

Bruinsma GM, van der Mei HC, Busscher HJ (2001) Bacterial adhesion to surface hydrophilic and hydrophobic contact lenses. Biomaterials 22:3217–3224. https://doi.org/10.1016/s0142-9612(01)00159-4

Buckner MMC, Ciusa ML, Piddock LJV (2018) Strategies to combat antimicrobial resistance: anti-plasmid and plasmid curing. FEMS Microbiol Rev 42:781–804. https://doi.org/10.1093/femsre/fuy031

Buttner H, Mack D, Rohde H (2015) Structural basis of Staphylococcus epidermidis biofilm formation: mechanisms and molecular interactions. Front Cell Infect Microbiol 5:14. https://doi.org/10.3389/fcimb.2015.00014

Castro-Sanchez E, Moore LS, Husson F et al (2016) What are the factors driving antimicrobial resistance? Perspectives from a public event in London, England. BMC Infect Dis 16:465. https://doi.org/10.1186/s12879-016-1810-x

Catalanotti P, Lanza M, Del Prete A et al (2005) Slime-producing Staphylococcus epidermidis and S. aureus in acute bacterial conjunctivitis in soft contact lens wearers. New Microbiol 28:345–354

Cavuoto K, Zutshi D, Karp CL et al (2008) Update on bacterial conjunctivitis in South Florida. Ophthalmology 115:51–56. https://doi.org/10.1016/j.ophtha.2007.03.076

Chaignon P, Sadovskaya I, Ragunah C et al (2007) Susceptibility of staphylococcal biofilms to enzymatic treatments depends on their chemical composition. Appl Microbiol Biotechnol 75: 125–132. https://doi.org/10.1007/s00253-006-0790-y

Chalita MR, Hofling-Lima AL, Paranhos A Jr et al (2004) Shifting trends in in vitro antibiotic susceptibilities for common ocular isolates during a period of 15 years. Am J Ophthalmol 137: 43–51. https://doi.org/10.1016/s0002-9394(03)00905-x

Characklis WG (1990) Biofilm processes. In: Characklis WG, Marshall K (eds) Biofilms. John Wiley & Sons, New York, NY, pp 195–231

Chaudhry NA, Flynn HW Jr, Murray TG et al (1999) Emerging ciprofloxacin-resistant Pseudomonas aeruginosa. Am J Ophthalmol 128:509–510. https://doi.org/10.1016/s0002-9394(99)00196-8

Chaudhry WN, Concepcion-Acevedo J, Park T et al (2017) Synergy and order effects of antibiotics and phages in killing Pseudomonas aeruginosa biofilms. PLoS One 12:e0168615. https://doi.org/10.1371/journal.pone.0168615

Cheng KH, Leung SL, Hoekman HW et al (1999) Incidence of contact-lens-associated microbial keratitis and its related morbidity. Lancet 354:181–185. https://doi.org/10.1016/S0140-6736(98)09385-4

Chilcott GS, Hughes KT (2000) Coupling of flagellar gene expression to flagellar assembly in Salmonella enterica serovar typhimurium and Escherichia coli. Microbiol Mol Biol Rev 64: 694–708. https://doi.org/10.1128/MMBR.64.4.694-708.2000

Clarke SR, Foster SJ (2006) Surface adhesins of Staphylococcus aureus. In: Poole RK (ed) Advances in microbial physiology, vol 51. Academic Press, London, pp 187–224. https://doi.org/10.1016/S0065-2911(06)51004-5

Conlon BP, Nakayasu ES, Fleck LE et al (2013) Activated ClpP kills persisters and eradicates a chronic biofilm infection. Nature 503:365–370. https://doi.org/10.1038/nature12790

Cordoba A, Graue-Hernandez EO, Bermudez-Magner JA et al (2019) Corneal biofilm plaques: a novel clinical presentation. Cornea 38:764–767. https://doi.org/10.1097/ICO.0000000000001923

Costerton JW, Irvin RT, Cheng KJ (1981) The bacterial glycocalyx in nature and disease. Annu Rev Microbiol 35:299–324. https://doi.org/10.1146/annurev.mi.35.100181.001503

Cui H, Ma C, Lin L (2016) Co-loaded proteinase K/thyme oil liposomes for inactivation of Escherichia coli O157:H7 biofilms on cucumber. Food Funct 7:4030–4040. https://doi.org/10.1039/c6fo01201a

Daneshvar H, MacInnis B, Hodge WG (2000) Nd:YAG laser corneal disruption as adjuvant treatment for infectious crystalline keratopathy. Am J Ophthalmol 129:800–801. https://doi.org/10.1016/s0002-9394(00)00468-2

Danis-Wlodarczyk K, Vandenheuvel D, Jang HB et al (2016) A proposed integrated approach for the preclinical evaluation of phage therapy in Pseudomonas infections. Sci Rep 6:28115. https://doi.org/10.1038/srep28115

Dantas G, Sommer M (2014) How to fight back against antibiotic resistance. Am Sci 102:42. https://doi.org/10.1511/2014.106.42

Darouiche RO (2004) Treatment of infections associated with surgical implants. N Engl J Med 350: 1422–1429. https://doi.org/10.1056/NEJMra035415

Dart JK, Seal DV (1988) Pathogenesis and therapy of Pseudomonas aeruginosa keratitis. Eye (Lond) 2(Suppl):S46–S55. https://doi.org/10.1038/eye.1988.133

Das S, Samantaray R, Mallick A et al (2019) Types of organisms and in-vitro susceptibility of bacterial isolates from patients with microbial keratitis: a trend analysis of 8 years. Indian J Ophthalmol 67:49–53. https://doi.org/10.4103/ijo.IJO_500_18

Dave A, Samarth A, Karolia R et al (2020) Characterization of ocular clinical isolates of pseudomonas aeruginosa from non-contact lens related keratitis patients from South India. Microorganisms 8:260. https://doi.org/10.3390/microorganisms8020260

Davies D (2003) Understanding biofilm resistance to antibacterial agents. Nat Rev Drug Discov 2: 114–122. https://doi.org/10.1038/nrd1008

Davies DG, Parsek MR, Pearson JP et al (1998) The involvement of cell-to-cell signals in the development of a bacterial biofilm. Science 280:295–298. https://doi.org/10.1126/science.280.5361.295

Davies SC, Fowler T, Watson J et al (2013) Annual report of the chief medical officer: infection and the rise of antimicrobial resistance. Lancet 381:1606–1609. https://doi.org/10.1016/S0140-6736(13)60604-2

De Kievit TR, Gillis R, Marx S et al (2001) Quorum-sensing genes in Pseudomonas aeruginosa biofilms: their role and expression patterns. Appl Environ Microbiol 67:1865–1873. https://doi.org/10.1128/AEM.67.4.1865-1873.2001

Desrousseaux C, Sautou V, Descamps S et al (2013) Modification of the surfaces of medical devices to prevent microbial adhesion and biofilm formation. J Hosp Infect 85:87–93. https://doi.org/10.1016/j.jhin.2013.06.015

Deziel E, Comeau Y, Villemur R (2001) Initiation of biofilm formation by Pseudomonas aeruginosa 57RP correlates with emergence of hyperpiliated and highly adherent phenotypic variants deficient in swimming, swarming, and twitching motilities. J Bacteriol 183:1195–1204. https://doi.org/10.1128/JB.183.4.1195-1204.2001

Ding W, Zhou Y, Qu Q et al (2018) Azithromycin inhibits biofilm formation by Staphylococcus xylosus and affects histidine biosynthesis pathway. Front Pharmacol 9:740. https://doi.org/10.3389/fphar.2018.00740

Doan T, Akileswaran L, Andersen D et al (2016) Paucibacterial microbiome and resident DNA virome of the healthy conjunctiva. Invest Ophthalmol Vis Sci 57:5116–5126. https://doi.org/10.1167/iovs.16-19803

Dobrynina OY, Bolshakova TN, Umyarov AM et al (2015) Disruption of bacterial biofilms using recombinant dispersin B. Microbiology 84:498–501. https://doi.org/10.1134/s0026261715040062

Dong YH, Zhang XF, An SW et al (2008) A novel two-component system BqsS-BqsR modulates quorum sensing-dependent biofilm decay in Pseudomonas aeruginosa. Commun Integr Biol 1: 88–96. https://doi.org/10.4161/cib.1.1.6717

Donlan RM, Costerton JW (2002) Biofilms: survival mechanisms of clinically relevant microorganisms. Clin Microbiol Rev 15:167–193. https://doi.org/10.1128/CMR.15.2.167-193.2002

Dow JM, Crossman L, Findlay K et al (2003) Biofilm dispersal in Xanthomonas campestris is controlled by cell-cell signaling and is required for full virulence to plants. Proc Natl Acad Sci U S A 100:10995–11000. https://doi.org/10.1073/pnas.1833360100

Doyle A, Beigi B, Early A et al (1995) Adherence of bacteria to intraocular lenses: a prospective study. Br J Ophthalmol 79:347–349. https://doi.org/10.1136/bjo.79.4.347

Duggirala A, Kenchappa P, Sharma S et al (2007) High-resolution genome profiling differentiated Staphylococcus epidermidis isolated from patients with ocular infections and normal individuals. Invest Ophthalmol Vis Sci 48:3239–3245. https://doi.org/10.1167/iovs.06-1365

Dutta D, Cole N, Willcox M (2012) Factors influencing bacterial adhesion to contact lenses. Mol Vis 18:14–21

Džidić S, Šušković J, Kos B (2008) Antibiotic resistance mechanisms in bacteria: biochemical and genetic aspects. Food Technol Biotechnol 46:11–21

Eguchi H, Kuwahara T, Miyamoto T et al (2008) High-level fluoroquinolone resistance in ophthalmic clinical isolates belonging to the species Corynebacterium macginleyi. J Clin Microbiol 46:527–532. https://doi.org/10.1128/JCM.01741-07

Elder MJ, Stapleton F, Evans E et al (1995) Biofilm-related infections in ophthalmology. Eye (Lond) 9(Pt 1):102–109. https://doi.org/10.1038/eye.1995.16

Elder MJ, Matheson M, Stapleton F et al (1996) Biofilm formation in infectious crystalline keratopathy due to Candida albicans. Cornea 15:301–304. https://doi.org/10.1097/00003226-199605000-00012

Fleiszig SM, Evans DJ (2010) Pathogenesis of contact lens-associated microbial keratitis. Optom Vis Sci 87:225–232. https://doi.org/10.1097/OPX.0b013e3181d408ee

Flemming HC, Wingender J (2010) The biofilm matrix. Nat Rev Microbiol 8:623–633. https://doi.org/10.1038/nrmicro2415

Flemming HC, Wingender J, Szewzyk U et al (2016) Biofilms: an emergent form of bacterial life. Nat Rev Microbiol 14:563–575. https://doi.org/10.1038/nrmicro.2016.94

Flores-Paez LA, Zenteno JC, Alcantar-Curiel MD et al (2015) Molecular and phenotypic characterization of Staphylococcus epidermidis isolates from healthy conjunctiva and a comparative analysis with isolates from ocular infection. PLoS One 10:e0135964. https://doi.org/10.1371/journal.pone.0135964

Foster TJ, Hook M (1998) Surface protein adhesins of Staphylococcus aureus. Trends Microbiol 6: 484–488. https://doi.org/10.1016/s0966-842x(98)01400-0

Fridkin SK, Hageman JC, Morrison M et al (2005) Methicillin-resistant Staphylococcus aureus disease in three communities. N Engl J Med 352:1436–1444. https://doi.org/10.1056/NEJMoa043252

de la Fuente-Nunez C, Reffuveille F, Haney EF et al (2014) Broad-spectrum anti-biofilm peptide that targets a cellular stress response. PLoS Pathog 10:e1004152. https://doi.org/10.1371/journal.ppat.1004152

Fulcher TP, Dart JKG, McLaughlin-Borlace L et al (2001) Demonstration of biofilm in infectious crystalline keratopathy using ruthenium red and electron microscopy. Ophthalmology 108: 1088–1092. https://doi.org/10.1016/s0161-6420(01)00561-9

Fux CA, Costerton JW, Stewart PS et al (2005) Survival strategies of infectious biofilms. Trends Microbiol 13:34–40. https://doi.org/10.1016/j.tim.2004.11.010

Gabriel MM, Ahearn DG, Chan KY et al (1998) In vitro adherence of Pseudomonas aeruginosa to four intraocular lenses. J Cataract Refract Surg 24:124

García-Sáenz MC, Arias-Puente A, Fresnadillo-Martinez MJ et al (2000) In vitro adhesion of Staphylococcus epidermidis to intraocular lenses. J Cataract Refract Surg 26:1673

Garty S, Shirakawa R, Warsen A et al (2011) Sustained antibiotic release from an intraocular lens-hydrogel assembly for cataract surgery. Invest Ophthalmol Vis Sci 52:6109–6116. https://doi.org/10.1167/iovs.10-6071

Geoghegan JA, Corrigan RM, Gruszka DT et al (2010) Role of surface protein SasG in biofilm formation by Staphylococcus aureus. J Bacteriol 192:5663–5673. https://doi.org/10.1128/JB.00628-10

Georgiou T, Qureshi SH, Chakrabarty A et al (2002) Biofilm formation and coccal organisms in infectious crystalline keratopathy. Eye (Lond) 16:89–92. https://doi.org/10.1038/sj.eye.6700020

Gerdes K, Maisonneuve E (2012) Bacterial persistence and toxin-antitoxin loci. Annu Rev Microbiol 66:103–123. https://doi.org/10.1146/annurev-micro-092611-150159

Gillis RJ, White KG, Choi KH et al (2005) Molecular basis of azithromycin-resistant Pseudomonas aeruginosa biofilms. Antimicrob Agents Chemother 49:3858–3867. https://doi.org/10.1128/AAC.49.9.3858-3867.2005

Gordon CA, Hodges NA, Marriott C (1988) Antibiotic interaction and diffusion through alginate and exopolysaccharide of cystic fibrosis-derived Pseudomonas aeruginosa. J Antimicrob Chemother 22:667–674. https://doi.org/10.1093/jac/22.5.667

Griffiths PG, Elliot TS, McTaggart L (1989) Adherence of Staphylococcus epidermidis to intraocular lenses. Br J Ophthalmol 73:402–406. https://doi.org/10.1136/bjo.73.6.402

Gross M, Cramton SE, Gotz F et al (2001) Key role of teichoic acid net charge in Staphylococcus aureus colonization of artificial surfaces. Infect Immun 69:3423–3426. https://doi.org/10.1128/IAI.69.5.3423-3426.2001

Guilhen C, Forestier C, Balestrino D (2017) Biofilm dispersal: multiple elaborate strategies for dissemination of bacteria with unique properties. Mol Microbiol 105:188–210. https://doi.org/10.1111/mmi.13698

Guiotti AM, da Silva EVF, Catanoze IA et al (2018) Microbiological analysis of conjunctival secretion in anophthalmic cavity, contralateral eye and ocular prosthesis of patients with maxillofacial abnormalities. Lett Appl Microbiol 66:104–109. https://doi.org/10.1111/lam.12830

Gupta SK (2011) Non-inferiority clinical trials: practical issues and current regulatory perspective. Indian J Pharm 43:371–374. https://doi.org/10.4103/0253-7613.83103

Hammersmith KM (2006) Diagnosis and management of Acanthamoeba keratitis. Curr Opin Ophthalmol 17:327–331. https://doi.org/10.1097/01.icu.0000233949.56229.7d

Han DP, Wisniewski SR, Wilson LA et al (1996) Spectrum and susceptibilities of microbiologic isolates in the Endophthalmitis Vitrectomy Study. Am J Ophthalmol 122:1–17. https://doi.org/10.1016/s0002-9394(14)71959-2

Hancock V, Klemm P (2007) Global gene expression profiling of asymptomatic bacteriuria Escherichia coli during biofilm growth in human urine. Infect Immun 75:966–976. https://doi.org/10.1128/IAI.01748-06

Harper T, Miller D, Flynn HW Jr (2007) In vitro efficacy and pharmacodynamic indices for antibiotics against coagulase-negative staphylococcus endophthalmitis isolates. Ophthalmology 114:871–875. https://doi.org/10.1016/j.ophtha.2007.01.007

Herrmann M, Vaudaux PE, Pittet D et al (1988) Fibronectin, fibrinogen, and laminin act as mediators of adherence of clinical staphylococcal isolates to foreign material. J Infect Dis 158:693–701. https://doi.org/10.1093/infdis/158.4.693

Herzberg M, Kaye IK, Peti W et al (2006) YdgG (TqsA) controls biofilm formation in Escherichia coli K-12 through autoinducer 2 transport. J Bacteriol 188:587–598. https://doi.org/10.1128/JB.188.2.587-598.2006

Hochbaum AI, Kolodkin-Gal I, Foulston L et al (2011) Inhibitory effects of D-amino acids on Staphylococcus aureus biofilm development. J Bacteriol 193:5616–5622. https://doi.org/10.1128/JB.05534-11

Holland SP, Pulido JS, Miller D et al (1991) Biofilm and scleral buckle-associated infections. A mechanism for persistence. Ophthalmology 98:933–938. https://doi.org/10.1016/s0161-6420(91)32199-7

Hooper DC, Wolfson JS, Ng EY et al (1987) Mechanisms of action of and resistance to ciprofloxacin. Am J Med 82:12–20

Hou W, Sun X, Wang Z et al (2012) Biofilm-forming capacity of Staphylococcus epidermidis, Staphylococcus aureus, and Pseudomonas aeruginosa from ocular infections. Invest Ophthalmol Vis Sci 53:5624–5631. https://doi.org/10.1167/iovs.11-9114

Hoyle BD, Alcantara J, Costerton JW (1992) Pseudomonas aeruginosa biofilm as a diffusion barrier to piperacillin. Antimicrob Agents Chemother 36:2054–2056. https://doi.org/10.1128/AAC.36.9.2054

Huang XD, Yao K, Zhang H et al (2007) Surface modification of silicone intraocular lens by 2-methacryloyloxyethyl phosphoryl-choline binding to reduce Staphylococcus epidermidis adherence. Clin Exp Ophthalmol 35:462–467. https://doi.org/10.1111/j.1442-9071.2007.01516.x

Hunts JH, Matoba AY, Osato MS et al (1993) Infectious crystalline keratopathy. The role of bacterial exopolysaccharide. Arch Ophthalmol 111:528–530. https://doi.org/10.1001/archopht.1993.01090040120044

Ibanez A, Trinidad A, Garcia-Berrocal JR et al (2011) Biofilm colonisation in nasolacrimal stents. B-ENT 7:7–10

Jackson DW, Suzuki K, Oakford L et al (2002) Biofilm formation and dispersal under the influence of the global regulator CsrA of Escherichia coli. J Bacteriol 184:290–301. https://doi.org/10.1128/JB.184.1.290-301.2002

Jassim SH, Sivaraman KR, Jimenez JC et al (2015) Bacteria colonizing the ocular surface in eyes with Boston type 1 keratoprosthesis: analysis of biofilm-forming capability and vancomycin tolerance. Invest Ophthalmol Vis Sci 56:4689–4696. https://doi.org/10.1167/iovs.15-17101

Jefferson KK, Goldmann DA, Pier GB (2005) Use of confocal microscopy to analyze the rate of vancomycin penetration through Staphylococcus aureus biofilms. Antimicrob Agents Chemother 49:2467–2473. https://doi.org/10.1128/AAC.49.6.2467-2473.2005

Jena S, Panda S, Nayak KC et al (2017) Identification of major sequence types among multidrug-resistant Staphylococcus epidermidis strains isolated from infected eyes and healthy conjunctiva. Front Microbiol 8:1430. https://doi.org/10.3389/fmicb.2017.01430

Joganathan V, Mehta P, Murray A et al (2010) Complications of intracanalicular plugs: a case series. Orbit 29:271–273. https://doi.org/10.3109/01676831003664376

Joo HS, Otto M (2012) Molecular basis of in vivo biofilm formation by bacterial pathogens. Chem Biol 19:1503–1513. https://doi.org/10.1016/j.chembiol.2012.10.022

Joseph J, Sontam B, Guda SJM et al (2019) Trends in microbiological spectrum of endophthalmitis at a single tertiary care ophthalmic hospital in India: a review of 25 years. Eye (Lond) 33:1090–1095. https://doi.org/10.1038/s41433-019-0380-8

Juarez-Verdayes MA, Ramon-Perez ML, Flores-Paez LA et al (2013) Staphylococcus epidermidis with the icaA(-)/icaD(-)/IS256(-) genotype and protein or protein/extracellular-DNA biofilm is frequent in ocular infections. J Med Microbiol 62:1579–1587. https://doi.org/10.1099/jmm.0.055210-0

Junker LM, Peters JE, Hay AG (2006) Global analysis of candidate genes important for fitness in a competitive biofilm using DNA-array-based transposon mapping. Microbiology (Reading) 152:2233–2245. https://doi.org/10.1099/mic.0.28767-0

Junker LM, Toba FA, Hay AG (2007) Transcription in Escherichia coli PHL628 biofilms. FEMS Microbiol Lett 268:237–243. https://doi.org/10.1111/j.1574-6968.2006.00585.x

Kaplan JB (2010) Biofilm dispersal: mechanisms, clinical implications, and potential therapeutic uses. J Dent Res 89:205–218. https://doi.org/10.1177/0022034509359403

Kapoor G, Saigal S, Elongavan A (2017) Action and resistance mechanisms of antibiotics: a guide for clinicians. J Anaesthesiol Clin Pharmacol 33:300–305. https://doi.org/10.4103/joacp.JOACP_349_15

Katiyar R, Vishwakarma A, Kaistha SD (2012) Analysis of biofilm formation and antibiotic resistance of microbial isolates from intraocular lens following conventional extracapsular cataract surgery. Int J Res Pure Appl Microbiol 2:20–24

Kaur G, Balamurugan P, Vasudevan S et al (2017) Antimicrobial and antibiofilm potential of acyclic amines and diamines against multi-drug resistant Staphylococcus aureus. Front Microbiol 8:1767. https://doi.org/10.3389/fmicb.2017.01767

Kazemzadeh-Narbat M, Lai BF, Ding C et al (2013) Multilayered coating on titanium for controlled release of antimicrobial peptides for the prevention of implant-associated infections. Biomaterials 34:5969–5977. https://doi.org/10.1016/j.biomaterials.2013.04.036

Kazi MS, Sharma VR, Kumar S et al (2015) Indications and outcomes of scleral buckle removal in a tertiary eye care center in South India. Oman J Ophthalmol 8:171–174. https://doi.org/10.4103/0974-620X.169891

Keay L, Edwards K, Naduvilath T et al (2006) Microbial keratitis predisposing factors and morbidity. Ophthalmology 113:109–116. https://doi.org/10.1016/j.ophtha.2005.08.013

Keay LJ, Gower EW, Iovieno A et al (2011) Clinical and microbiological characteristics of fungal keratitis in the United States, 2001-2007: a multicenter study. Ophthalmology 118:920–926. https://doi.org/10.1016/j.ophtha.2010.09.011

Khan W, Bernier SP, Kuchma SL et al (2010) Aminoglycoside resistance of Pseudomonas aeruginosa biofilms modulated by extracellular polysaccharide. Int Microbiol 13:207–212. https://doi.org/10.2436/20.1501.01.127

Khater TT, Jones DB, Wilhelmus KR (1997) Infectious crystalline keratopathy caused by gram-negative bacteria. Am J Ophthalmol 124:19–23. https://doi.org/10.1016/s0002-9394(14)71638-1

de Kievit TR (2009) Quorum sensing in Pseudomonas aeruginosa biofilms. Environ Microbiol 11:279–288. https://doi.org/10.1111/j.1462-2920.2008.01792.x

Kim SJ, Toma HS (2011) Ophthalmic antibiotics and antimicrobial resistance a randomized, controlled study of patients undergoing intravitreal injections. Ophthalmology 118:1358–1363. https://doi.org/10.1016/j.ophtha.2010.12.014

Kim T, Lee J, Kim KS (2013) Escherichia coli YmdB regulates biofilm formation independently of its role as an RNase III modulator. BMC Microbiol 13:266. https://doi.org/10.1186/1471-2180-13-266

Kirchhoff L, Arweiler-Harbeck D, Arnolds J et al (2020) Imaging studies of bacterial biofilms on cochlear implants-bioactive glass (BAG) inhibits mature biofilm. PLoS One 15:e0229198. https://doi.org/10.1371/journal.pone.0229198

Klotz SA, Butrus SI, Misra RP et al (1989) The contribution of bacterial surface hydrophobicity to the process of adherence of Pseudomonas aeruginosa to hydrophilic contact lenses. Curr Eye Res 8:195–202. https://doi.org/10.3109/02713688908995192

Kobayakawa S, Jett BD, Gilmore MS (2005) Biofilm formation by Enterococcus faecalis on intraocular lens material. Curr Eye Res 30:741–745. https://doi.org/10.1080/02713680591005959

Kodjikian L, Burillon C, Roques C et al (2003) Bacterial adherence of Staphylococcus epidermidis to intraocular lenses: a bioluminescence and scanning electron microscopy study. Invest Ophthalmol Vis Sci 44:4388–4394. https://doi.org/10.1167/iovs.03-0186

Kolodkin-Gal I, Romero D, Cao S et al (2010) D-amino acids trigger biofilm disassembly. Science 328:627–629. https://doi.org/10.1126/science.1188628

de Kraker ME, Stewardson AJ, Harbarth S (2016) Will 10 million people die a year due to antimicrobial resistance by 2050? PLoS Med 13:e1002184. https://doi.org/10.1371/journal.pmed.1002184

Kresloff MS, Castellarin AA, Zarbin MA (1998) Endophthalmitis. Surv Ophthalmol 43:193–224. https://doi.org/10.1016/s0039-6257(98)00036-8

Kulshrestha S, Khan S, Hasan S et al (2016) Calcium fluoride nanoparticles induced suppression of Streptococcus mutans biofilm: an in vitro and in vivo approach. Appl Microbiol Biotechnol 100:1901–1914. https://doi.org/10.1007/s00253-015-7154-4

Kumar Shukla S, Rao TS (2013) Dispersal of Bap-mediated Staphylococcus aureus biofilm by proteinase K. J Antibiot (Tokyo) 66:55–60. https://doi.org/10.1038/ja.2012.98

Kumar S, Singh B (2013) An overview of mechanisms and emergence of antimicrobial drug resistance. Adv Anim Vet Sci 1:7–14

Kumarasamy KK, Toleman MA, Walsh TR et al (2010) Emergence of a new antibiotic resistance mechanism in India, Pakistan, and the UK: a molecular, biological, and epidemiological study. Lancet Infect Dis 10:597–602. https://doi.org/10.1016/S1473-3099(10)70143-2

Kunimoto DY, Sharma S, Garg P et al (1999) In vitro susceptibility of bacterial keratitis pathogens to ciprofloxacin. Ophthalmology 106:80–85. https://doi.org/10.1016/s0161-6420(99)90008-8

Lasa I, Penades JR (2006) Bap: a family of surface proteins involved in biofilm formation. Res Microbiol 157:99–107. https://doi.org/10.1016/j.resmic.2005.11.003

Lee AE, Niruttan K, Rawson TM et al (2019) Antibacterial resistance in ophthalmic infections: a multi-centre analysis across UK care settings. BMC Infect Dis 19:768. https://doi.org/10.1186/s12879-019-4418-0

Lefebvre E, Vighetto C, Di Martino P et al (2016) Synergistic antibiofilm efficacy of various commercial antiseptics, enzymes and EDTA: a study of Pseudomonas aeruginosa and Staphylococcus aureus biofilms. Int J Antimicrob Agents 48:181–188. https://doi.org/10.1016/j.ijantimicag.2016.05.008

Levy SB (2002) Active efflux, a common mechanism for biocide and antibiotic resistance. J Appl Microbiol 92:65S–S71. https://doi.org/10.1046/j.1365-2672.92.5s1.4.x

Li B, Webster TJ (2018) Bacteria antibiotic resistance: new challenges and opportunities for implant-associated orthopedic infections. J Orthop Res 36:22–32. https://doi.org/10.1002/jor.23656

Li L, Mendis N, Trigui H et al (2014) The importance of the viable but non-culturable state in human bacterial pathogens. Front Microbiol 5:258. https://doi.org/10.3389/fmicb.2014.00258

Lichtinger A, Yeung SN, Kim P et al (2012) Shifting trends in bacterial keratitis in Toronto: an 11-year review. Ophthalmology 119:1785–1790. https://doi.org/10.1016/j.ophtha.2012.03.031

Lister JL, Horswill AR (2014) Staphylococcus aureus biofilms: recent developments in biofilm dispersal. Front Cell Infect Microbiol 4:178. https://doi.org/10.3389/fcimb.2014.00178

Lönn-Stensrud J, Landin MA, Benneche T, Petersen FC, Scheie AA (2009) Furanones, potential agents for preventing Staphylococcus epidermidis biofilm infections? J Antimicrob Chemother 63(2):309–316. https://doi.org/10.1093/jac/dkn501. Epub 2008 Dec 20

Lu TK, Collins JJ (2007) Dispersing biofilms with engineered enzymatic bacteriophage. Proc Natl Acad Sci U S A 104:11197–11202. https://doi.org/10.1073/pnas.0704624104

Lu TK, Collins JJ (2009) Engineered bacteriophage targeting gene networks as adjuvants for antibiotic therapy. Proc Natl Acad Sci U S A 106:4629–4634. https://doi.org/10.1073/pnas.0800442106

Ma L, Conover M, Lu H et al (2009) Assembly and development of the Pseudomonas aeruginosa biofilm matrix. PLoS Pathog 5:e1000354. https://doi.org/10.1371/journal.ppat.1000354

Ma Q, Yang Z, Pu M et al (2011) Engineering a novel c-di-GMP-binding protein for biofilm dispersal. Environ Microbiol 13:631–642. https://doi.org/10.1111/j.1462-2920.2010.02368.x

Mack D, Riedewald J, Rohde H et al (1999) Essential functional role of the polysaccharide intercellular adhesin of Staphylococcus epidermidis in hemagglutination. Infect Immun 67:1004–1008. https://doi.org/10.1128/IAI.67.2.1004-1008.1999

Mah TF, O'Toole GA (2001) Mechanisms of biofilm resistance to antimicrobial agents. Trends Microbiol 9:34–39. https://doi.org/10.1016/s0966-842x(00)01913-2

Makki AR, Sharma S, Duggirala A et al (2011) Phenotypic and genotypic characterization of coagulase negative staphylococci (CoNS) other than Staphylococcus epidermidis isolated from ocular infections. Invest Ophthalmol Vis Sci 52:9018–9022. https://doi.org/10.1167/iovs.11-7777

Mann EE, Rice KC, Boles BR et al (2009) Modulation of eDNA release and degradation affects Staphylococcus aureus biofilm maturation. PLoS One 4:e5822. https://doi.org/10.1371/journal.pone.0005822

Masselos K, Tsang HH, Ooi JL et al (2009) Laser corneal biofilm disruption for infectious crystalline keratopathy. Clin Exp Ophthalmol 37:177–180. https://doi.org/10.1111/j.1442-9071.2008.01912.x

Maxe I, Ryden C, Wadstrom T et al (1986) Specific attachment of Staphylococcus aureus to immobilized fibronectin. Infect Immun 54:695–704. https://doi.org/10.1128/iai.54.3.695-704.1986

May RM, Hoffman MG, Sogo MJ et al (2014) Micro-patterned surfaces reduce bacterial colonization and biofilm formation in vitro: potential for enhancing endotracheal tube designs. Clin Transl Med 3:8. https://doi.org/10.1186/2001-1326-3-8

Mayo MS, Cook WL, Schlitzer RL et al (1986) Antibiograms, serotypes, and plasmid profiles of Pseudomonas aeruginosa associated with corneal ulcers and contact lens wear. J Clin Microbiol 24:372–376. https://doi.org/10.1128/jcm.24.3.372-376.1986

McElroy MC, Cain DJ, Tyrrell C et al (2002) Increased virulence of a fibronectin-binding protein mutant of Staphylococcus aureus in a rat model of pneumonia. Infect Immun 70:3865–3873. https://doi.org/10.1128/IAI.70.7.3865-3873.2002

McGavin MH, Krajewska-Pietrasik D, Ryden C et al (1993) Identification of a Staphylococcus aureus extracellular matrix-binding protein with broad specificity. Infect Immun 61:2479–2485. https://doi.org/10.1128/iai.61.6.2479-2485.1993

Meisler DM, Langston RH, Naab TJ et al (1984) Infectious crystalline keratopathy. Am J Ophthalmol 97:337–343. https://doi.org/10.1016/0002-9394(84)90633-0

Mihara E, Shimizu M, Touge C et al (2004) Case of a large, movable bacterial concretion with biofilm formation on the ocular surface. Cornea 23:513–515. https://doi.org/10.1097/01.ico.0000114838.91149.6d

Miller DM, Vedula AS, Flynn HW Jr et al (2007) Endophthalmitis caused by staphylococcus epidermidis: in vitro antibiotic susceptibilities and clinical outcomes. Ophthalmic Surg Lasers Imaging 38:446–451. https://doi.org/10.3928/15428877-20071101-01

Miller KV, Eisley KM, Shanks RM et al (2011) Recurrent enterococcal endophthalmitis seeded by an intraocular lens biofilm. J Cataract Refract Surg 37:1355–1359. https://doi.org/10.1016/j.jcrs.2011.04.014

Misba L, Khan AU (2018) Enhanced photodynamic therapy using light fractionation against Streptococcus mutans biofilm: type I and type II mechanism. Future Microbiol 13:437–454. https://doi.org/10.2217/fmb-2017-0207

Misba L, Kulshrestha S, Khan AU (2016) Antibiofilm action of a toluidine blue O-silver nanoparticle conjugate on Streptococcus mutans: a mechanism of type I photodynamic therapy. Biofouling 32:313–328. https://doi.org/10.1080/08927014.2016.1141899

Misba L, Zaidi S, Khan AU (2017) A comparison of antibacterial and antibiofilm efficacy of phenothiazinium dyes between Gram positive and Gram negative bacterial biofilm. Photodiagn Photodyn Ther 18:24–33. https://doi.org/10.1016/j.pdpdt.2017.01.177

Moreno A, Dos Santos DM, Neto CL d MM et al (2020) In vitro evaluation of the effect of different disinfectants on the biofilm of Staphylococcus epidermidis and Staphylococcus aureus formed on acrylic ocular prostheses. PLoS One 15:e0240116. https://doi.org/10.1371/journal.pone.0240116

Morrissey I, Burnett R, Viljoen L et al (2004) Surveillance of the susceptibility of ocular bacterial pathogens to the fluoroquinolone gatifloxacin and other antimicrobials in Europe during 2001/2002. J Infect 49:109–114. https://doi.org/10.1016/j.jinf.2004.03.007

Muloi D, Kiiru J, Ward MJ et al (2019) Epidemiology of antimicrobial-resistant Escherichia coli carriage in sympatric humans and livestock in a rapidly urbanizing city. Int J Antimicrob Agents 54:531–537. https://doi.org/10.1016/j.ijantimicag.2019.08.014

Munoz-Egea MC, Garcia-Pedrazuela M, Mahillo-Fernandez I et al (2016) Effect of antibiotics and antibiofilm agents in the ultrastructure and development of biofilms developed by nonpigmented rapidly growing mycobacteria. Microb Drug Resist 22:1–6. https://doi.org/10.1089/mdr.2015.0124

Niba ET, Naka Y, Nagase M et al (2007) A genome-wide approach to identify the genes involved in biofilm formation in E. coli. DNA Res 14:237–246. https://doi.org/10.1093/dnares/dsm024

Nickel JC, Downey JA, Costerton JW (1989) Ultrastructural study of microbiologic colonization of urinary catheters. Urology 34:284–291. https://doi.org/10.1016/0090-4295(89)90327-0

O'Toole GA, Kolter R (1998) Flagellar and twitching motility are necessary for Pseudomonas aeruginosa biofilm development. Mol Microbiol 30:295–304. https://doi.org/10.1046/j.1365-2958.1998.01062.x

O'Toole G, Kaplan HB, Kolter R (2000) Biofilm formation as microbial development. Annu Rev Microbiol 54:49–79. https://doi.org/10.1146/annurev.micro.54.1.49

Okajima Y, Kobayakawa S, Tsuji A et al (2006) Biofilm formation by Staphylococcus epidermidis on intraocular lens material. Invest Ophthalmol Vis Sci 47:2971–2975. https://doi.org/10.1167/iovs.05-1172

Oluyombo O, Penfold CN, Diggle SP (2019) Competition in biofilms between cystic fibrosis isolates of Pseudomonas aeruginosa is shaped by R-pyocins. MBio 10:e01828. https://doi.org/10.1128/mBio.01828-18

Ormerod LD, Hertzmark E, Gomez DS et al (1987) Epidemiology of microbial keratitis in southern California. A multivariate analysis. Ophthalmology 94:1322–1333. https://doi.org/10.1016/s0161-6420(87)80019-2

Osakabe Y, Yaguchi C, Miyai T et al (2006) Detection of Streptococcus species by polymerase chain reaction in infectious crystalline keratopathy. Cornea 25:1227–1230. https://doi.org/10.1097/01.ico.0000240083.31432.70

Otto M (2008) Staphylococcal biofilms. Curr Top Microbiol Immunol 322:207–228. https://doi.org/10.1007/978-3-540-75418-3_10

Paharik AE, Horswill AR (2016) The staphylococcal biofilm: adhesins, regulation, and host response. Microbiol Spectr 4. https://doi.org/10.1128/microbiolspec.VMBF-0022-2015

Pakkulnan R, Anutrakunchai C, Kanthawong S et al (2019) Extracellular DNA facilitates bacterial adhesion during Burkholderia pseudomallei biofilm formation. PLoS One 14:e0213288. https://doi.org/10.1371/journal.pone.0213288

Papenfort K, Bassler BL (2016) Quorum sensing signal-response systems in Gram-negative bacteria. Nat Rev Microbiol 14:576–588. https://doi.org/10.1038/nrmicro.2016.89

Paranhos RM, Batalhao CH, Semprini M et al (2007) Evaluation of ocular prosthesis biofilm and anophthalmic cavity contamination after use of three cleansing solutions. J Appl Oral Sci 15:33–38. https://doi.org/10.1590/s1678-77572007000100008

Parsa K, Schaudinn C, Gorur A et al (2010) Demonstration of bacterial biofilms in culture-negative silicone stent and jones tube. Ophthalmic Plast Reconstr Surg 26:426–430. https://doi.org/10.1097/IOP.0b013e3181cff61f

Passos RM, Cariello AJ, Yu MC et al (2010) Microbial keratitis in the elderly: a 32-year review. Arq Bras Oftalmol 73:315–319. https://doi.org/10.1590/s0004-27492010000400002

Pathengay A, Karosekar S, Raju B et al (2004) Microbiologic spectrum and susceptibility of isolates in scleral buckle infection in India. Am J Ophthalmol 138:663–664. https://doi.org/10.1016/j.ajo.2004.04.056

Pathengay A, Flynn HW Jr, Isom RF et al (2012) Endophthalmitis outbreaks following cataract surgery: causative organisms, etiologies, and visual acuity outcomes. J Cataract Refract Surg 38:1278–1282. https://doi.org/10.1016/j.jcrs.2012.04.021

Pei L, Palma M, Nilsson M et al (1999) Functional studies of a fibrinogen binding protein from Staphylococcus epidermidis. Infect Immun 67:4525–4530. https://doi.org/10.1128/IAI.67.9.4525-4530.1999

Pelgrift RY, Friedman AJ (2013) Nanotechnology as a therapeutic tool to combat microbial resistance. Adv Drug Deliv Rev 65:1803–1815. https://doi.org/10.1016/j.addr.2013.07.011

Pichi F, Nucci P, Baynes K et al (2014) Acute and chronic Staphylococcus epidermidis post-operative endophthalmitis: the importance of biofilm production. Int Ophthalmol 34:1267–1270. https://doi.org/10.1007/s10792-014-0011-0

Pilchová T, Hernould M, Prévost H et al (2014) Influence of food processing environments on structure initiation of static biofilm of Listeria monocytogenes. Food Control 35:366–372. https://doi.org/10.1016/j.foodcont.2013.07.021

Pinweha P, Pumirat P, Cuccui J et al (2018) Inactivation of bpsl1039-1040 ATP-binding cassette transporter reduces intracellular survival in macrophages, biofilm formation and virulence in the murine model of Burkholderia pseudomallei infection. PLoS One 13:e0196202. https://doi.org/10.1371/journal.pone.0196202

Pogliano J, Pogliano N, Silverman JA (2012) Daptomycin-mediated reorganization of membrane architecture causes mislocalization of essential cell division proteins. J Bacteriol 194:4494–4504. https://doi.org/10.1128/JB.00011-12

Portolés M, Refojo MF, Leong F-L (1993) Reduced bacterial adhesion to heparin-surface-modified intraocular lenses. J Cataract Refract Surg 19:755

Prigent-Combaret C, Prensier G, Le Thi TT et al (2000) Developmental pathway for biofilm formation in curli-producing Escherichia coli strains: role of flagella, curli and colanic acid. Environ Microbiol 2:450–464. https://doi.org/10.1046/j.1462-2920.2000.00128.x

Ramappa M, Majji AB, Murthy SI et al (2012) An outbreak of acute post-cataract surgery Pseudomonas sp. endophthalmitis caused by contaminated hydrophilic intraocular lens solution. Ophthalmology 119:564–570. https://doi.org/10.1016/j.ophtha.2011.09.031

Ramphal R, McNiece MT, Polack FM (1981) Adherence of Pseudomonas aeruginosa to the injured cornea: a step in the pathogenesis of corneal infections. Ann Ophthalmol 13:421–425

Ranjith K, Arunasri K, Reddy GS et al (2017) Global gene expression in Escherichia coli, isolated from the diseased ocular surface of the human eye with a potential to form biofilm. Gut Pathog 9:15. https://doi.org/10.1186/s13099-017-0164-2

Ranjith K, Ramchiary J, Prakash JSS et al (2019) Gene targets in ocular pathogenic Escherichia coli for mitigation of biofilm formation to overcome antibiotic resistance. Front Microbiol 10:1308. https://doi.org/10.3389/fmicb.2019.01308

Ranjith K, SaiAbhilash CR, Sai Prashanthi G et al (2020) Phylogenetic grouping of human ocular Escherichia coli based on whole-genome sequence analysis. Microorganisms 8:422. https://doi.org/10.3390/microorganisms8030422

Ranjith K, Saiabhilash CR, Shivaji S (2021a) Biofilm-forming potential of ocular fluid Staphylococcus aureus and Staphylococcus epidermidis on ex vivo human corneas from attachment to dispersal phase. Microorganisms 9:1124. https://doi.org/10.3390/microorganisms9061124

Ranjith K, Sharma S, Shivaji S (2021b) Microbes of the human eye: microbiome, antimicrobial resistance and biofilm formation. Exp Eye Res 205:108476. https://doi.org/10.1016/j.exer.2021.108476

Raskin EM, Speaker MG, McCormick SA et al (1993) Influence of haptic materials on the adherence of staphylococci to intraocular lenses. Arch Ophthalmol 111:250–253. https://doi.org/10.1001/archopht.1993.01090020104032

Rau G, Seedor JA, Shah MK et al (2008) Incidence and clinical characteristics of enterococcus keratitis. Cornea 27:895–899. https://doi.org/10.1097/ICO.0b013e31816f633b

Reiss GR, Campbell RJ, Bourne WM (1986) Infectious crystalline keratopathy. Surv Ophthalmol 31:69–72. https://doi.org/10.1016/0039-6257(86)90053-6

Ren DH, Petroll WM, Jester JV et al (1999) The relationship between contact lens oxygen permeability and binding of Pseudomonas aeruginosa to human corneal epithelial cells after overnight and extended wear. CLAO J 25:80–100

Ren D, Bedzyk LA, Thomas SM et al (2004) Gene expression in Escherichia coli biofilms. Appl Microbiol Biotechnol 64:515–524. https://doi.org/10.1007/s00253-003-1517-y

Ren CP, Beatson SA, Parkhill J et al (2005) The Flag-2 locus, an ancestral gene cluster, is potentially associated with a novel flagellar system from Escherichia coli. J Bacteriol 187: 1430–1440. https://doi.org/10.1128/JB.187.4.1430-1440.2005

Reygaert WC (2018) An overview of the antimicrobial resistance mechanisms of bacteria. AIMS Microbiol 4:482–501. https://doi.org/10.3934/microbiol.2018.3.482

Rhem MN, Wilhelmus KR, Font RL (1996) Infectious crystalline keratopathy caused by Candida parapsilosis. Cornea 15:543–545

Rohde H, Burandt EC, Siemssen N et al (2007) Polysaccharide intercellular adhesin or protein factors in biofilm accumulation of Staphylococcus epidermidis and Staphylococcus aureus isolated from prosthetic hip and knee joint infections. Biomaterials 28:1711–1720. https://doi.org/10.1016/j.biomaterials.2006.11.046

Romero D, Vlamakis H, Losick R et al (2011) An accessory protein required for anchoring and assembly of amyloid fibres in B. subtilis biofilms. Mol Microbiol 80:1155–1168. https://doi.org/10.1111/j.1365-2958.2011.07653.x

Rose WE, Otto DP, Aucamp ME et al (2015) Prevention of biofilm formation by methacrylate-based copolymer films loaded with rifampin, clarithromycin, doxycycline alone or in combination. Pharm Res 32:61–73. https://doi.org/10.1007/s11095-014-1444-x

Rubini D, Banu SF, Nisha P et al (2018) Essential oils from unexplored aromatic plants quench biofilm formation and virulence of Methicillin resistant Staphylococcus aureus. Microb Pathog 122:162–173. https://doi.org/10.1016/j.micpath.2018.06.028

Rutherford ST, Bassler BL (2012) Bacterial quorum sensing: its role in virulence and possibilities for its control. Cold Spring Harb Perspect Med 2:a012427. https://doi.org/10.1101/cshperspect.a012427

Sadaka A, Durand ML, Gilmore MS (2012) Bacterial endophthalmitis in the age of outpatient intravitreal therapies and cataract surgeries: host-microbe interactions in intraocular infection. Prog Retin Eye Res 31:316–331. https://doi.org/10.1016/j.preteyeres.2012.03.004

Salwiczek M, Qu Y, Gardiner J et al (2014) Emerging rules for effective antimicrobial coatings. Trends Biotechnol 32:82–90. https://doi.org/10.1016/j.tibtech.2013.09.008

Samimi DB, Bielory BP, Miller D et al (2013) Microbiologic trends and biofilm growth on explanted periorbital biomaterials: a 30-year review. Ophthalmic Plast Reconstr Surg 29:376–381. https://doi.org/10.1097/IOP.0b013e31829a7313

Saraswathi P, Beuerman RW (2015) Corneal biofilms: from planktonic to microcolony formation in an experimental keratitis infection with pseudomonas aeruginosa. Ocul Surf 13:331–345. https://doi.org/10.1016/j.jtos.2015.07.001

Sauer K, Camper AK, Ehrlich GD et al (2002) Pseudomonas aeruginosa displays multiple phenotypes during development as a biofilm. J Bacteriol 184:1140–1154. https://doi.org/10.1128/jb.184.4.1140-1154.2002

Schaeffer CR, Woods KM, Longo GM et al (2015) Accumulation-associated protein enhances Staphylococcus epidermidis biofilm formation under dynamic conditions and is required for infection in a rat catheter model. Infect Immun 83:214–226. https://doi.org/10.1128/IAI.02177-14

Schauersberger J, Amon M, Aichinger D et al (2003) Bacterial adhesion to rigid and foldable posterior chamber intraocular lenses: in vitro study. J Cataract Refract Surg 29:361–366. https://doi.org/10.1016/s0886-3350(02)01741-8

Scheie HG, Grayson MC (1971) Ocular manifestations of systemic diseases. Dis Mon 17:1–51. https://doi.org/10.1016/s0011-5029(71)80006-8

Schembri MA, Kjaergaard K, Klemm P (2003) Global gene expression in Escherichia coli biofilms. Mol Microbiol 48:253–267. https://doi.org/10.1046/j.1365-2958.2003.03432.x

Schlag S, Nerz C, Birkenstock TA et al (2007) Inhibition of staphylococcal biofilm formation by nitrite. J Bacteriol 189:7911–7919. https://doi.org/10.1128/JB.00598-07

Schmelcher M, Shen Y, Nelson DC et al (2015) Evolutionarily distinct bacteriophage endolysins featuring conserved peptidoglycan cleavage sites protect mice from MRSA infection. J Antimicrob Chemother 70:1453–1465. https://doi.org/10.1093/jac/dku552

Shimizu K, Kobayakawa S, Tsuji A et al (2006) Biofilm formation on hydrophilic intraocular lens material. Curr Eye Res 31:989–997. https://doi.org/10.1080/02713680601038816

Shimizu Y, Toshida H, Honda R et al (2013) Prevalence of drug resistance and culture-positive rate among microorganisms isolated from patients with ocular infections over a 4-year period. Clin Ophthalmol 7:695–702. https://doi.org/10.2147/OPTH.S43323

Shirodkar AR, Pathengay A, Flynn HW Jr et al (2012) Delayed- versus acute-onset endophthalmitis after cataract surgery. Am J Ophthalmol 153:391–398.e392. https://doi.org/10.1016/j.ajo.2011.08.029

Silver LL (2016) Appropriate targets for antibacterial drugs. Cold Spring Harb Perspect Med 6: a030239. https://doi.org/10.1101/cshperspect.a030239

Smiddy WE, Miller D, Flynn HW (1993) Scleral buckle removal following retinal reattachment surgery: clinical and microbiologic aspects. Ophthal Surg Las Imag Ret 24:440–445. https://doi.org/10.3928/1542-8877-19930701-04

Smith HW, Huggins MB (1982) Successful treatment of experimental Escherichia coli infections in mice using phage: its general superiority over antibiotics. J Gen Microbiol 128:307–318. https://doi.org/10.1099/00221287-128-2-307

Soto SM (2013) Role of efflux pumps in the antibiotic resistance of bacteria embedded in a biofilm. Virulence 4:223–229. https://doi.org/10.4161/viru.23724

Stapleton F, Dart J (1995) Pseudomonas keratitis associated with biofilm formation on a disposable soft contact lens. Br J Ophthalmol 79:864–865. https://doi.org/10.1136/bjo.79.9.864

Stewart PS, Franklin MJ (2008) Physiological heterogeneity in biofilms. Nat Rev Microbiol 6:199–210. https://doi.org/10.1038/nrmicro1838

Stover CK, Pham XQ, Erwin AL et al (2000) Complete genome sequence of Pseudomonas aeruginosa PAO1, an opportunistic pathogen. Nature 406:959–964. https://doi.org/10.1038/35023079

Sugita J, Yokoi N, Fullwood NJ et al (2001) The detection of bacteria and bacterial biofilms in punctal plug holes. Cornea 20:362–365. https://doi.org/10.1097/00003226-200105000-00005

Suzuki T, Kawamura Y, Uno T et al (2005) Prevalence of Staphylococcus epidermidis strains with biofilm-forming ability in isolates from conjunctiva and facial skin. Am J Ophthalmol 140:844–850. https://doi.org/10.1016/j.ajo.2005.05.050

Tack KJ, Sabath LD (1985) Increased minimum inhibitory concentrations with anaerobiasis for tobramycin, gentamicin, and amikacin, compared to latamoxef, piperacillin, chloramphenicol, and clindamycin. Chemotherapy 31:204–210. https://doi.org/10.1159/000238337

Tam C, Mun JJ, Evans DJ et al (2010) The impact of inoculation parameters on the pathogenesis of contact lens-related infectious keratitis. Invest Ophthalmol Vis Sci 51:3100–3106. https://doi.org/10.1167/iovs.09-4593

Tang J, Zhou R, Shi X et al (2008) Two thermostable nucleases coexisted in Staphylococcus aureus: evidence from mutagenesis and in vitro expression. FEMS Microbiol Lett 284:176–183. https://doi.org/10.1111/j.1574-6968.2008.01194.x

Teichmann KD (2000) Propionibacterium acnes endophthalmitis requiring intraocular lens removal after failure of medical therapy. J Cataract Refract Surg 26:1085

Tetz GV, Artemenko NK, Tetz VV (2009) Effect of DNase and antibiotics on biofilm characteristics. Antimicrob Agents Chemother 53:1204–1209. https://doi.org/10.1128/AAC.00471-08

Tiller JC, Liao CJ, Lewis K et al (2001) Designing surfaces that kill bacteria on contact. Proc Natl Acad Sci U S A 98:5981–5985. https://doi.org/10.1073/pnas.111143098

Torelli R, Cacaci M, Papi M et al (2017) Different effects of matrix degrading enzymes towards biofilms formed by E. faecalis and E. faecium clinical isolates. Colloids Surf B: Biointerfaces 158:349–355. https://doi.org/10.1016/j.colsurfb.2017.07.010

Torres-Barcelo C, Arias-Sanchez FI, Vasse M et al (2014) A window of opportunity to control the bacterial pathogen Pseudomonas aeruginosa combining antibiotics and phages. PLoS One 9:e106628. https://doi.org/10.1371/journal.pone.0106628

Van Leeuwenhoek A (1684) Microscopical observations about animals in the scurf of the teeth the substance call'd worms in the nose, the cuticula consisting of scales. Philos Trans R Soc 14:568–574

Vianney A, Jubelin G, Renault S et al (2005) Escherichia coli tol and rcs genes participate in the complex network affecting curli synthesis. Microbiology (Reading) 151:2487–2497. https://doi.org/10.1099/mic.0.27913-0

Watnick P, Kolter R (2000) Biofilm, city of microbes. J Bacteriol 182:2675–2679. https://doi.org/10.1128/JB.182.10.2675-2679.2000

Watters CM, Burton T, Kirui DK et al (2016) Enzymatic degradation of in vitro Staphylococcus aureus biofilms supplemented with human plasma. Infect Drug Resist 9:71–78. https://doi.org/10.2147/IDR.S103101

Wehrli W (1983) Rifampin: mechanisms of action and resistance. Rev Infect Dis 5(Suppl 3):S407–S411. https://doi.org/10.1093/clinids/5.supplement_3.s407

Weigel LM, Clewell DB, Gill SR et al (2003) Genetic analysis of a high-level vancomycin-resistant isolate of Staphylococcus aureus. Science 302:1569–1571. https://doi.org/10.1126/science.1090956

Whiteley M, Bangera MG, Bumgarner RE et al (2001) Gene expression in Pseudomonas aeruginosa biofilms. Nature 413:860–864. https://doi.org/10.1038/35101627

Willcox MD, Carnt N, Diec J et al (2010) Contact lens case contamination during daily wear of silicone hydrogels. Optom Vis Sci 87:456–464. https://doi.org/10.1097/OPX.0b013e3181e19eda

Wilton M, Charron-Mazenod L, Moore R et al (2016) Extracellular DNA acidifies biofilms and induces aminoglycoside resistance in Pseudomonas aeruginosa. Antimicrob Agents Chemother 60:544–553. https://doi.org/10.1128/AAC.01650-15

Witzky A, Tollerson R II, Ibba M (2019) Translational control of antibiotic resistance. Open Biol 9:190051. https://doi.org/10.1098/rsob.190051

Wu H, Moser C, Wang HZ et al (2015) Strategies for combating bacterial biofilm infections. Int J Oral Sci 7:1–7. https://doi.org/10.1038/ijos.2014.65

Xu KD, McFeters GA, Stewart PS (2000) Biofilm resistance to antimicrobial agents. Microbiology (Reading) 146(Pt 3):547–549. https://doi.org/10.1099/00221287-146-3-547

Xu Z, Liang Y, Lin S et al (2016) Crystal violet and XTT assays on Staphylococcus aureus biofilm quantification. Curr Microbiol 73:474–482. https://doi.org/10.1007/s00284-016-1081-1

Yokoi N, Okada K, Sugita J et al (2000) Acute conjunctivitis associated with biofilm formation on a punctal plug. Jpn J Ophthalmol 44:559–560. https://doi.org/10.1016/s0021-5155(00)00214-8

Zegans ME, Becker HI, Budzik J et al (2002) The role of bacterial biofilms in ocular infections. DNA Cell Biol 21:415–420. https://doi.org/10.1089/10445490260099700

Zhu H, Thuruthyil SJ, Willcox MD (2001) Production of N-acyl homoserine lactones by gram-negative bacteria isolated from contact lens wearers. Clin Exp Ophthalmol 29:150–152. https://doi.org/10.1046/j.1442-9071.2001.00397.x

Zhu H, Thuruthyil SJ, Willcox MDP (2002) Determination of quorum-sensing signal molecules and virulence factors of Pseudomonas aeruginosa isolates from contact lens-induced microbial keratitis. J Med Microbiol 51:1063–1070. https://doi.org/10.1099/0022-1317-51-12-1063

Zuberi A, Ahmad N, Khan AU (2017a) CRISPRi induced suppression of fimbriae gene (fimH) of a uropathogenic Escherichia coli: an approach to inhibit microbial biofilms. Front Immunol 8: 1552. https://doi.org/10.3389/fimmu.2017.01552

Zuberi A, Misba L, Khan AU (2017b) CRISPR interference (CRISPRi) inhibition of luxS gene expression in E. coli: an approach to inhibit biofilm. Front Cell Infect Microbiol 7:214. https://doi.org/10.3389/fcimb.2017.00214

Chapter 5
The Cultivable Fungi of the Human Eye

Abstract The mycobiome (fungal microbiome) of the human body has been studied over the years to understand their role in the health of the host. In this context, the gut mycobiome has been studied but comparatively less is known about the mycobiome of the eye. This is important since the eye is prone to several fungal diseases and is important to establish the healthy viable mycobiome of the healthy and diseased eyes. The focus of this chapter is to enumerate the cultivable fungi of the human eye by the conventional culture-based methods and changes in the diversity and abundance of the viable fungi in the diseased eye.

Keywords Cultivable fungi · Eye · Diversity · Abundance · Ocular fungal disease

5.1 Introduction

Microbes (bacteria, fungi, and viruses) are ubiquitous in their distribution and are present in every possible habitat (Rajilic-Stojanovic et al. 2007; Qin et al. 2010; Clemente et al. 2012; Hamad et al. 2016). The human body serves as a unique habitat that harbours. microbes both within and on the surface. These microbes may be commensals and may play a role in the health of the host (Huseyin et al. 2017). Between the bacteria, fungi and viruses, the bacteria have been extensively analysed for their abundance, diversity and function in some niches of the human body such as the gut, so as to unravel their role in various human diseases (Huseyin et al. 2017; Shivaji 2017, 2019, 2020). In contrast, fungi and viruses have not been analysed as much. Among the various niches of the human body, the eye has been least studied. We have a reasonable estimate of the abundance and diversity of bacteria associated with the eye, but data on fungi and viruses that constitute a minor fraction (1%) of the microbiota of the eye (Wen et al. 2017) are less studied.

This chapter would focus on the cultivable fungi of the human eye as determined by conventional culture-based method and changes in the viable fungi in the diseased state of the eye.

S. Shivaji, *Human Ocular Microbiome*,
https://doi.org/10.1007/978-981-19-1754-7_5

5.2 Collection and Processing of Clinical Samples for Ocular Fungi

The ultimate determination of the causative agent of an infection depends on the precise collection of the sample followed by identification of the microorganism. The clinical samples to be analysed for ocular infections should be collected from the site of infection and transported to the laboratory within the shortest period of time. As far as possible, the samples are collected without any transport/collection media (except for virus or chlamydia isolation) and processed directly. However, Amies transport medium without charcoal is recommended for collection of corneal scrapings for culture of bacteria and fungi (McLeod et al. 2005). Sharma (2012a) has compiled a list of samples that are collected along with the suitable collection media for the diagnosis of various ocular infections (Table 5.1).

Clinical samples are processed at the earliest without delay. But, in the event that the samples need to be transported to another facility, the slides, media and samples (contact lenses, contact lens cases, contact lens solutions, etc.) are transported in secure boxes for processing. A few of the samples such as contact lenses, contact lens solutions, intraocular lenses, corneal biopsy/buttons and iris tissues are directly processed for culture of the microorganisms in appropriate media without prior

Table 5.1 Collection of samples for the diagnosis of various ocular infections

Ocular infection	Sample to be collected	Sample collection device and procedure
Blepharitis	Scales/discharge from lid margin	Forceps/cotton swab
Conjunctivitis, dacryocystitis	Fluid/discharge from lower conjunctival sac	Calcium alginate/cotton swab
Keratitis	Corneal scraping	Kimura spatula, no. 15 surgical blade, bent needle
Uveitis	Anterior chamber fluid	Paracentesis (anterior chamber tap) with tuberculin syringe
Endophthalmitis	Anterior chamber fluid Vitreous aspirate Vitreous biopsy	Paracentesis (anterior chamber tap) with tuberculin syringe, tuberculin syringe, vitrectomy
Panophthalmitis	Vitreous biopsy Evisceration contents	Vitrectomy, evisceration
Deep-seated stromal infiltrate in keratitis	Corneal biopsy	Lamellar biopsy
Non-healing keratitis requiring keratoplasty	Corneal buttons	Penetrating keratoplasty
Contact lens-associated keratitis	Contact lenses, lens cases and lens solution	Aseptically removed, aseptically collected
Postoperative endophthalmitis	Intraocular lens	Surgical removal
Eye injury with iris prolapse/incarceration	Iris tissue	Surgical removal

Modified from Sharma (2012a)

Table 5.2 Smear examination methods used for the diagnosis of fungal eye infections

Sample	Staining method	Microorganism
Conjunctival swabs/ scrapings	Gram, Giemsa, KOH + calcofluor white, direct/ indirect immunofluorescence or immunoperoxidase	Bacteria, fungi, parasites (microsporidia), viral antigens
Corneal scrapings	Gram, Giemsa, KOH + calcofluor white, lactophenol cotton blue, Gomori methenamine silver, Ziehl-Neelsen stain, direct/indirect immunofluorescence or immunoperoxidase	Bacteria, fungi, parasites (*Acanthamoeba*, Microsporidia), viral antigens
Aqueous/vitreous fluids/ biopsy	Gram, Giemsa, calcofluor white, Gomori methenamine silver, direct/indirect immunofluorescence or immunoperoxidase	Bacteria, fungi Viral antigens

Modified from Sharma (2012a, b)

preparation of smears. Table 5.2 summarizes the direct microscopy methods that are used for detection of various organisms from ocular samples.

5.3 Direct Microscopy Examination of Clinical Samples

Microscopy examination of smears of clinical samples is rapid and often provides etiological diagnosis that does not require further culture. A few of the methods that are routinely used include:

1. Gram staining that demonstrates the presence of bacteria, fungi, and parasites. It has also been observed that the sensitivity and specificity of Gram stain in the identification of bacterial and fungal elements in corneal scrapings is reported to be affected by the stage of the disease (Sharma et al. 2002).
2. Giemsa staining has also been used to identify fungi in ocular infected samples. Zhang et al. (2010) demonstrated that in corneal scrapings from patients with clinically suspected fungal keratitis, Giemsa stain positivity was 39.7% but addition of calcofluor white increased the sensitivity to 98.3%. Giemsa stain has also been reported to detect microsporidia spores in in 73.3% of corneal scrapings (Joseph et al. 2006).
3. KOH + calcofluor white (CFW) stain has been used for the detection of fungi in corneal ulcers. Evaluation of 143 consecutive patients with corneal ulcers indicated that CFW stain detected fungi in 95.2% patients in comparison with 71.4% patients by the KOH method (Chander et al. 1993). But the sensitivity of KOH wet mount was higher (99.3%) in the detection of fungi than that of Gram-stained smears (89.2%) (Bharathi et al. 2006).
4. Lactophenol cotton blue (LPCB) is used to examine fungal elements following either a tape preparation or a scraping. This stain contains phenol, which kills the organisms, lactic acid which preserves fungal structures and cotton blue which stains the chitin found in the fungal cell walls. LPCB mounts of corneal ulcer scrapings were assessed by direct microscopic examination, and fungi were

detected in 78% of the culture-positive patients compared to 73% detected by the Gram-staining method (Thomas et al. 1991).

5. Gomori's methenamine silver impregnation technique is a highly reliable and achievable method of detecting fungal filaments (Vemuganti et al. 2002; Mukherjee et al. 2013).
6. Ziehl-Neelsen stain is a fungal stain and reported to be efficient in diagnosing microsporidiosis (Joseph et al. 2006).

5.4 Molecular Methods in the Diagnosis of Ocular Infections

Polymerase chain reaction (PCR) using primers specific to 18S rRNA or 28S rRNA have been routinely employed to detect fungal infections of the eye (Font et al. 2002; Embong et al. 2008; Anand 2001; Bagyalakshmi et al. 2006; Ferrer and Alio 2011; Sharma 2012a, b; Minkus et al. 2019). These studies indicated that PCR was an effective, rapid and sensitive method for the diagnosis of fungal infections in keratitis, endophthalmitis and uveitis compared to staining and culture methods. Ten cases of ocular mycotic infections were evaluated using histopathology, microbial culturing and PCR using a pan-fungal PCR and a *Candida albicans*-specific PCR. The PCR results correlated accurately with histopathology and culture results in both *Candida* (*C. albicans* and *C. parasilopsis)* and non-*Candida* fungal infections caused by *Aspergillus flavus*, *Fusarium* sp., *Scedosporium apiospermum* and *Phoma* spp. (Font et al. 2002). Bagyalakshmi et al. (2006) compared the results of uniplex PCR with a multiplex PCR (mPCR) for the detection of eubacterial genome, *Pseudomonas acnes* genome and pan fungal genomes in intraocular fluids from patients clinically diagnosed to have infective endophthalmitis. mPCR detected all the three genomes, and the results correlated with those of uniplex PCR (Bagyalakshmi et al. 2006). In another study (Balne et al. 2012), three phylogenetic markers of fungi, namely, 18S rRNA, 28S rRNA, and ITS regions, were amplified by PCR for the detection of fungi in corneal scrapings of patients with mycotic keratitis, and the positivity for fungal DNA was 40, 32 and 43 by 18S rRNA PCR, 28S rRNA PCR and ITS PCR, respectively, thus indicating that ITS PCR was significantly better than 28S rRNA PCR. Recently, a Direct Strip PCR method, which skips the DNA extraction step, was developed, and it increased pathogen detection in the intraocular fluids of uveitis patients (Sugita et al. 2021).

5.5 Confocal Microscopy in the Diagnosis of Ocular Infections

Corneal cultures are considered to be the gold standard diagnostic tool for the detection of the causative agent. Nevertheless, routine culture methods yield positive results in only half the cases. In vivo confocal microscopy (IVCM) has been demonstrated as useful in the diagnosis and management of *Acanthamoeba* and filamentous fungal keratitis (Labbé et al. 2009; Vaddavalli et al. 2011). Hau et al. (2010) reported that the diagnostic accuracy of microbial keratitis by confocal microscopy is dependent on observer experience. Wang et al. (2019) investigated the utility of IVCM in the diagnosis of infectious keratitis (IK) which included bacterial keratitis (BK), *Acanthamoeba* keratitis (AK), fungal keratitis (FK) and HSV viral keratitis (VK) and observed that the sensitivity and specificity of IVCM in identifying atypical IK (AK and FK cases combined) were 85.3% and 100%, respectively, and the sensitivity and specificity of IVCM in identifying BK and VK were also comparably high. Additionally, in FK subgroup IVCM was as accurate as clinical assessment (Wang et al. 2019). IVCM and anterior segment optical coherence tomography have also been used in the diagnosis of micro-cystic keratitis (Dembski et al. 2020).

5.6 Experimental In Vivo Models for Ocular Infections

Several animal models have also been used to monitor fungal infections and evaluate risk factors affecting the development of fungal infections (Kalkanci and Ozdek 2011). The animal models should be highly reproducible and suitable to evaluate effects of trauma, infection and co-infection and most importantly display clinical features comparable to the human counterpart and permit isolation of fungi from the diseased tissues for growth and analysis (Kalkanci and Ozdek 2011). Pigs, rabbits, hamsters, rats and mice are most widely used animals for experimental ocular infections. These animals are easy to breed and maintain, cheap and suitable for research which is facilitated by the availability of antibodies suitable to study response to infection by monitoring host factors and defence mechanisms (Ma et al. 2009; Yuan and Wilhelmus 2010). A few of the popular models used for studying ocular infections are described below.

5.6.1 Mouse Models

Outbred NIH Swiss and inbred BALB/C mice have been used as models for keratitis. For suitable response the mice are subjected to immunosuppression using either methylprednisolone (100 mg/kg body weight) or cyclophosphamide (180 mg/kg body weight) at 5, 3, and 1 day(s) prior to corneal inoculation. On the inoculation day, the mice are anesthetized by administering ketamine (37.5 mg/kg) and xylazine (1.9 mg/kg), and a superficial wound is generated on the corneal surface by scratching with a needle. Subsequently, fungi (1×10^3) CFU are inoculated on the scratched cornea for the keratitis model. The mice are killed at the end of the experiment, and their eyes are enucleated and used for histological examination, quantitative microbial culturing and molecular analysis (Wu et al. 2003a, b) to understand expression of inflammatory cytokines, such as MIP2, KC, IL-1 and IL-6, and the pathogenesis of ocular infection (Mitchell et al. 2007; Wu et al. 2004; Zhong et al. 2009; Yuan et al. 2010). Blood may also be collected for monitoring various parameters related to immunity, infection, etc. Contact lens-induced keratitis has also been studied using *Fusarium* grown as a biofilm on contact lenses to induce keratitis on injured corneas (Sun et al. 2010). Mice are easy to handle but mice have a certain death rate even with surface infections.

5.6.2 Rabbit Models

New Zealand albino rabbits were most frequently used animal models for ocular infections like contact lens-induced keratitis (Velpandian et al. 2006), keratomycosis by injection of *Candida* cells into the central corneal stroma (Ozturk et al. 2007) and fungal endophthalmitis (Kocak et al. 2009). For keratitis studies animals were first anaesthetized with intramuscular ketamine (35 mg/kg) and xylazine (5 mg/kg). In addition, corneal anaesthesia is performed with topical preparations such as 0.2% novocaine. Corneal epithelium was then removed and fungal inoculum transferred to the denuded cornea and covered with a contact lens. The lids are closed with sutures to prevent contact lens extrusion (Goldblum et al. 2005). In Chinchilla rabbits keratomycosis was established by the injection of *Candida* cells into the central corneal stroma. Clinical scoring system was used to grade conjunctival hyperaemia, corneal clouding, diameter of corneal neovascularization and hypopyon level (Ozturk et al. 2007). Fungal endophthalmitis has also been created in immunocompetent and immune-suppressed rabbits (McGuire et al. 1991) and different antifungal treatment regimens have been evaluated (Kocak et al. 2009).

5.6.3 Rat Models

Rats are considered superior models because of their suitable size and immune response and also because the size of their eyes makes controlled surgical procedures easier (Ren and Wu 2010). Wistar rats or Lewis rats are used as experimental animals for the keratitis model. The procedure is performed under an operation microscope. Initially, a half thickness linear blade incision is made 2 mm from the centre of cornea. A microliter syringe and 30 G needle are used for the incision. Fungal inoculum is injected into the stroma.

5.6.4 Other Models

In pigs, the large size of their eyes which eases the fitting of soft contact lenses made this a promising host for development as an animal model for contact lens-associated keratitis. Female Yucatan micropigs weighing approximately 25 kg were used for an experimental study (He et al. 1992).

5.7 Cultivable Fungi of the Normal Human Eye

Fungi were discovered on the human eye as early as the nineteenth century (Köllner 1906; Von Graefe and Mittheilungen 1857). However, the conventional culture-based method to identify fungi from the ocular surface using conjunctival swabs has always been a challenge, and often the culture-based and the PCR-based methods did not yield any fungi (Ozkan et al. 2017; Segal et al. 1977; Wu et al. 2003a, b). The frequency of positive cultures of fungi ranged from 3% to 32%, and most conjunctival swabs yielded only one culture (Segal et al. 1977; Wu et al. 2003a, b; Gopinathan et al. 2009; Sehgal et al. 1981; Wang et al. 2020; Williamson et al. 1968; Behrens-Baumann and Rüchel 1999). The success of culturing fungi from ocular samples also depends on the choice of the culture medium, the conditions used for growth, age of the subject (Williamson et al. 1968; Behrens-Baumann and Rüchel 1999) and the geographical region (Williamson et al. 1968; Marchlewitz Ba 1966). Further, longitudinal studies indicated that the fungi detected from an individual varied with time (1–4 weeks) implying that the fungi are not residents of the eye but are probably acquired from the surroundings (Williamson et al. 1968; Williamson and Haynes 2018). It has also been observed that the lid margins (80%) are more abundantly colonized than the conjunctivae (5%) (Williamson et al. 1968; Olson 1969). Filamentous fungi such as *Aspergillus* spp., *Aspergillus flavus*, *A. niger* and *Fusarium* spp. and the non-filamentous *Candida* spp. are the most predominant fungi on the ocular surface (Karsten and Foster 2012). These fungi also exhibit a geographical preference with *Penicillium* spp., *Aspergillus* spp. and *Candida* spp. dominating in central Europe (Williamson et al. 1968; Marchlewitz Ba 1966) and *Aspergillus flavus* and *A. niger* in the tropics (Wang et al. 2020; Zegans and Van Gelder 2014; Sharma 2012b). Other fungi associated with the normal human eye are listed in Table 5.3.

Table 5.3 Cultivable fungi of the normal (healthy) human eye

Fungi	Reference
Alternaria spp., *Aspergillus niger*, *A. flavus*, *Botrytis* spp., *Curvularia* spp., *Candida* spp., *Candida albicans*, *C. guilliermondii*, *C. parapsilosis*, *Cladosporium* spp., *Fusarium* spp., *Geotrichum* spp., *Gliocladium* spp., *Helminthosporium* spp., *Hormodendrum* spp., *Isaria* spp., *Mucor* spp., *Nigrospora* spp. *Papulospora* spp., *Penicillium* spp., *P. citrinum*, *Phialophora* spp., *Rhizoctonia* spp., *Rhizopus* spp., *Rhodotorula* spp., *Rhodotorula rubra*, *Saccharomyces* spp., *S. cerevisiae*, *Scopulariopsis* spp., *Stachybotrys* spp. and *Trichoderma* spp.	Ando and Takatori 1982; Sehgal et al. 1981; Wang et al. 2020; Hammeke and Ellis 1960; Ainley and Smith 1965; Mitsui and Hanabusa 1955; Nema et al. 1966; Williamson et al. 1968

5.8 Cultivable Fungi of the Diseased Human Eye

Fungi are the causative agents of several ocular diseases, such as keratitis, endophthalmitis, blepharitis, conjunctivitis, scleritis, retinitis, etc. (Klotz et al. 2000) and are considered important cause of vision impairment. Thus, it is important to identify and catalogue the ocular surface mycobiome of the diseased eye (Ozkan et al. 2017; Zegans and Van Gelder 2014) such as in individuals with keratitis, blepharitis, conjunctivitis, scleritis, endophthalmitis, post-fever retinitis, orbital cellulitis and several other diseases. In Chap. 2, a brief description of all the ocular diseases is given with respect to whether it is an inflammatory or non-inflammatory disease; its effect on vision; the etiological agent; the influence of trauma; the predisposing factors; the confounding factors such as age, sex and geographical location of the individuals; and the incidence of the disease on different parts of the ocular surface. These details are not provided in this chapter but the cultivable fungi associated with various ocular diseases are highlighted.

5.8.1 Fungi Associated with Fungal Keratitis

Fungal keratitis (FK) is an inflammatory disease of the cornea that could lead to serious complications including loss of vision (Rautaraya et al. 2011). Invariably, FK occurs after an injury to the eye, in people wearing contact lenses, by the presence of a foreign body in the eye and in immunocompromised patients (Fig. 5.1a). The risk factors include injury with vegetative and animal debris, soil, dust and contaminated water that facilitate fungal conidial invasion of the corneal stroma or the epithelium (Gopinathan et al. 2002; Thomas 2003a, b; Nath et al. 2011). The incidence of fungal keratitis is higher in the tropical countries, as high as 50% of all microbial keratitis patients (Bharathi et al. 2007) and less common in

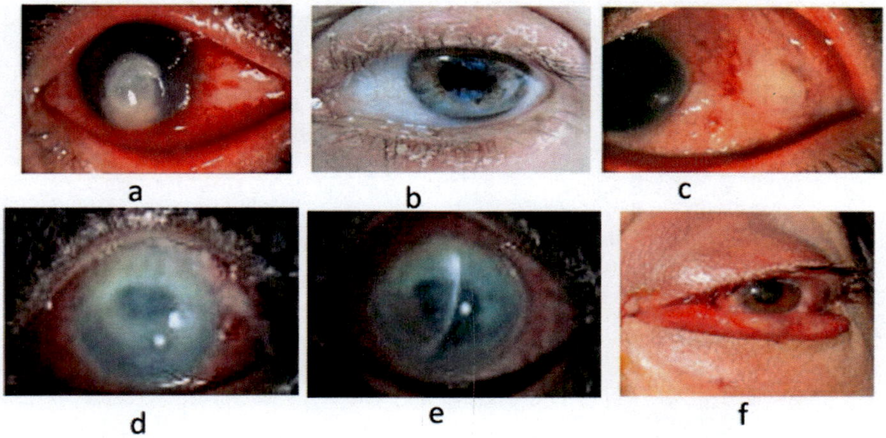

Fig. 5.1 Fungal diseases of the eye. (**a**) Keratitis caused by *Aspergillus fumigatus*; (**b**) blepharitis caused by *Trichophyton ajelloi*; (**c**) scleritis caused by *Histoplasma capsulatum*; (**d**) endophthalmitis caused by a fungus; (**e**) slit lamp photograph of fungal endophthalmitis; (**f**) orbital inflammatory involvement of fungal sinusitis, diagnosed as mucormycosis. Except (**d**) and (**e**) which were retrieved from EMR, LVPEI, Hyderabad, the remaining photomicrographs (**a–c** and **f**) were from Zagora et al. (2017) published as an open access article distributed under the Creative Commons Attribution License

more temperate climates (Rautaraya et al. 2011; Lalitha et al. 2015). A higher incidence is also reported in paediatric population, in males and during the monsoon and winter seasons than summer in India (Satpathy et al. 2019). Several fungal taxa (approximately 65–92) have been identified from the cornea of individuals with fungal keratitis (Karsten and Foster 2012; Thomas 2003a, b; Maharana et al. 2016). Each of these studies reported one to three genera and generally included members of the 'core' genera (*Aspergillus* spp., *Cladosporium* spp., *Malassezia* spp. and *Fusarium* spp.) (Shivaji et al. 2019). The other four core genera (*Setosphaeria* spp., *Hematonectria* spp., *Emericella* spp., and *Choiromyces* spp.) were not detected in FK, and the possible reason could be that these four 'core' genera, more often, are only healthy conjunctival commensals while not discounting the differences between the cultivable and the NGS method (Brooks et al. 2015; Huang et al. 2016).

Thomas (2003a, b) categorized the fungi associated with fungal keratitis into four broad groups: hyaline filamentous fungi, dematiaceous fungi, yeasts and *Zygomycetes* fungi. A decade later, in 2013, the same group reorganized the fungi involved in mycotic keratitis (Thomas and Kaliamurthy 2013) into seven groups, viz., (1) hyaline filamentous fungi, (2) phaeohyphomycetes, (3) phaeoid sphaerosidales, (4) yeasts and yeast like fungi, (5) dimorphic fungi, (6) other fungi and (7) newly reported fungi (Table 5.4). Comparison of the above studies clearly indicated a significant increase in the number of fungi identified in FK patients in all categories, but more prominently in the filamentous group. Many other studies have also reported higher proportion of filamentous fungi in FK patients (Jhanji et al. 2009; Gaujoux et al. 2012; Motley et al. 2011; Sangwan et al. 2013). Among the

Table 5.4 Cultivable fungi associated with the eye of individuals with keratitis

Fungi	References
Hyaline filamentous fungi *Acremonium atrogriseum*, *A. curvum*, *A. kiliense*, *A. potroni*, *A. recifei*, *Apophysomyces elegans*, *Arthrographis kalrae*, *Aspergillus clavatus*, *A. fischerianus*, *A. flavipes*, *A. flavus*, *A. fumigatus*, *A. glaucus*, *A. janus*, *A. niger*, *A. terreus*, *A. nidulans*, *A. oryzae*, *A. viridinutans*, *A. wentii*, *Aspergillus* spp., *Beauveria bassiana*, *Cephaliophora irregularis*, *Chrysonilia sitophila*, *Chrysosporium parvum*, *Cylindrocarpon lichenicola*, *Diplosporium* spp., *Engyodontium alba*, *Epidermophyton* spp., *Fusarium aquaeductum*, *F. dimerum*, *F. nivale*, *F. oxysporum*, *F. solani*, *F. subglutinans*, *F. ventricosum*, *F. verticillioides*, *Fusarium* spp., *Glenospora graphii*, *Metarhizium anisopliae*, *Microsporum* spp., *M. canis*, *Mucor ramosissimus*, *Myrathecum* spp., *Ovadendron sulphureo-ochraceum*, *Paecilomyces farinosus*, *P. lilacinus*, *P. variotii*, *Paecilomyces* spp., *Penicillium citrinum*, *P. expansum*, *Rhizoctonia* spp., *Rhizomucor pusillus*, *Rhizopus arrhizus*, *Sarcopodium oculorum*, *Saksenaea vasiformis*, *Scedosporium apiospermum*, *S. prolificans*, *Tritirachium oryzae*, *Ustilago* spp., *Verticillium searrae* and *Verticillium* spp.	Thomas 2003a, b; Thomas and Kaliamurthy 2013; Jhanji et al. 2009; Gaujoux et al. 2012; Motley et al. 2011; Yohai et al. 1994; Chang and Chodosh 2011; Khairallah et al. 1992; Krishnan et al. 2009; Yuan et al. 2009
Phaeohyphomycetes *Alternaria alternata*, *A. infectoria*, *Aureobasidium pullulans*, *Bipolaris hawaiiensis*, *B. spicifera*, *Cladosporium cladosporioides*, *Curvularia brachyspora*, *C. geniculata*, *C. lunatus*, *C. pallescens*, *C. senegalensis*, *C. verruculosa*, *Dichotomophthoropsis nymphearum*, *D. portulacae*, *Doratomyces stemonitis*, *Exophiala jeanselmei*, *E. dermatitidis*, *Exserohilum rostratum*, *E. longirostratum*, *Fonsecaea pedrosoi*, *Lecythophora mutabilis*, *L. hoffmannii*, *Phaeoisaria clematidis*, *Phaeotrichoconis crotalariae*, *Phialophora bubakii*, *P. verrucose* and *Tetraploa aristate*	Thomas and Kaliamurthy 2013; Sangwan et al. 2013; Revankar and Sutton 2010
Phaeoid sphaerosidales *Colletotrichum capsici*, *C. coccodes*, *C. dematium*, *C. graminicola*, *C. gloeosporioides*, *Lasiodiplodia theobromae*, *Microsphaeropsis olivacea*, *Phoma oculo-hominis*, *Phoma* spp. and *Sphaeropsis subglobosa*	Thomas and Kaliamurthy 2013

(continued)

Table 5.4 (continued)

Fungi	References
Yeast and yeast-like fungi *Candida albicans, C. famata, C. fermentati, C. glabrata, C. guilliermondii, C. krusei, C. parapsilosis, C. tropicalis, Cryptococcus laurentii, C. neoformans, Geotrichum candidum, Malassezia furfur, Rhodotorula glutinis, R. rubra, Rhodotorula* spp. and *Rhodosporidium toruloides*	Thomas and Kaliamurthy 2013; Sun et al. 2007; Gopinathan et al. 2002
Dimorphic fungi *Blastomyces dermatitidis, Coccidioides immitis, Paracoccidioides brasiliensis* and *Sporothrix schenckii*	Thomas and Kaliamurthy 2013
Other fungi *Absidia corymbifera, Blastoschizomyces capitatus, Chlamydoabsidia padenii, Pythium insidiosum, Scytalidium* spp., *Ulocladium atrum* and *Thielavia subthermophila*	Thomas and Kaliamurthy 2013; Sharma et al. 2015
Newly reported fungi *Aspergillus viridinutans, Candida fermentati* and *Thelavia subthermophila*	Thomas and Kaliamurthy 2013

yeasts, *Candida albicans* and other yeasts are common in temperate regions (Sun et al. 2007). In rare instances, mixed infections with bacteria and fungi were detected in FK patients and varied from 6 to 17% (Rautaraya et al. 2011; Bharathi et al. 2003). Fungal pathogens like *Fonsecaea pedrosoi* (Sangwan et al. 2013), *Alternaria* spp., *Acremonium* spp., *Colletotrichum* spp., *Lasiodiplodia theobromae* (Lekhanont et al. 2015), *Cylindrocarpon lichenicola* (Gaujoux et al. 2012), *Metarhizium anisopliae* (Motley et al. 2011) and *Pythium insidiosum* (Sharma et al. 2015) have been reported in FK patients (Thomas 2003b). Obviously, one needs a more comprehensive study to enlist all fungi associated with keratitis.

5.8.2 Fungi Associated with Blepharitis

Blepharitis is an inflammatory condition of the eyelid margin (Putnam 2016) and is less often caused by fungi than bacteria, virus and parasites (Jackson 2008) (Fig. 5.1b). Table 5.5 lists the fungi cultivated from eyelash of patients with blepharitis. Among the fungi, *Malassezia globosa* was more abundant than *C. albicans* (Fierro-Arias et al. 2014). Direct microscopic examination has revealed *Demodex* mites, such as *Microsporum canis*, in 42.1% of the blepharitis group (Creach et al. 1995). Though NGS analysis on ocular bacterial communities of eyelash and tear samples of individuals with blepharitis has been attempted, similar studies related to the fungal community have not been reported (Lee et al. 2012).

Table 5.5 Cultivable fungi associated with the diseased human eye

Disease	Fungi	References
Blepharitis	*Aspergillus flavus*, *Candida* spp., *C. albicans*, *Malassezia globosa*, *M. furfur*, *Penicillium* spp., and *Trichophyton verrucosum*	Klotz et al. 2000; Fierro-Arias et al. 2014; Jackson 2008; Anane et al. 2007; Dadaci et al. 2015; Lee et al. 2016; Sahin et al. 2014
Conjunctivitis	*Candida albicans*, *C. parapsilosis*, *C. tropicalis*, *Sporotrichum schenckii*, *Paracoccidioides brasiliensis*, *Coccidioides immitis*, *Blastomyces dermatitidis*, *Rhinosporidium seeberi*, *Sporothrix schenckii*, etc.	Lupetti et al. 2002; Forte et al. 2009: Sarkar et al. 2016
Scleritis	*Acremonium* spp., *Aspergillus* spp., *A. flavus*, *A. oryzae*, *A. terreus*, *Candida parapsilosis*, *Cephalosporium* spp., *Cladosporium* spp., *Colletotrichum dematium*, *Fusarium* spp., *Metarhizium anisopliae*, *Paecilomyces* spp., *P. lilacinus*, *Penicillium* spp., Petriellidium *boydii*, Rhizopus spp., *Scedosporium spp.*, *S. apiospermum*, *Sporotrichum* spp. and *Trichophyton* spp.	Jain et al. 2009; Jeang et al. 2017; Reddy et al. 2015; Bernauer et al. 1998; Huang et al. 2000; Fincher and Fulcher 2007; Todokoro et al. 2018
Exogenous endophthalmitis	*Acremonium kiliense*, *Aspergillus* spp., A. *niger*, *Candida* spp., *C. famata*, *C. glabrata*, *C. parapsilosis*, *Exophiala jeanselmei*, *Fusarium* spp., *F. moniliforme*, *Helminthosporium* spp., *L. theobromae*, *Paecilomyces lilacinus*, *P. boydii*, *Penicillium chrysogenum*, *Scytalidium dimidiatum* and *S. schenckii*	Anand 2001; Behera et al. 2018; Gupta et al. 2003; Das et al. 1994; Stern et al. 1985; Pettit et al. 1980; Pflugfelder et al. 1988; Hammer et al. 1983; Carney et al. 1996; Al-Rajhi et al. 1993; Borderie et al. 1997
Endogenous endophthalmitis	*Aspergillus flavus*, *A. fumigatus*, *A. Glaucus*, *A. nidulans*, *A. niger*, *A. terreus*, *Blastomyces dermatitidis*, *Candida* spp., *Coccidioides immitis*, *Cryptococcus neoformans*, *Fusarium* spp., *Paecilomyces* spp., *Histoplasma capsulatum*, *Penicillium* spp. and Sporothrix *schenckii*	Klotz et al. 2000; Bannerman et al. 1997; Kurosawa et al. 1988; Louie et al. 1994; Speaker et al. 1991; Swan et al. 1985; Van Buren 1958; Weishaar et al. 1998: Sarkar et al. 2016
Retinitis	*Aspergillus* spp., *Candida* spp., *Histoplasma capsulatum* and *H. duboisii*.	Invernizzi et al. 2018
Chorioretinitis	*Candida* spp., *C. albicans*, *Coccidioides immitis*, *Cryptococcus* spp., and *Pneumocystis carinii*	Sheu et al. 1998; Lu et al. 2017; Toomey et al. 2019; Fowler et al. 2011

(continued)

Table 5.5 (continued)

Disease	Fungi	References
Post-fever retinitis	*C. albicans*	Menia et al. 2019; Bartolini and Duker 2012; Song et al. 2002; Kose 2020
Orbital cellulitis	*Aspergillus* spp., *Bipolaris* spp., *Candida* spp., *Klebsiella* spp., *Mucor* spp., *Rhizopus* spp., *Scedosporium* spp. and *Scopulariopsis* spp.	Gamaletsou et al. 2012; Levin et al. 1996; Ajayi et al. 1986; Kammoun et al. 2018; Jacobson et al. 1992; Mukherjee et al. 2016; Pandian et al. 2011; Vida and Moel 1974; Johnson 2000; Pornpanich et al. 2011
Dry eye disease	*Candida* spp.	Klotz et al. 2000
Meibomian gland dysfunction	*Malassezia* spp.	Wang and Deng 2020
Uveitis	*Aspergillus fumigatus*, *C. albicans* and *Fusarium* spp.	Barisani-Asenbauer et al. 2012; Amin et al. 2015; Deshmukh et al. 2018
Stevens-Johnson syndrome and toxic epidermal necrolysis	*Coccidioides* spp., *Epidermophyton* spp., *Histoplasma microsporum* and *Trichophyton* spp.	Hazin et al. 2008
Sjögren's syndrome	*C. albicans*, *C. parapsilosis* and *H. capsulatum*	Williamson et al. 1968

5.8.3 Fungi Associated with Conjunctivitis

Conjunctivitis results from an infection or allergy that affects exclusively the conjunctiva. It may be unilateral or bilateral and could be contagious (Azari and Barney 2013). The usual etiological agents are either a bacterium or a virus; but in rare cases fungi have been detected (Azari and Barney 2013) (Table 5.5). *Candida albicans*, *C. parapsilosis*, and *C. tropicalis* with primary infections in the oral mucosa or vagina have been implicated in conjunctivitis in newborns and school children (Lupetti et al. 2002; Sarkar et al. 2016) and in adults (Forte et al. 2009). Conjunctivitis by filamentous fungus is very rare. The dimorphic fungus *Sporotrichum schenckii* is reported from the conjunctiva of patients from Japan and Mexico (Vanzzini Zago et al. 2012). Protists like rhinosporidiosis of tarsal conjunctiva were reported in 8-year-olds from India and Sri Lanka; whereas sporadic cases were seen in Europe and North America (Nair et al. 2015).

5.8.4 Fungi Associated with Scleritis

Scleritis is a chronic, painful inflammation of the sclera and (or) adjacent tissues and could lead to vision impairment (Fig. 5.1c). The symptoms include decreased vision and moderate to severe eye pain (Azari and Barney 2013). Scleritis could be idiopathic, autoimmune or infectious (Jain et al. 2009). The common predisposing factors are ocular trauma, spread from the cornea of microbial keratitis or infectious endophthalmitis, ocular surgery, use of antimetabolites (such as mitomycin C), local radiotherapy, immunosuppressed conditions (such as HIV infection, diabetes, systemic immunosuppression), debilitating ocular diseases caused by herpes simplex virus or varicella zoster virus, recurrent keratitis, dry eye and chronic use of topical corticosteroids (Murthy et al. 2020). In the vast majority of cases, infectious scleritis follows accidental or surgical trauma and rarely can be attributed to hematogenous spread (Jeang et al. 2017; Reddy et al. 2015; Stenson et al. 1982).

The reported incidence of scleritis due to infectious aetiology is 4.6–18% of all scleritis cases in adults; and herpes zoster virus and tuberculosis have been implicated as the most common causes of infectious scleritis (Hemady et al. 1992; Watson and Hayreh 1976). Fungal scleritis is less common than bacterial infection (Jain et al. 2009; Reddy et al. 2015), and fungal scleritis is less often reported from temperate climate zones such as in Europe than in tropical countries like India. Several filamentous and non-filamentous fungi have been reported from patients with scleritis (Table 5.5). The mycobiome of scleritis patients is yet to be described.

5.8.5 *Fungi Associated with Endophthalmitis*

Endophthalmitis is inflammation of the vitreous caused by bacteria or fungi after intraocular surgery, trauma (both exogenous) or hematogenous spread (endogenous) (Chee and Jap 2001; Connell et al. 2011; Durand 2013; Ramakrishnan et al. 2009). Irrespective of the cause, endophthalmitis presents with conjunctival hyperaemia, exudates in the anterior chamber and vitreous opacities (Fig. 5.1d, e). Patients have reduced vision which leads to severe vision loss and even blindness (Das et al. 1995). In general, bacterial endophthalmitis is more common than fungal endophthalmitis and is more often reported from Asia (Das 2017).

5.8.5.1 Exogenous Endophthalmitis

In tropical countries like India, 90% of fungal endophthalmitis are exogenous (Chakrabarti et al. 2008; Lalitha et al. 2015). The incidence of exogenous endophthalmitis in India is between 0.04–0.15% (Lalitha et al. 2015) after intraocular surgery and 2.4–17% following trauma (Das et al. 2012; Levison et al. 2014). The predominant causative fungi are *Aspergillus* spp., *Fusarium* spp. and *Candida* spp., the former two are detected in exogenous endophthalmitis, and the latter in endogenous endophthalmitis (Das et al. 2020). *Aspergillus* spp. is the most common infecting fungus, reported both in India and the USA (Anand 2001; Behera et al. 2018; Gupta et al. 2003; Sarkar et al. 2016). Table 5.5 lists the candidate genera associated with exogenous endophthalmitis.

5.8.5.2 Endogenous Endophthalmitis

In general, the incidence of endogenous endophthalmitis is much lower than exogenous endophthalmitis and manifests in patients with systemic debilitating diseases like cancer and often after intravenous drug therapy, chemotherapy and systemic antibiotics, in individuals in long-term steroid usage (Modjtahedi et al. 2019), in immunosuppressed individuals and in users of intravenous drugs (Modjtahedi et al. 2017) and antimetabolites like 5-fluorouracil (105) and mitomycin C (Greenfield et al. 1996). Fungi are more often isolated in endogenous than exogenous endophthalmitis and include several different genera (Table 5.5). Studies have also indicated that 37% of endogenous endophthalmitis are caused by fungi (Modjtahedi et al. 2019). *Candida* spp. are the most common isolated fungus in Europe, the USA and Korea (Durand 2017; Christian et al. 2019). Using a murine model, it has been demonstrated that *Aspergillus* spp. infection leads to influx of polymorphonuclear cells and production of various pro-inflammatory cytokines and chemokines which cause retinal tissue damage (Gupta et al. 2019). The fungal microbiome of the vitreous in endophthalmitis patients has been described recently (Gandhi et al. 2019). Four genera, namely, Aspergillus spp., Fusarium spp., *Exserohilum* spp. and Candida spp., were the most commonly involved fungi in the culture-negative endophthalmitis cases, and these organisms were taxonomically similar to the species identified by conventional culture methods (Gandhi et al. 2019).

5.8.6 Fungi Associated with Retinitis

Retinitis is an inflammation of the retina that could cause blindness. The prevalence is 1 in 2500–7000 people and affects people in the age range of 20–60 years. Retinitis is common in immunocompromised patients (Agarwal et al. 2014) and is caused by viruses, bacteria, fungi, ticks and parasites (*Toxoplasma gondii*) (Ahuja et al. 2008; Williamson and Haynes 2018). The fungi commonly causing retinitis include *Candida* spp., *Aspergillus* spp., *Histoplasma capsulatum* and *H. duboisii* (Invernizzi et al. 2018). The mycobiome of retinitis patients is not yet described.

5.8.7 Fungi Associated with Chorioretinitis

Chorioretinitis (CR) is an inflammation of both choroid and retina. Similar to retinitis, chorioretinitis occurs frequently in immunodeficient subjects such as people with HIV/AIDS or in patients on immunosuppressant drugs (Kasper et al. 2005). The other possible causes are syphilis, sarcoidosis, tuberculosis, Behcet's disease, onchocerciasis, West Nile virus (Shaikh and Trese 2004) and ocular histoplasmosis (Diaz et al. 2015; Nielsen et al. 2012). It is also known that chorioretinitis is usually the result of antecedent fungemia (Klotz et al. 2000). The fungi associated with CR are *Cryptococcus* spp. (Sheu et al. 1998), *Candida* spp. (Sheu et al. 1998), *C. albicans* (Lu et al. 2017), coccidioidal infection (Toomey et al. 2019) and *Pneumocystis carinii* (Foster et al. 1991). Disseminated chorioretinitis with confirmed presence of yeast cells is seen in immunocompromised patients (Table 5.5).

5.8.8 Fungi Associated with Post-fever Retinitis (PFR)

Post-fever retinitis manifests with various uveo-retinal features 2–4 weeks after the onset of fever, and very often the symptoms of the original illness may no longer be obvious. Patients generally present with positive serology for bacteria, viruses or protozoa. The observed effects may be the result of a direct invasion by the pathogen or by indirect mechanism mediated through the immune system (Vishwanath et al. 2014). Ocular symptoms include sudden, painless diminution of vision, floaters, flashes of light, loss of one-half of visual field and central blackouts. Post-fever retinitis is an infectious ocular disease involving bacteria, fungi, viruses and protozoans (Vishwanath et al. 2014) and is more common in tropical countries (Vishwanath et al. 2014; Kawali et al. 2019). PFR has been observed after typhoid fever (Prabhushanker et al. 2017), rickettsioses (Sinha et al. 2012) and malaria (Beare et al. 2011). Post-fever retinitis has also been observed in patients after a variety of viral fevers, such as, dengue, Chikungunya (Khairallah et al. 2013), Zika (Zhao et al. 2017), Ebola (Shantha et al. 2016), influenza (Faridi et al. 2010), West Nile (Garg and Jampol 2005) and corona viruses (Marinho et al. 2020), and also occurs in immunocompetent individuals (Menia et al. 2019). Comparatively, there are very few reports on fungal infection. Fungal retinitis has been reported in a patient with acute lymphocytic leukaemia (Song et al. 2002) (*C. albicans* has been detected in retinitis following influenza virus type A infection (Menia et al. 2019)) and was also seen in a patient after liver transplantation (Kose 2020). In a recent review, Sanjay et al. (2020) concluded that PFR presents with a similar morphological pattern irrespective of the aetiology and follows a preset natural course before resolution; treatment may not be required.

5.8.9 Fungi Associated with Orbital cellulitis

Orbital cellulitis (OC) is an inflammation of the eye behind the orbital septum accompanied by pain, edema, chemosis, decreased ocular motility and proptosis (Chaudhry et al. 2007). OC includes pre-septal cellulitis (72.2%), orbital cellulitis (22.2%) and abscess (5.6%) and apparently more prevalent in children than in adults (Hegde and Sundar 2017). The major causes are sinusitis (58%) (Fig. 5.1f), lid or face infection (28%), foreign body (11%), hematogenous (4%) and odontogenic

(2–5%). Infection in OC is caused by bacteria, ocular parasites and fungi (Partamian et al. 1983; Howe and Jones 2004). Fungal OC is associated with a high mortality despite aggressive antifungal treatment and orbital exenteration (Farooq et al. 2015; Bedwell and Bauman 2011). Several factors like renal disease, leukaemia, hemiparesis, diabetes mellitus and bilateral orbital involvement are associated with the high mortality in OC patients (Yohai et al. 1994). Invasive aspergillosis due to *Aspergillus* spp. also causes OC which initially infects the sinuses and then spreads to the orbit (Levin et al. 1996). Other fungal species that cause OC are *Candida* spp., *Scedosporium* spp., *Scopulariopsis* spp., *Bipolaris* spp., Klebsiella spp., *Mucor* spp., *Rhizopus* spp. and *Aspergillus* spp. (Pandian et al. 2011; Mukherjee et al. 2016; Johnson 2000; Ajayi et al. 1986; Pornpanich et al. 2011; Vida and Moel 1974; Kammoun et al. 2018; Jacobson et al. 1992; Gamaletsou et al. 2012; Levin et al. 1996). Fungal species that cause OC are listed in Table 5.5. Rarely, fungal infections and granulomas are seen in immunocompetent patients (Hegde and Sundar 2017).

5.8.10 Fungi Associated with Dry Eye Disease

Dry eye disease (DED) is a multifactorial disease of the tears and ocular surface, and patients present with increased osmolarity of the tear film and inflammation of the ocular surface (Helmick et al. 2008; Yamaguchi 2018). The common symptoms of DED are ocular burning, itching, foreign body sensation, photophobia, redness, and reduced visual acuity (Kassan and Moutsopoulos 2004). Several physiological conditions such as allergic inflammatory diseases, hormonal imbalance, advanced age, gender, exposure to allergens, cigarette smoke, dry climate, high altitudes, excessive coffee consumption, contact lens use, multivitamins and systemic disorders, such as diabetes mellitus, thyroid disease, rheumatoid arthritis and systemic lupus erythematosus, can also lead to DED (Javadi and Feizi 2011; Kuo et al. 2019). In addition, previous corneal transplantation and cataract and refractive surgery could also predispose to DED (Solomon et al. 2001). It is more prevalent in women and elderly individuals (Smith et al. 2007). Several studies indicated an association between DED and ocular surface bacterial infection (Albietz and Lenton 2006; Graham et al. 2007; Hori et al. 2008; Zilliox et al. 2020). Microbiome analysis also confirmed distinct differences in the ocular microbiome of DED patients with a near complete substitution of *Staphylococcus* spp. with *Pseudomonas* spp. (Graham et al. 2007); *Corynebacterium* spp. is one of the most frequent organisms in DED (Zilliox et al. 2020, p. 181). Fungal associations with DED are limited. Chronic dry eye could be caused by yeasts, usually *Candida* species (Klotz et al. 2000) (Table 5.5). Mycobiome of the ocular surface of dry eye patients is not yet reported.

5.8.11 Fungi Associated with Meibomian Gland Dysfunction

Meibum secreted by the meibomian glands (MG) along with water and mucus forms the tears. Meibum protects the water from evaporating in the tears. Obviously, when the duct of the meibomian gland is blocked or altered, less meibum reaches the eye surface leading to dry eye disease (DED). Several bacteria, which included the gram-positive *Staphylococcus* spp., *Lysinibacillus* spp., *Microbacteriaceae* spp., *Corynebacterium* spp., *C. macginleyi*, *Paenibacillus* spp. and *Bacillus* spp. and the gram-negative *Moraxella osloensis*, *Sphingomonas* spp. and *Xanthomonadaceae* are dominant in the eyes with MGD (Jiang et al. 2018; Dong et al. 2019; Nattis et al. 2019). In a more recent study, Zhao et al. (2020) observed that MGD meibum contained an abundance of ocular pathogens, namely, *Campylobacter coli*, *Campylobacter jejuni* and *Enterococcus faecium* (Janssen et al. 2008; Kuriyan et al. 2014) which were almost absent in the unaffected individuals. It has been reported that the microbiomes of the meibum, conjunctival sac and the skin samples were distinctly different (Suzuki et al. 2020). Fungi do not seem to be associated with MGD though there is a recent suggestion of a possible role of *Malassezia* spp., in the pathogenesis of MGD (Wang and Deng 2020) (Table 5.5).

5.8.12 Fungi Associated with Uveitis

Uveitis (UVT) is an inflammation of the iris, ciliary body and choroid. Photophobia, pain and blurred vision are the common symptoms. It is often bilateral, and depending on the involved tissue, it could be anterior uveitis or iritis (predominantly the iris is affected), intermediate uveitis (predominantly affects tissues behind the iris and lens) and posterior uveitis (predominantly the choroid is affected). Anterior uveitis is the most common type accounting for ~85% of all cases of UVT (Gritz and Wong 2004) and is associated with human leukocyte antigen B27 (HLA-B27) throughout the world (Rosenbaum 1989; Zagora et al. 2017; Yang et al. 2018). The prevalence of UVT is around 38,200 and 730 per 100,000 in France, the USA and India, respectively (Sangwan 2010). All kinds of spondyloarthritis including ankylosing spondylitis, reactive arthritis, psoriatic arthritis and arthritis have known association with uveitis (Rosenbaum and Asquith 2018). Infectious uveitis could be caused by viruses, bacteria, fungi or parasites (Lin 2015). A few of the bacteria that have been identified from UVT patients include *Mycobacterium tuberculosis*, *Chlamydia trachomatis*, *Treponema pallidum* and *Borrelia burgdorferi* (Barisani-Asenbauer et al. 2012). The fungi associated with UVT include *Fusarium* spp., *Aspergillus fumigatus* and *C. albicans* (Barisani-Asenbauer et al. 2012; Amin et al. 2015; Deshmukh et al. 2018). We have shown changes in the gut microbiomes and mycobiomes that are associated with auto-immune and idiopathic uveitis (Jayasudha et al. 2019; Kalyana Chakravarthy et al. 2018).

5.8.13 Fungi Associated with Stevens-Johnson Syndrome and Toxic Epidermal Necrolysis

Ocular involvement has been reported in 35–90% of the Stevens-Johnson syndrome (SJS) and toxic epidermal necrolysis (TEN) patients (De Rojas et al. 2007; Sotozono et al. 2007; Van Zyl et al. 2014). Patients present with hyperaemia, pseudomembrane formation, extensive exfoliation of the epidermis, mucosal tissue conjunctivitis and corneal abrasion (Kim et al. 2015). Incidence of SJS is 1.2–6 per million persons per year and the mortality rate is 1–5% (Abrol et al. 2020). SJS/TEN could be either 'acute' or 'chronic' (Chang et al. 2016). Systemic infections by viruses (herpes simplex virus, AIDS, coxsackievirus, influenza, hepatitis, mumps virus, Epstein-Barr virus and enteroviruses), bacteria (group A beta-hemolytic streptococci, diphtheria, brucellosis, *Mycobacteria*, *Mycoplasma pneumoniae* and *Salmonella typhi*), parasites and fungi have been implicated in SJS (Frizon et al. 2014). Recent studies indicated that the ocular microbiomes in SJS patients exhibited a greater diversity of bacteria compared to the healthy controls (Frizon et al. 2014) with a preponderance of few pathogenic bacteria (*Enterobacter* spp., *Serratia nonliquefaciens*, *E. coli*, *Morganella morganii*, *Proteus mirabilis*, and *Haemophilus* spp.) and a few commensals (*Micrococcus* spp., and *Streptococcus* spp.) (Frizon et al. 2014; Kittipibul et al. 2020; Kohanim et al. 2016; Venugopal et al. 2016). Fungal infections with coccidioidomycosis (caused by *Coccidioides*) and dermatophytosis (caused by fungi from three genera: *Microsporum*, *Trichophyton* and *Epidermophyton*) have been reported on the ocular surface of SJS patients (Hazin et al. 2008) (Table 5.5).

5.8.14 Fungi Associated with Sjögren's Syndrome

Sjögren's syndrome (SS) is a slow, progressive, chronic, autoimmune disease destroying the body's moisture-producing glands like the lacrimal and salivary glands, thus causing oral and ocular dryness (Tsigalou et al. 2018). Sjögren's syndrome affects approximately 2% of the adult population and remains undiagnosed in a vast majority of the people. Patients with Sjögren's syndrome frequently present with xerostomia, keratoconjunctivitis sicca and parotid gland enlargement. Symptoms also include itchiness of the eyes, reduced tear secretion, chronic irritation and destruction of the corneal and bulbar conjunctival epithelium. The other symptoms include dry skin, vaginal dryness, chronic cough, numbness in arms and legs, fatigue, muscle and joint pain and thyroid problems. Sjögren's syndrome could develop at any age and is more common in perimenopausal women (Brito-Zeron et al. 2016). Ocular surface microbiome has been studied in patients with Sjögren's syndrome (de Paiva et al. 2016; Moon et al. 2020). Williamson et al. (1968) reported higher yield of fungi from the conjunctival sacs of patients with Sjögren's syndrome. The isolates included the intraocular pathogens *C. albicans* and *C. parapsilosis* (Williamson et al. 1968). Histoplasmosis caused by saprophytic dimorphic *H. capsulatum* has also been documented in patients with SS. Patients with dry eye showed compositional changes of gut microbiome somewhere in between Sjögren's syndrome and controls (Moon et al. 2020) (Table 5.5).

5.8.15 Fungi Associated with Lax Eyelid Syndrome

Eyelid laxity can result from a number of involutional, local and systemic diseases, but it is frequently of unknown aetiology. When eyelid laxity is associated with papillary conjunctivitis and dry eyes, it is referred to as lax eyelid syndrome (LES) and patients typically present with 'floppy' eyelids (Fowler and Dutton 2010). LES affects patients of all ages and patients have systemic elevation of inflammatory markers like matrix metalloproteinases (Sward et al. 2018). *Corynebacterium* spp., *Hydrotalea* spp., *Ralstonia* spp., *Clostridium* XI, *Staphylococcus* spp. and *Rothia* spp. were detected in the eyes of LES patients. The involvement of fungi in LES is not yet established. A subset of LES is known as the floppy eyelid syndrome (Culbertson and Ostler 1981); it occurs in males and characterized by a combination of eyelid laxity, papillary conjunctivitis, obesity and obstructive sleep apnea and is associated with keratoconus (Fowler and Dutton 2010; Sward et al. 2018; Mastrota 2008).

5.8.16 Fungi Associated with Ocular Graft Vs Host Disease

Ocular graft vs host disease (oGVHD) is an immune-mediated disease caused by complication of hematopoietic stem cell transplantation and is a leading cause of morbidity and mortality (Mirza et al. 2016). These patients present with keratoconjunctivitis sicca, cataract, blepharitis, ocular hypertension, filamentary keratitis, dry eye, conjunctivitis and punctate keratopathy (Franklin et al. 1983; Lin and Cavanagh 2015). Shimizu et al. (2019) analysed the conjunctiva of oGVHD patients and identified several Gram-positive and Gram-negative bacteria compared to the non-GVHD patients and the control group. Zilliox et al. (2020) recently analysed the bacterial microbiome of the ocular surface of patients with oGVHD and observed that oGVHD eyes shared *Lactobacillus/Streptococcus* spp. with the healthy eyes. Zilliox et al. (2020) also observed higher relative abundances of *Corynebacterium* spp. in DED patients compared to oGVHD. Association of fungi with oGVHD is not known.

5.9 Treatment of Ocular Infectious Diseases

Treatment and management of fungal eye infections is disease-dependent and ranges from use of appropriate topical antifungal agents to eye surgery and occasionally removal of the eye. Several reviews are available on this aspect (Maharana et al. 2016; Klotz et al. 2000; Thomas 2003a, b; Hegde and Sundar 2017; Kuo et al. 2019; Ozdemir et al. 2012; Tsirouki et al. 2018*). In brief,* conjunctivitis, keratitis, cellulitis, etc. respond to topical medications (Maharana et al. 2016); endophthalmitis to vitrectomy and intravitreal antifungal agents (Das et al. 2020); blepharitis to systemic medication along with topical corticosteroid and lubricating eye drops; and uveitis to topical and systemic therapy (Wilhelmus and Jones 2001). Most often one needs treatment for prolonged periods of several weeks to months. Further, many patients need systemic treatment for a prolonged period. However, one must exercise caution in long-term systemic use of antifungal agents since these are not without toxicity, e.g. amphotericin B is nephrotoxic, voriconazole and ketoconazole are hepatotoxic and fluconazole and itraconazole have gastrointestinal side effects. It may also be mentioned that several eye infections like conjunctivitis resolve without treatment.

5.10 The Gap Areas

Studies confirm the existence of fungi on the ocular surface. Further, the cultivable fungi of the ocular surface and the ocular fluids of the eye would contribute significantly to our understanding of the role of fungi in the diseased eye vis-à-vis the healthy eye. Despite these studies, several aspects of the mycobiome are not yet addressed. Few of the gap areas include:

1. Several ocular fungi (both surface and fluid) are not identified up to the species level.
2. Characterization of ocular cultivable fungi (conjunctiva, corneal scrapings, ocular fluids, lacrimal glands, etc.) in different ethnicities, climates, geographical regions, etc. are required to unravel the confounding factors that influence the diversity and abundance.
3. One needs to characterize ocular fungi under different physiological conditions including various ocular diseases.
4. Longitudinal studies would establish the stability of the ocular cultivable fungi with age.
5. The interaction between different genera and species of fungi in the eye is not completely understood.
6. It is also imperative that we understand whether the ocular mycobiome is modulated by changes in the gut mycobiome.

5.11 Conclusions

Fungal infections of the eye are rarer than bacterial infections, but they are more serious. Research on the cultivable fungi of the ocular surface will help build a targeted and individualized treatment. This chapter has dealt with the cultivable fungi in various infections of the eye and has shown that the abundance of commensal and pathological fungi in different ophthalmic conditions could form the basis for therapy to overcome fungal infections.

References

Abrol A, Gulanikar A, Thakre S et al (2020) Study of ocular manifestations of Stevens-Johnson syndrome/toxic epidermal necrolysis. Indian Dermatol Online J 11:570–574. https://doi.org/10.4103/idoj.IDOJ_377_19

Agarwal A, Kumari N, Trehan A et al (2014) Outcome of cytomegalovirus retinitis in immunocompromised patients without human immunodeficiency virus treated with intravitreal ganciclovir injection. Graefes Arch Clin Exp Ophthalmol 252:1393–1401. https://doi.org/10.1007/s00417-014-2587-5

Ahuja Y, Couch SM, Razonable RR et al (2008) Infectious retinitis: a review. Retin Physician 2008:17–25

Ainley R, Smith B (1965) Fungal flora of the conjunctival sac in healthy and diseased eyes. Br J Ophthalmol 49:505–515. https://doi.org/10.1136/bjo.49.10.505

Ajayi BG, Osuntokun B, Olurin O et al (1986) Orbital histoplasmosis due to Histoplasma capsulatum var. duboisii: successful treatment with Septrin. J Trop Med Hyg 89:179–187

Albietz JM, Lenton LM (2006) Effect of antibacterial honey on the ocular flora in tear deficiency and meibomian gland disease. Cornea 25:1012–1019. https://doi.org/10.1097/01.ico.0000225716.85382.7b

Al-Rajhi AA, Awad AH, Al-Hedaithy SS et al (1993) Scytalidium dimidiatum fungal endophthalmitis. Br J Ophthalmol 77:388–390. https://doi.org/10.1136/bjo.77.6.388

Amin RM, Hamdy I, Osman IM (2015) Chronic candida endophthalmitis as a cause of intermediate uveitis. BMJ Case Rep 2015. https://doi.org/10.1136/bcr-2014-208847

Anand A (2001) Use of polymerase chain reaction in the diagnosis of fungal endophthalmitis. Ophthalmology 108:326–330. https://doi.org/10.1016/s0161-6420(00)00517-0

Anane S, Anane Touzri R, Malouche N et al (2007) Which is the role of parasites and yeasts in the genesis of chronic blepharitis? Pathol Biol 55:323–327. https://doi.org/10.1016/j.patbio.2007.03.005

Ando N, Takatori K (1982) Fungal flora of the conjunctival sac. Am J Ophthalmol 94:67–74. https://doi.org/10.1016/0002-9394(82)90193-3

Azari AA, Barney NP (2013) Conjunctivitis: a systematic review of diagnosis and treatment. JAMA 310:1721–1729. https://doi.org/10.1001/jama.2013.280318

Bagyalakshmi R, Madhavan HN, Therese KL (2006) Development and application of multiplex polymerase chain reaction for the etiological diagnosis of infectious endophthalmitis. J Postgrad Med 52:179–182

Balne PK, Reddy AK, Kodiganti M et al (2012) Evaluation of three PCR assays for the detection of fungi in patients with mycotic keratitis. Br J Ophthalmol 96:911–912. https://doi.org/10.1136/bjophthalmol-2011-300207

Bannerman TL, Rhoden DL, McAllister SK et al (1997) The source of coagulase-negative staphylococci in the Endophthalmitis Vitrectomy Study. A comparison of eyelid and intraocular isolates using pulsed-field gel electrophoresis. Arch Ophthalmol 115:357–361. https://doi.org/10.1001/archopht.1997.01100150359008

Barisani-Asenbauer T, Maca SM, Mejdoubi L et al (2012) Uveitis-a rare disease often associated with systemic diseases and infections- a systematic review of 2619 patients. Orphanet J Rare Dis 7:57. https://doi.org/10.1186/1750-1172-7-57

Bartolini CE, Duker JS (2012) Man presents with unilateral loss of vision after fever. Ocular Surg News 17:7

Beare NA, Lewallen S, Taylor TE et al (2011) Redefining cerebral malaria by including malaria retinopathy. Future Microbiol 6:349–355. https://doi.org/10.2217/fmb.11.3

Bedwell J, Bauman NM (2011) Management of pediatric orbital cellulitis and abscess. Curr Opin Otolaryngol Head Neck Surg 19:467–473. https://doi.org/10.1097/MOO.0b013e32834cd54a

Behera UC, Budhwani M, Das T et al (2018) Role of early vitrectomy in the treatment of fungal endophthalmitis. Retina 38:1385–1392. https://doi.org/10.1097/IAE.0000000000001727

Behrens-Baumann W, Rüchel R (1999) Mycosis of the eye and its adnexa. Karger Medical and Scientific Publishers, Basel, p 32

Bernauer W, Allan BD, Dart JK (1998) Successful management of Aspergillus scleritis by medical and surgical treatment. Eye 12(2):311–316. https://doi.org/10.1038/eye.1998.71

Bharathi MJ, Ramakrishnan R, Vasu S et al (2003) Epidemiological characteristics and laboratory diagnosis of fungal keratitis. A three-year study. Indian J Ophthalmol 51:315–321

Bharathi MJ, Ramakrishnan R, Meenakshi R et al (2006) Microbiological diagnosis of infective keratitis: comparative evaluation of direct microscopy and culture results. Br J Ophthalmol 90: 1271–1276. https://doi.org/10.1136/bjo.2006.096230

Bharathi MJ, Ramakrishnan R, Meenakshi R et al (2007) Microbial keratitis in South India: influence of risk factors, climate, and geographical variation. Ophthalmic Epidemiol 14:61–69. https://doi.org/10.1080/09286580601001347

Borderie VM, Bourcier TM, Poirot JL et al (1997) Endophthalmitis after Lasiodiplodia theobromae corneal abscess. Graefes Arch Clin Exp Ophthalmol 235:259–261. https://doi.org/10.1007/BF00941769

Brito-Zeron P, Baldini C, Bootsma H et al (2016) Sjogren syndrome. Nat Rev Dis Primers 2:16047. https://doi.org/10.1038/nrdp.2016.47

Brooks JP, Edwards DJ, Harwich MD Jr et al (2015) The truth about metagenomics: quantifying and counteracting bias in 16S rRNA studies. BMC Microbiol 15:66. https://doi.org/10.1186/s12866-015-0351-6

Carney MD, Tabassian A, Guerry RK (1996) Pseudo-Allescheria boydii endophthalmitis. Retina 16:263–264

Chakrabarti A, Shivaprakash MR, Singh R et al (2008) Fungal endophthalmitis: fourteen years' experience from a center in India. Retina 28:1400–1407. https://doi.org/10.1097/iae.0b013e318185e943

Chander J, Chakrabarti A, Sharma A et al (1993) Evaluation of Calcofluor staining in the diagnosis of fungal corneal ulcer. Mycoses 36:243–245. https://doi.org/10.1111/j.1439-0507.1993.tb00758.x

Chang HY, Chodosh J (2011) Diagnostic and therapeutic considerations in fungal keratitis. Int Ophthalmol Clin 51:33–42. https://doi.org/10.1097/IIO.0b013e31822d64dc

Chang VS, Chodosh J, Papaliodis GN (2016) Chronic ocular complications of Stevens-Johnson Syndrome and toxic epidermal necrolysis: the role of systemic immunomodulatory therapy. Semin Ophthalmol 31:178–187. https://doi.org/10.3109/08820538.2015.1114841

Chaudhry IA, Shamsi FA, Elzaridi E et al (2007) Outcome of treated orbital cellulitis in a tertiary eye care center in the middle East. Ophthalmology 114:345–354. https://doi.org/10.1016/j.ophtha.2006.07.059

Chee SP, Jap A (2001) Endogenous endophthalmitis. Curr Opin Ophthalmol 12:464–470. https://doi.org/10.1097/00055735-200112000-00012

Christian M, Loos D, Feucht N et al (2019) Endogenous endophthalmitis: epidemiology, clinic, therapy and visual outcome. Klin Monatsbl Augenheilkd 236:1208–1214. https://doi.org/10.1055/s-0044-101152

Clemente JC, Ursell LK, Parfrey LW et al (2012) The impact of the gut microbiota on human health: an integrative view. Cell 148:1258–1270. https://doi.org/10.1016/j.cell.2012.01.035

Connell PP, O'Neill EC, Fabinyi D et al (2011) Endogenous endophthalmitis: 10-year experience at a tertiary referral centre. Eye 25:66–72. https://doi.org/10.1038/eye.2010.145

Creach P, Auffret N, Buot G et al (1995) Microsporum canis tinea ciliaris and blepharitis. Ann Dermatol Venereol 122:773–774

Culbertson WW, Ostler HB (1981) The floppy eyelid syndrome. Am J Ophthalmol 92:568–575. https://doi.org/10.1016/0002-9394(81)90652-8

Dadaci Z, Kilinc F, Ozer TT et al (2015) Periodic acid-Schiff staining demonstrates fungi in chronic anterior blepharitis. Eye 29:1522–1527. https://doi.org/10.1038/eye.2015.144

Das T (2017) Redefining evidence in the management of acute post-cataract surgery endophthalmitis in India - The 2014 Adenwalla Oration, All India Ophthalmological Society. Indian J Ophthalmol 65:1403–1406. https://doi.org/10.4103/ijo.IJO_755_17

Das T, Gopinathan U, Sharma S (1994) Exogenous helminthosporium endophthalmitis. Br J Ophthalmol 78:492–493. https://doi.org/10.1136/bjo.78.6.492

Das T, Dogra MR, Gopal L et al (1995) Postsurgical endophthalmitis: diagnosis and management. Indian J Ophthalmol 43:103–116

Das T, Hussain A, Naduvilath T et al (2012) Case control analyses of acute endophthalmitis after cataract surgery in South India associated with technique, patient care, and socioeconomic status. J Ophthalmol 2012:298459. https://doi.org/10.1155/2012/298459

Das T, Agarwal M, Behera U et al (2020) Diagnosis and management of fungal endophthalmitis: India perspective. Expert Rev Ophthalmol 15:355–365. https://doi.org/10.1080/17469899.2020.1820322

de Paiva CS, Jones DB, Stern ME et al (2016) Altered mucosal microbiome diversity and disease severity in sjogren syndrome. Sci Rep 6:23561. https://doi.org/10.1038/srep23561

De Rojas MV, Dart JK, Saw VP (2007) The natural history of Stevens Johnson syndrome: patterns of chronic ocular disease and the role of systemic immunosuppressive therapy. Br J Ophthalmol 91:1048–1053. https://doi.org/10.1136/bjo.2006.109124

Dembski M, Nowinska A, Ulfik K et al (2020) In vivo confocal microscopy and anterior segment optical coherence tomography analysis of the microcystic keratitis. J Ophthalmol 2020: 8871904. https://doi.org/10.1155/2020/8871904

Deshmukh S, Das D, Medhi J et al (2018) Fungal hyphae in the anterior chamber in a case of anterior uveitis: a case report. Adv Ophthalmol Vis Syst 8:231–232. https://doi.org/10.15406/aovs.2018.08.00309

Diaz RI, Sigler EJ, Rafieetary MR et al (2015) Ocular histoplasmosis syndrome. Surv Ophthalmol 60:279–295. https://doi.org/10.1016/j.survophthal.2015.02.005

Dong X, Wang Y, Wang W et al (2019) Composition and diversity of bacterial community on the ocular surface of patients with meibomian gland dysfunction. Invest Ophthalmol Vis Sci 60: 4774–4783. https://doi.org/10.1167/iovs.19-27719

Durand ML (2013) Endophthalmitis. Clin Microbiol Infect 19:227–234. https://doi.org/10.1111/1469-0691.12118

Durand ML (2017) Endophthalmitis. Mandell, douglas, and Bennett's infectious disease essentials. Elsevier, Philadelphia

Embong Z, Wan Hitam WH, Yean CY et al (2008) Specific detection of fungal pathogens by 18S rRNA gene PCR in microbial keratitis. BMC Ophthalmol 8:7. https://doi.org/10.1186/1471-2415-8-7

Faridi OS, Ranchod TM, Ho LY et al (2010) Pandemic 2009 influenza A H1N1 retinopathy. Can J Ophthalmol 45:286–287. https://doi.org/10.3129/i10-030

Farooq AV, Patel RM, Lin AY et al (2015) Fungal orbital cellulitis: presenting features, management and outcomes at a referral center. Orbit 34:152–159. https://doi.org/10.3109/01676830.2015.1014512

Ferrer C, Alio JL (2011) Evaluation of molecular diagnosis in fungal keratitis. Ten years of experience. J Ophthalmic Inflamm Infect 1:15–22. https://doi.org/10.1007/s12348-011-0019-9

Fierro-Arias L, Araiza J, Cortés D et al (2014) Blepharitis associated to opportunistic fungi. Dermatol Rev Mex 58:134–141

Fincher T, Fulcher SF (2007) Diagnostic and therapeutic challenge of Aspergillus flavus scleritis. Cornea 26:618–620. https://doi.org/10.1097/ICO.0b013e318033de67

Font RL, Wu TG, Wilhelmus KR et al (2002) Polymerase chain reaction (PCR) in the diagnosis of fungal infections of the eye. Invest Ophthalmol Vis Sci 43:1609–1609

Forte R, Cennamo G, Del Prete S et al (2009) Allergic conjunctivitis and latent infections. Cornea 28:839–842. https://doi.org/10.1097/ICO.0b013e318197711cf

Foster RE, Lowder CY, Meisler DM et al (1991) Presumed Pneumocystis carinii choroiditis. Unifocal presentation, regression with intravenous pentamidine, and choroiditis recurrence. Ophthalmology 98:1360–1365. https://doi.org/10.1016/S0161-6420(91)32125-0

Fowler AM, Dutton JJ (2010) Floppy eyelid syndrome as a subset of lax eyelid conditions: relationships and clinical relevance (an ASOPRS thesis). Ophthalmic Plast Reconstr Surg 26: 195–204. https://doi.org/10.1097/IOP.0b013e3181b9e37e

Fowler B, Shen C, Mastellone J et al (2011) Acute systemic histoplasmosis associated with chorioretinitis in an immunocompetent adolescent. Digit J Ophthalmol 17:31–35. https://doi.org/10.5693/djo.02.2011.07.002

Franklin RM, Kenyon KR, Tutschka PJ et al (1983) Ocular manifestations of graft-vs-host disease. Ophthalmology 90:4–13. https://doi.org/10.1016/s0161-6420(83)34604-2

Frizon L, Araujo MC, Andrade L et al (2014) Evaluation of conjunctival bacterial flora in patients with Stevens-Johnson Syndrome. Clinics 69:168–172. https://doi.org/10.6061/clinics/2014(03)04

Gamaletsou MN, Sipsas NV, Roilides E et al (2012) Rhino-orbital-cerebral mucormycosis. Curr Infect Dis Rep 14:423–434. https://doi.org/10.1007/s11908-012-0272-6

Gandhi J, Jayasudha R, Naik P et al (2019) Targeted high-throughput sequencing identifies predominantly fungal pathogens in patients with clinically infectious, culture-negative endophthalmitis in South India. Microorganisms 7:100411. https://doi.org/10.3390/microorganisms7100411

Garg S, Jampol LM (2005) Systemic and intraocular manifestations of West Nile virus infection. Surv Ophthalmol 50:3–13. https://doi.org/10.1016/j.survophthal.2004.10.001

Gaujoux T, Borsali E, Gavrilov JC et al (2012) Fungal keratitis caused by Cylindrocarpon lichenicola. J Fr Ophtalmol 35(356):351–355. https://doi.org/10.1016/j.jfo.2011.06.005

Goldblum D, Frueh BE, Sarra GM et al (2005) Topical caspofungin for treatment of keratitis caused by Candida albicans in a rabbit model. Antimicrob Agents Chemother 49:1359–1363. https://doi.org/10.1128/AAC.49.4.1359-1363.2005

Gopinathan U, Garg P, Fernandes M et al (2002) The epidemiological features and laboratory results of fungal keratitis: a 10-year review at a referral eye care center in South India. Cornea 21:555–559. https://doi.org/10.1097/00003226-200208000-00004

Gopinathan U, Sharma S, Garg P et al (2009) Review of epidemiological features, microbiological diagnosis and treatment outcome of microbial keratitis: experience of over a decade. Indian J Ophthalmol 57:273–279. https://doi.org/10.4103/0301-4738.53051

Graham JE, Moore JE, Jiru X et al (2007) Ocular pathogen or commensal: a PCR-based study of surface bacterial flora in normal and dry eyes. Invest Ophthalmol Vis Sci 48:5616–5623. https://doi.org/10.1167/iovs.07-0588

Greenfield DS, Suner IJ, Miller MP et al (1996) Endophthalmitis after filtering surgery with mitomycin. Arch Ophthalmol 114:943–949. https://doi.org/10.1001/archopht.1996.01100140151007

Gritz DC, Wong IG (2004) Incidence and prevalence of uveitis in Northern California; the Northern California Epidemiology of Uveitis Study. Ophthalmology 111(491-500):discussion 500. https://doi.org/10.1016/j.ophtha.2003.06.014
Gupta A, Gupta V, Gupta A et al (2003) Spectrum and clinical profile of post cataract surgery endophthalmitis in north India. Indian J Ophthalmol 51:139–145
Gupta N, Singh PK, Revankar SG et al (2019) Pathobiology of Aspergillus fumigatus endophthalmitis in immunocompetent and immunocompromised mice. Microorganisms 7. https://doi.org/10.3390/microorganisms7090297
Hamad I, Raoult D, Bittar F (2016) Repertory of eukaryotes (eukaryome) in the human gastrointestinal tract: taxonomy and detection methods. Parasite Immunol 38:12–36. https://doi.org/10.1111/pim.12284
Hammeke JC, Ellis PP (1960) Mycotic flora of the conjunctiva. Am J Ophthalmol 49:1174–1178. https://doi.org/10.1016/0002-9394(60)91632-9
Hammer ME, Harding S, Wynn P (1983) Post-traumatic fungal endophthalmitis caused by Exophiala jeanselmei. Ann Ophthalmol 15:853–855
Hau SC, Dart JK, Vesaluoma M et al (2010) Diagnostic accuracy of microbial keratitis with in vivo scanning laser confocal microscopy. Br J Ophthalmol 94:982–987. https://doi.org/10.1136/bjo.2009.175083
Hazin R, Ibrahimi OA, Hazin MI et al (2008) Stevens-Johnson syndrome: pathogenesis, diagnosis, and management. Ann Med 40:129–138. https://doi.org/10.1080/07853890701753664
He YG, McCulley JP, Alizadeh H et al (1992) A pig model of Acanthamoeba keratitis: transmission via contaminated contact lenses. Invest Ophthalmol Vis Sci 33:126–133
Hegde R, Sundar G (2017) Orbital cellulitis- a review. TNOA J Ophthalmic Sci Res 55:211–219. https://doi.org/10.4103/tjosr.tjosr_9_18
Helmick CG, Felson DT, Lawrence RC et al (2008) Estimates of the prevalence of arthritis and other rheumatic conditions in the United States. Part I. Arthritis Rheum 58:15–25. https://doi.org/10.1002/art.23177
Hemady R, Sainz de la Maza M, Raizman MB et al (1992) Six cases of scleritis associated with systemic infection. Am J Ophthalmol 114:55–62. https://doi.org/10.1016/s0002-9394(14)77413-6
Hori Y, Maeda N, Sakamoto M et al (2008) Bacteriologic profile of the conjunctiva in the patients with dry eye. Am J Ophthalmol 146:729–734. https://doi.org/10.1016/j.ajo.2008.06.003
Howe L, Jones NS (2004) Guidelines for the management of periorbital cellulitis/abscess. Clin Otolaryngol Allied Sci 29:725–728. https://doi.org/10.1111/j.1365-2273.2004.00889.x
Huang FC, Huang SP, Tseng SH (2000) Management of infectious scleritis after pterygium excision. Cornea 19:34–39. https://doi.org/10.1097/00003226-200001000-00008
Huang Y, Yang B, Li W (2016) Defining the normal core microbiome of conjunctival microbial communities. Clin Microbiol Infect 22:643. https://doi.org/10.1016/j.cmi.2016.04.008
Huseyin CE, O'Toole PW, Cotter PD et al (2017) Forgotten fungi-the gut mycobiome in human health and disease. FEMS Microbiol Rev 41:479–511. https://doi.org/10.1093/femsre/fuw047
Invernizzi A, Symes R, Miserocchi E et al (2018) Spectral domain optical coherence tomography findings in endogenous candida endophthalmitis and their clinical relevance. Retina 38:1011–1018. https://doi.org/10.1097/IAE.0000000000001630
Jackson WB (2008) Blepharitis: current strategies for diagnosis and management. Can J Ophthalmol 43:170–179. https://doi.org/10.1139/i08-016
Jacobson M, Galetta SL, Atlas SW et al (1992) Bipolaris-induced orbital cellulitis. J Clin Neuroophthalmol 12:250–256
Jain V, Garg P, Sharma S (2009) Microbial scleritis-experience from a developing country. Eye 23:255–261. https://doi.org/10.1038/sj.eye.6703099
Janssen R, Krogfelt KA, Cawthraw SA et al (2008) Host-pathogen interactions in Campylobacter infections: the host perspective. Clin Microbiol Rev 21:505–518. https://doi.org/10.1128/CMR.00055-07
Javadi MA, Feizi S (2011) Dry eye syndrome. J Ophthalmic Vis Res 6:192–198

Jayasudha R, Kalyana Chakravarthy S, Sai Prashanthi G et al (2019) Implicating dysbiosis of the gut fungal microbiome in uveitis, an inflammatory disease of the eye. Invest Ophthalmol Vis Sci 60:1384–1393. https://doi.org/10.1167/iovs.18-26426

Jeang LJ, Davis A, Madow B et al (2017) Occult fungal scleritis. Ocul Oncol Pathol 3:41–44. https://doi.org/10.1159/000449103

Jhanji V, Yohendran J, Constantinou M et al (2009) Scedosporium scleritis or keratitis or both: case series. Eye Contact Lens 35:312–315. https://doi.org/10.1097/ICL.0b013e3181be722e

Jiang X, Deng A, Yang J et al (2018) Pathogens in the Meibomian gland and conjunctival sac: microbiome of normal subjects and patients with Meibomian gland dysfunction. Infect Drug Resist 11:1729–1740. https://doi.org/10.2147/IDR.S162135

Johnson TE (2000) Fungal disease of the orbit. Ophthalmol Clin N Am 13:643–656

Joseph J, Murthy S, Garg P et al (2006) Use of different stains for microscopic evaluation of corneal scrapings for diagnosis of microsporidial keratitis. J Clin Microbiol 44:583–585. https://doi.org/10.1128/JCM.44.2.583-585.2006

Kalkanci A, Ozdek S (2011) Ocular fungal infections. Curr Eye Res 36:179–189. https://doi.org/10.3109/02713683.2010.533810

Kalyana Chakravarthy S, Jayasudha R, Sai Prashanthi G et al (2018) Dysbiosis in the gut bacterial microbiome of patients with uveitis, an inflammatory disease of the eye. Indian J Microbiol 58:457–469. https://doi.org/10.1007/s12088-018-0746-9

Kammoun S, Rekik M, Trabelsi H et al (2018) Orbital cellulitis secondary to a fungal sinusitis caused by Scopulariopsis: the first case in Tunisia. J Mycol Med 28:384–386. https://doi.org/10.1016/j.mycmed.2018.04.006

Karsten E, Watson SL, Foster LJ (2012) Diversity of microbial species implicated in keratitis: a review. Open Ophthalmol J 6:110–124. https://doi.org/10.2174/1874364101206010110

Kasper DL, Braunwald E, Fauci AS, Hauser SL, Longo DL, Jameson JL (eds) (2005) Harrison's principles of internal medicine, 16th edn. McGraw-Hill Medical Publishing Division, New York

Kassan SS, Moutsopoulos HM (2004) Clinical manifestations and early diagnosis of Sjogren syndrome. Arch Intern Med 164:1275–1284. https://doi.org/10.1001/archinte.164.12.1275

Kawali A, Mahendradas P, Mohan A et al (2019) Epidemic retinitis. Ocul Immunol Inflamm 27:571–577. https://doi.org/10.1080/09273948.2017.1421670

Khairallah SH, Byrne KA, Tabbara KF (1992) Fungal keratitis in Saudi Arabia. Doc Ophthalmol 79:269–276. https://doi.org/10.1007/BF00158257

Khairallah M, Kahloun R, Ben Yahia S et al (2013) New infectious etiologies for posterior uveitis. Ophthalmic Res 49:66–72. https://doi.org/10.1159/000344009

Kim DH, Yoon KC, Seo KY et al (2015) The role of systemic immunomodulatory treatment and prognostic factors on chronic ocular complications in Stevens-Johnson syndrome. Ophthalmology 122:254–264. https://doi.org/10.1016/j.ophtha.2014.08.013

Kittipibul T, Puangsricharern V, Chatsuwan T (2020) Comparison of the ocular microbiome between chronic Stevens-Johnson syndrome patients and healthy subjects. Sci Rep 10:4353. https://doi.org/10.1038/s41598-020-60794-w

Klotz SA, Penn CC, Negvesky GJ et al (2000) Fungal and parasitic infections of the eye. Clin Microbiol Rev 13:662–685. https://doi.org/10.1128/CMR.13.4.662

Kocak N, Kaynak S, Kirdar S et al (2009) Comparison of different antifungal treatment regimens for experimental Candida endophthalmitis in rabbit models. Mikrobiyol Bul 43:619–626

Kohanim S, Palioura S, Saeed HN et al (2016) Acute and chronic ophthalmic involvement in Stevens-Johnson syndrome/toxic epidermal necrolysis - a comprehensive review and guide to therapy. II. Ophthalmic disease. Ocul Surf 14:168–188. https://doi.org/10.1016/j.jtos.2016.02.001

Köllner H (1906) Schimmelpilzerkrankung der Sklera. Ophthalmologica 16:441–447. https://doi.org/10.1159/000291135

Kose A (2020) Candida retinitis in a liver transplantation recipient. Int J Transplant Res Med 6:1510047. https://doi.org/10.23937/2572-4045.1510047

Krishnan S, Manavathu EK, Chandrasekar PH (2009) Aspergillus flavus: an emerging non-fumigatus Aspergillus species of significance. Mycoses 52:206–222. https://doi.org/10.1111/j.1439-0507.2008.01642.x

Kuo YK, Lin IC, Chien LN et al (2019) Dry eye disease: a review of epidemiology in Taiwan, and its clinical treatment and merits. J Clin Med 8. https://doi.org/10.3390/jcm8081227

Kuriyan AE, Sridhar J, Flynn HW Jr et al (2014) Endophthalmitis caused by Enterococcus faecalis: clinical features, antibiotic sensitivities, and outcomes. Am J Ophthalmol 158:1018–1023. https://doi.org/10.1016/j.ajo.2014.07.038

Kurosawa A, Pollock SC, Collins MP et al (1988) Sporothrix schenckii endophthalmitis in a patient with human immunodeficiency virus infection. Arch Ophthalmol 106:376–380. https://doi.org/10.1001/archopht.1988.01060130402030

Labbé A, Khammari C, Dupas B et al (2009) Contribution of in vivo confocal microscopy to the diagnosis and management of infectious keratitis. Ocul Surf 7:41–52. https://doi.org/10.1016/s1542-0124(12)70291-4

Lalitha P, Prajna NV, Manoharan G et al (2015) Trends in bacterial and fungal keratitis in South India, 2002-2012. Br J Ophthalmol 99:192–194. https://doi.org/10.1136/bjophthalmol-2014-305000

Lee SH, Oh DH, Jung JY et al (2012) Comparative ocular microbial communities in humans with and without blepharitis. Invest Ophthalmol Vis Sci 53:5585–5593. https://doi.org/10.1167/iovs.12-9922

Lee CY, Ho YJ, Sun CC et al (2016) Recurrent fungal keratitis and blepharitis caused by Aspergillus flavus. Am J Trop Med Hyg 95:1216–1218. https://doi.org/10.4269/ajtmh.16-0453

Lekhanont K, Nonpassopon M, Nimvorapun N et al (2015) Treatment with intrastromal and intracameral voriconazole in 2 eyes with Lasiodiplodia theobromae keratitis: case reports. Medicine 94:e541. https://doi.org/10.1097/MD.0000000000000541

Levin LA, Avery R, Shore JW et al (1996) The spectrum of orbital aspergillosis: a clinicopathological review. Surv Ophthalmol 41:142–154. https://doi.org/10.1016/s0039-6257(96)80004-x

Levison AL, Mendes TS, Bhisitkul R (2014) Postprocedural endophthalmitis: a review. Expert Rev Ophthalmol 8:45–62. https://doi.org/10.1586/eop.12.77

Lin P (2015) Infectious uveitis. Curr Ophthalmol Rep 3:170–183. https://doi.org/10.1007/s40135-015-0076-6

Lin X, Cavanagh HD (2015) Ocular manifestations of graft-versus-host disease: 10 years' experience. Clin Ophthalmol 9:1209–1213. https://doi.org/10.2147/OPTH.S84704

Louie T, el Baba F, Shulman M et al (1994) Endogenous endophthalmitis due to Fusarium: case report and review. Clin Infect Dis 18:585–588. https://doi.org/10.1093/clinids/18.4.585

Lu CW, Tao Y, Li XH et al (2017) Fungal chorioretinitis with systemic candidiasis in an infant following treatment with broad spectrum antibiotics: a case report. Exp Ther Med 14:286–288. https://doi.org/10.3892/etm.2017.4507

Lupetti A, Tavanti A, Davini P et al (2002) Horizontal transmission of Candida parapsilosis candidemia in a neonatal intensive care unit. J Clin Microbiol 40:2363–2369. https://doi.org/10.1128/JCM.40.7.2363-2369.2002

Ma L, Xie L, Dong X et al (2009) Role of extracellular phospholipase B of Candida albicans as a virulent factor in experimental keratomycosis. Curr Eye Res 34:761–768. https://doi.org/10.1080/02713680903056391

Maharana PK, Sharma N, Nagpal R et al (2016) Recent advances in diagnosis and management of Mycotic Keratitis. Indian J Ophthalmol 64:346–357. https://doi.org/10.4103/0301-4738.185592

Marchlewitz Ba MM (1966) Die Pilzflora gesunder Augen. Z Militarmed 3:170–174

Marinho PM, Marcos AAA, Romano AC et al (2020) Retinal findings in patients with COVID-19. Lancet 395:1610. https://doi.org/10.1016/s0140-6736(20)31014-x

Mastrota KM (2008) Impact of floppy eyelid syndrome in ocular surface and dry eye disease. Optom Vis Sci 85:814–816. https://doi.org/10.1097/OPX.0b013e3181852777

McGuire TW, Bullock JD, Bullock JD Jr et al (1991) Fungal endophthalmitis. An experimental study with a review of 17 human ocular cases. Arch Ophthalmol 109:1289–1296. https://doi.org/10.1001/archopht.1991.01080090115034

McLeod SD, Kumar A, Cevallos V et al (2005) Reliability of transport medium in the laboratory evaluation of corneal ulcers. Am J Ophthalmol 140:1027–1031. https://doi.org/10.1016/j.ajo.2005.06.042

Menia NK, Sharma SP, Bansal R (2019) Fungal retinitis following influenza virus type A (H1N1) infection. Indian J Ophthalmol 67:1483–1484. https://doi.org/10.4103/ijo.IJO_1691_18

Minkus CL, Bispo PJM, Papaliodis GN et al (2019) Real-time multiplex PCR analysis in infectious uveitis. Semin Ophthalmol 34:252–255. https://doi.org/10.1080/08820538.2019.1620803

Mirza N, Zierhut M, Korn A et al (2016) Graft versus self (GvS) against T-cell autoantigens is a mechanism of graft-host interaction. Proc Natl Acad Sci U S A 113:13827–13832. https://doi.org/10.1073/pnas.1609118113

Mitchell BM, Wu TG, Chong EM et al (2007) Expression of matrix metalloproteinases 2 and 9 in experimental corneal injury and fungal keratitis. Cornea 26:589–593. https://doi.org/10.1097/ICO.0b013e318033b504

Mitsui Y, Hanabusa J (1955) Corneal infections after cortisone therapy. Br J Ophthalmol 39:244–250. https://doi.org/10.1136/bjo.39.4.244

Modjtahedi BS, Finn AP, Papakostas TD et al (2017) Intravenous Drug Use-associated endophthalmitis. Ophthalmol Retina 1:192–199. https://doi.org/10.1016/j.oret.2016.10.013

Modjtahedi BS, Finn AP, Barb SM et al (2019) Characteristics and outcomes of endogenous endophthalmitis: eight-year experience at a tertiary care center. Ophthalmol Retina 3:61–72. https://doi.org/10.1016/j.oret.2018.08.009

Moon J, Choi SH, Yoon CH et al (2020) Gut dysbiosis is prevailing in Sjogren's syndrome and is related to dry eye severity. PLoS One 15:e0229029. https://doi.org/10.1371/journal.pone.0229029

Motley WW, Melson AT, Mortensen JE (2011) Pediatric Metarrhizium anisopliae keratitis. J AAPOS 15:101–103. https://doi.org/10.1016/j.jaapos.2010.12.003

Mukherjee B, Chatterjee R, Biswas J (2013) Reverse masquerade syndrome: fungal adnexal infection mimicking carcinoma in a HIV-positive patient. Indian J Ophthalmol 61:521–523. https://doi.org/10.4103/0301-4738.119454

Mukherjee B, Raichura ND, Alam MS (2016) Fungal infections of the orbit. Indian J Ophthalmol 64:337–345. https://doi.org/10.4103/0301-4738.185588

Murthy SI, Sabhapandit S, Balamurugan S et al (2020) Scleritis: differentiating infectious from non-infectious entities. Indian J Ophthalmol 68:1818–1828. https://doi.org/10.4103/ijo.IJO_2032_20

Nair AG, Ali MJ, Kaliki S et al (2015) Rhinosporidiosis of the tarsal conjunctiva. Indian J Ophthalmol 63:462–463. https://doi.org/10.4103/0301-4738.159897

Nath R, Baruah S, Saikia L et al (2011) Mycotic corneal ulcers in upper Assam. Indian J Ophthalmol 59:367–371. https://doi.org/10.4103/0301-4738.83613

Nattis A, Perry HD, Rosenberg ED et al (2019) Influence of bacterial burden on meibomian gland dysfunction and ocular surface disease. Clin Ophthalmol 13:1225–1234. https://doi.org/10.2147/OPTH.S215071

Nema HV, Ahuja OP, Mohapatra LN (1966) Mycotic flora of the conjunctiva. Am J Ophthalmol 62:968–970. https://doi.org/10.1016/0002-9394(66)91928-3

Nielsen JS, Fick TA, Saggau DD et al (2012) Intravitreal anti-vascular endothelial growth factor therapy for choroidal neovascularization secondary to ocular histoplasmosis syndrome. Retina 32:468–472. https://doi.org/10.1097/IAE.0b013e318229b220

Olson CL (1969) Fungal contamination of conjunctiva and lid margin. Arch Ophthalmol 81:351–355. https://doi.org/10.1001/archopht.1969.00990010353009

Ozdemir HG, Oz Y, Ilkit M et al (2012) Antifungal susceptibility of ocular fungal pathogens recovered from around the world against itraconazole, voriconazole, amphotericin B, and caspofungin. Med Mycol 50:130–135. https://doi.org/10.3109/13693786.2011.584328

Ozkan J, Nielsen S, Diez-Vives C et al (2017) Temporal stability and composition of the ocular surface microbiome. Sci Rep 7:9880. https://doi.org/10.1038/s41598-017-10494-9

Ozturk F, Yavas GF, Kusbeci T et al (2007) Efficacy of topical caspofungin in experimental fusarium keratitis. Cornea 26:726–728. https://doi.org/10.1097/ICO.0b013e3180553b9d

Pandian DG, Babu RK, Chaitra A et al (2011) Nine years' review on preseptal and orbital cellulitis and emergence of community-acquired methicillin-resistant Staphylococus aureus in a tertiary hospital in India. Indian J Ophthalmol 59:431–435. https://doi.org/10.4103/0301-4738.86309

Partamian LG, Jay WM, Fritz KJ (1983) Anaerobic orbital cellulitis. Ann Ophthalmol 15:123–126

Pettit TH, Olson RJ, Foos RY et al (1980) Fungal endophthalmitis following intraocular lens implantation. A surgical epidemic. Arch Ophthalmol 98:1025–1039. https://doi.org/10.1001/archopht.1980.01020031015002

Pflugfelder SC, Flynn HW Jr, Zwickey TA et al (1988) Exogenous fungal endophthalmitis. Ophthalmology 95:19–30. https://doi.org/10.1016/s0161-6420(88)33229-x

Pornpanich K, Uiprasertkul M, Luemsamran P et al (2011) Entomophthoramycosis: a rare fungal orbital infection presenting with dacryocystitis. Orbit 30:21–23. https://doi.org/10.3109/01676830.2010.535640

Prabhushanker M, Topiwalla TT, Ganesan G et al (2017) Bilateral retinitis following typhoid fever. Int J Retina Vitreous 3:11. https://doi.org/10.1186/s40942-017-0065-z

Putnam CM (2016) Diagnosis and management of blepharitis: an optometrist's perspective. Clin Optom 8:71–78. https://doi.org/10.2147/OPTO.S84795

Qin J, Li R, Raes J et al (2010) A human gut microbial gene catalogue established by metagenomic sequencing. Nature 464:59–65. https://doi.org/10.1038/nature08821

Rajilic-Stojanovic M, Smidt H, de Vos WM (2007) Diversity of the human gastrointestinal tract microbiota revisited. Environ Microbiol 9:2125–2136. https://doi.org/10.1111/j.1462-2920.2007.01369.x

Ramakrishnan R, Bharathi MJ, Shivkumar C et al (2009) Microbiological profile of culture-proven cases of exogenous and endogenous endophthalmitis: a 10-year retrospective study. Eye 23:945–956. https://doi.org/10.1038/eye.2008.197

Rautaraya B, Sharma S, Kar S et al (2011) Diagnosis and treatment outcome of mycotic keratitis at a tertiary eye care center in eastern India. BMC Ophthalmol 11:39. https://doi.org/10.1186/1471-2415-11-39

Reddy JC, Murthy SI, Reddy AK et al (2015) Risk factors and clinical outcomes of bacterial and fungal scleritis at a tertiary eye care hospital. Middle East Afr J Ophthalmol 22:203–211. https://doi.org/10.4103/0974-9233.150634

Ren M, Wu X (2010) Evaluation of three different methods to establish animal models of Acanthamoeba keratitis. Yonsei Med J 51:121–127. https://doi.org/10.3349/ymj.2010.51.1.121

Revankar SG, Sutton DA (2010) Melanized fungi in human disease. Clin Microbiol Rev 23:884–928. https://doi.org/10.1128/CMR.00019-10

Rosenbaum JT (1989) Uveitis. An internist's view. Arch Intern Med 149:1173–1176. https://doi.org/10.1001/archinte.149.5.1173

Rosenbaum JT, Asquith M (2018) The microbiome and HLA-B27-associated acute anterior uveitis. Nat Rev Rheumatol 14:704–713. https://doi.org/10.1038/s41584-018-0097-2

Sahin GO, Dadaci Z, Ozer TT (2014) Two cases of tinea ciliaris with blepharitis due to Microsporum audouinii and Trichophyton verrucosum and review of the literature. Mycoses 57:577–580. https://doi.org/10.1111/myc.12201

Sangwan VS (2010) Treatment of uveitis: beyond steroids. Indian J Ophthalmol 58:1–2. https://doi.org/10.4103/0301-4738.58466

Sangwan J, Lathwal S, Juyal D et al (2013) Fonsecaea pedrosoi: a rare etiology in fungal keratitis. J Clin Diagn Res 7:2272–2273. https://doi.org/10.7860/JCDR/2013/6627.3491

Sanjay S, Agrawal S, Mahendradas P et al (2020) Post fever uveoretinal manifestations in an immunocompetent individual. EMJ Allergy Immunol 5:91–105

Sarkar S, Bhattacharya TK, Roy A et al (2016) A review of ocular fungal infections other than keratomycosis: the global and indian scenario. J Evol Med Dent Sci 5:3168–3172. https://doi.org/10.14260/jemds/2016/734

Satpathy G, Ahmed NH, Nayak N et al (2019) Spectrum of mycotic keratitis in north India: sixteen years study from a tertiary care ophthalmic centre. J Infect Public Health 12:367–371. https://doi.org/10.1016/j.jiph.2018.12.005

Segal E, Romano A, Eylan E et al (1977) Fungal flora of the normal conjunctival sac. Mykosen 20: 9–14. https://doi.org/10.1111/j.1439-0507.1977.tb01475.x

Sehgal SC, Dhawan S, Chhiber S et al (1981) Frequency and significance of fungal isolations from conjunctival sac and their role in ocular infections. Mycopathologia 73:17–19. https://doi.org/10.1007/BF00443007

Shaikh S, Trese MT (2004) West Nile virus chorioretinitis. Br J Ophthalmol 88:1599–1560. https://doi.org/10.1136/bjo.2004.049460

Shantha JG, Yeh S, Nguyen QD (2016) Ebola virus disease and the eye. Curr Opin Ophthalmol 27: 538–544. https://doi.org/10.1097/ICU.0000000000000313

Sharma S (2012a) Diagnosis of fungal keratitis: current options. Expert Opin Med Diagn 6:449–455. https://doi.org/10.1517/17530059.2012.679656

Sharma S (2012b) Diagnosis of infectious diseases of the eye. Eye 26:177–184. https://doi.org/10.1038/eye.2011.275

Sharma S, Kunimoto DY, Gopinathan U et al (2002) Evaluation of corneal scraping smear examination methods in the diagnosis of bacterial and fungal keratitis: a survey of eight years of laboratory experience. Cornea 21:643–647. https://doi.org/10.1097/00003226-200210000-00002

Sharma S, Balne PK, Motukupally SR et al (2015) Pythium insidiosum keratitis: clinical profile and role of DNA sequencing and zoospore formation in diagnosis. Cornea 34:438–442. https://doi.org/10.1097/ICO.0000000000000349

Sheu SJ, Chen YC, Kuo NW et al (1998) Endogenous cryptococcal endophthalmitis. Ophthalmology 105:377–381. https://doi.org/10.1016/s0161-6420(98)93679-x

Shimizu E, Ogawa Y, Saijo Y et al (2019) Commensal microflora in human conjunctiva; characteristics of microflora in the patients with chronic ocular graft-versus-host disease. Ocul Surf 17: 265–271. https://doi.org/10.1016/j.jtos.2019.02.001

Shivaji S (2017) We are not alone: a case for the human microbiome in extra intestinal diseases. Gut Pathog 9:13. https://doi.org/10.1186/s13099-017-0163-3

Shivaji S (2019) Connect between gut microbiome and diseases of the human eye. J Biosci 44:0110

Shivaji S (2020) A systematic review of gut microbiome and ocular inflammatory diseases: are they associated? Indian J Ophthalmol 69:535–542. https://doi.org/10.4103/ijo.IJO_1362_20

Shivaji S, Jayasudha R, Sai Prashanthi G et al (2019) The human ocular surface fungal microbiome. Invest Ophthalmol Vis Sci 60:451–459. https://doi.org/10.1167/iovs.18-26076

Sinha MK, Jalali S, Nalamada S (2012) Review of endogenous endophthalmitis caused by Salmonella species including delayed onset Salmonella typhi endophthalmitis. Semin Ophthalmol 27:94–98. https://doi.org/10.3109/08820538.2012.680642

Smith JA, Albenz J, Begley C et al (2007) The epidemiology of dry eye disease: report of the epidemiology subcommittee of the International Dry Eye WorkShop. Ocul Surf 5:93–107. https://doi.org/10.1016/s1542-0124(12)70082-4

Solomon A, Dursun D, Liu Z et al (2001) Pro- and anti-inflammatory forms of interleukin-1 in the tear fluid and conjunctiva of patients with dry-eye disease. Invest Ophthalmol Vis Sci 42:2283–2292

Song A, Dubovy SR, Berrocal AM et al (2002) Endogenous fungal retinitis in a patient with acute lymphocytic leukemia manifesting as uveitis and optic nerve lesion. Arch Ophthalmol 120: 1754–1756

Sotozono C, Ang LP, Koizumi N et al (2007) New grading system for the evaluation of chronic ocular manifestations in patients with Stevens-Johnson syndrome. Ophthalmology 114:1294–1302. https://doi.org/10.1016/j.ophtha.2006.10.029

Speaker MG, Milch FA, Shah MK et al (1991) Role of external bacterial flora in the pathogenesis of acute postoperative endophthalmitis. Ophthalmology 98:639–649. https://doi.org/10.1016/s0161-6420(91)32239-5

Stenson S, Brookner A, Rosenthal S (1982) Bilateral endogenous necrotizing scleritis due to Aspergillus oryzae. Ann Ophthalmol 14:67–72

Stern WH, Tamura E, Jacobs RA et al (1985) Epidemic postsurgical Candida parapsilosis endophthalmitis. Clinical findings and management of 15 consecutive cases. Ophthalmology 92:1701–1709. https://doi.org/10.1016/s0161-6420(85)34095-2

Sugita S, Takase H, Nakano S (2021) Practical use of multiplex and broad-range PCR in ophthalmology. Jpn J Ophthalmol 65:155–168. https://doi.org/10.1007/s10384-020-00794-5

Sun RL, Jones DB, Wilhelmus KR (2007) Clinical characteristics and outcome of Candida keratitis. Am J Ophthalmol 143:1043–1045. https://doi.org/10.1016/j.ajo.2007.02.016

Sun Y, Chandra J, Mukherjee P et al (2010) A murine model of contact lens-associated fusarium keratitis. Invest Ophthalmol Vis Sci 51:1511–1516. https://doi.org/10.1167/iovs.09-4237

Suzuki T, Sutani T, Nakai H et al (2020) The microbiome of the meibum and ocular surface in healthy subjects. Invest Ophthalmol Vis Sci 61:18. https://doi.org/10.1167/iovs.61.2.18

Swan SK, Wagner RA, Myers JP et al (1985) Mycotic endophthalmitis caused by Penicillium sp. after parenteral drug abuse. Am J Ophthalmol 100:408–410. https://doi.org/10.1016/0002-9394(85)90502-1

Sward M, Kirk C, Kumar S et al (2018) Lax eyelid syndrome (LES), obstructive sleep apnea (OSA), and ocular surface inflammation. Ocul Surf 16:331–336. https://doi.org/10.1016/j.jtos.2018.04.003

Thomas PA (2003a) Current perspectives on ophthalmic mycoses. Clin Microbiol Rev 16:730–797. https://doi.org/10.1128/CMR.16.4.730-797.2003

Thomas PA (2003b) Fungal infections of the cornea. Eye 17:852–862. https://doi.org/10.1038/sj.eye.6700557

Thomas PA, Kaliamurthy J (2013) Mycotic keratitis: epidemiology, diagnosis and management. Clin Microbiol Infect 19:210–220. https://doi.org/10.1111/1469-0691.12126

Thomas PA, Kuriakose T, Kirupashanker MP et al (1991) Use of lactophenol cotton blue mounts of corneal scrapings as an aid to the diagnosis of mycotic keratitis. Diagn Microbiol Infect Dis 14: 219–224. https://doi.org/10.1016/0732-8893(91)90035-e

Todokoro D, Hoshino J, Yo A et al (2018) Scedosporium apiospermum infectious scleritis following posterior subtenon triamcinolone acetonide injection: a case report and literature review. BMC Ophthalmol 18:40. https://doi.org/10.1186/s12886-018-0707-4

Toomey CB, Gross A, Lee J et al (2019) A case of unilateral coccidioidal chorioretinitis in a patient with HIV-associated meningoencephalitis. Case Rep Ophthalmol Med 2019:1475628. https://doi.org/10.1155/2019/1475628

Tsigalou C, Stavropoulou E, Bezirtzoglou E (2018) Current insights in microbiome shifts in Sjogren's syndrome and possible therapeutic interventions. Front Immunol 9:1106. https://doi.org/10.3389/fimmu.2018.01106

Tsirouki T, Dastiridou AI, Ibanez Flores N et al (2018) Orbital cellulitis. Surv Ophthalmol 63:534–553. https://doi.org/10.1016/j.survophthal.2017.12.001

Vaddavalli PK, Garg P, Sharma S et al (2011) Role of confocal microscopy in the diagnosis of fungal and acanthamoeba keratitis. Ophthalmology 118:29–35. https://doi.org/10.1016/j.ophtha.2010.05.018

Van Buren JM (1958) Septic retinitis due to Candida albicans. AMA Arch Pathol 65:137–146

Van Zyl L, Carrara H, Lecuona K (2014) Prevalence of chronic ocular complications in Stevens-Johnson syndrome and toxic epidermal necrolysis. Middle East Afr J Ophthalmol 21:332–335. https://doi.org/10.4103/0974-9233.142272

Vanzzini Zago V, Alcantara Castro M, Naranjo Tackman R (2012) Support of the laboratory in the diagnosis of fungal ocular infections. Int J Inflamm 2012:643104. https://doi.org/10.1155/2012/643104

Velpandian T, Narayanan K, Nag TC et al (2006) Retinal toxicity of intravitreally injected plain and liposome formulation of fluconazole in rabbit eye. Indian J Ophthalmol 54:237–240. https://doi.org/10.4103/0301-4738.27947

Vemuganti GK, Naidu C, Gopinathan U (2002) Rapid detection of fungal filaments in corneal scrapings by microwave heating-assisted Grocott's methenamine silver staining. Indian J Ophthalmol 50:326–328

Venugopal R, Satpathy G, Sangwan S et al (2016) Conjunctival microbial flora in ocular Stevens-Johnson syndrome sequelae patients at a tertiary eye care center. Cornea 35:1117–1121. https://doi.org/10.1097/ICO.0000000000000857

Vida L, Moel SA (1974) Systemic North American blastomycosis with orbital involvement. Am J Ophthalmol 77:240–242. https://doi.org/10.1016/0002-9394(74)90680-1

Vishwanath S, Badami K, Sriprakash KS et al (2014) Post-fever retinitis: a single center experience from south India. Int Ophthalmol 34:851–857. https://doi.org/10.1007/s10792-013-9891-7

Von Graefe A, Mittheilungen K (1857) Verschwarung der sclera nach einer schieloperation. Arch Ophthalmol 3:409–411

Wang L, Deng Y (2020) Malassezia species may play potential roles in the pathogenesis of meibomian gland dysfunction. Med Hypotheses 144:110137. https://doi.org/10.1016/j.mehy.2020.110137

Wang YE, Tepelus TC, Vickers LA et al (2019) Role of in vivo confocal microscopy in the diagnosis of infectious keratitis. Int Ophthalmol 39:2865–2874. https://doi.org/10.1007/s10792-019-01134-4

Wang Y, Chen H, Xia T et al (2020) Characterization of fungal microbiota on normal ocular surface of humans. Clin Microbiol Infect 26:123. https://doi.org/10.1016/j.cmi.2019.05.011

Watson PG, Hayreh SS (1976) Scleritis and episcleritis. Br J Ophthalmol 60:163–191. https://doi.org/10.1136/bjo.60.3.163

Weishaar PD, Flynn HW Jr, Murray TG et al (1998) Endogenous Aspergillus endophthalmitis. Clinical features and treatment outcomes. Ophthalmology 105:57–65. https://doi.org/10.1016/s0161-6420(98)71225-3

Wen X, Miao L, Deng Y et al (2017) The influence of age and sex on ocular surface microbiota in healthy adults. Invest Ophthalmol Vis Sci 58:6030–6037. https://doi.org/10.1167/iovs.17-22957

Wilhelmus KR, Jones DB (2001) Curvularia keratitis. Trans Am Ophthalmol Soc 99:111–130

Williamson J, Haynes J (2018) The bugs behind infectious retinitis. Rev Optom 2018:15

Williamson J, Gordon AM, Wood R et al (1968) Fungal flora of the conjunctival sac in health and disease. Influence of topical and systemic steroids. Br J Ophthalmol 52:127–137. https://doi.org/10.1136/bjo.52.2.127

Wu T, Mitchell B, Carothers T et al (2003a) Molecular analysis of the pediatric ocular surface for fungi. Curr Eye Res 26:33–36. https://doi.org/10.1076/ceyr.26.1.33.14253

Wu TG, Wilhelmus KR, Mitchell BM (2003b) Experimental keratomycosis in a mouse model. Invest Ophthalmol Vis Sci 44:210–216. https://doi.org/10.1167/iovs.02-0446

Wu TG, Keasler VV, Mitchell BM et al (2004) Immunosuppression affects the severity of experimental Fusarium solani keratitis. J Infect Dis 190:192–198. https://doi.org/10.1086/421300

Yamaguchi T (2018) Inflammatory response in dry eye. Invest Ophthalmol Vis Sci 59:192–199. https://doi.org/10.1167/iovs.17-23651

Yang P, Wan W, Du L et al (2018) Clinical features of HLA-B27-positive acute anterior uveitis with or without ankylosing spondylitis in a Chinese cohort. Br J Ophthalmol 102:215–219. https://doi.org/10.1136/bjophthalmol-2016-309499

Yohai RA, Bullock JD, Aziz AA et al (1994) Survival factors in rhino-orbital-cerebral mucormycosis. Surv Ophthalmol 39:3–22. https://doi.org/10.1016/s0039-6257(05)80041-4

Yuan X, Wilhelmus KR (2010) Toll-like receptors involved in the pathogenesis of experimental Candida albicans keratitis. Invest Ophthalmol Vis Sci 51:2094–2100. https://doi.org/10.1167/iovs.09-4330

Yuan X, Wilhelmus KR, Matoba AY et al (2009) Pathogenesis and outcome of Paecilomyces keratitis. Am J Ophthalmol 147:691–696. https://doi.org/10.1016/j.ajo.2008.11.016
Yuan X, Hua X, Wilhelmus KR (2010) Proinflammatory chemokines during Candida albicans keratitis. Exp Eye Res 90:413–419. https://doi.org/10.1016/j.exer.2009.12.001
Zagora SL, Symes R, Yeung A et al (2017) Etiology and clinical features of ocular inflammatory diseases in a Tertiary Referral Centre in Sydney. Ocul Immunol Inflamm 25:107–114. https://doi.org/10.1080/09273948.2016.1247871
Zegans ME, Van Gelder RN (2014) Considerations in understanding the ocular surface microbiome. Am J Ophthalmol 158:420–422. https://doi.org/10.1016/j.ajo.2014.06.014
Zhang W, Yang H, Jiang L et al (2010) Use of potassium hydroxide, Giemsa and calcofluor white staining techniques in the microscopic evaluation of corneal scrapings for diagnosis of fungal keratitis. J Int Med Res 38:1961–1967. https://doi.org/10.1177/147323001003800609
Zhao Z, Yang M, Azar SR et al (2017) Viral retinopathy in experimental models of zika infection. Invest Ophthalmol Vis Sci 58:4355–4365. https://doi.org/10.1167/iovs.17-22016
Zhao F, Zhang D, Ge C et al (2020) Metagenomic profiling of ocular surface microbiome changes in meibomian gland dysfunction. Invest Ophthalmol Vis Sci 61:22. https://doi.org/10.1167/iovs.61.8.22
Zhong W, Yin H, Xie L (2009) Expression and potential role of major inflammatory cytokines in experimental keratomycosis. Mol Vis 15:1303–1311
Zilliox MJ, Gange WS, Kuffel G et al (2020) Assessing the ocular surface microbiome in severe ocular surface diseases. Ocul Surf 18:706–712. https://doi.org/10.1016/j.jtos.2020.07.007

Chapter 6
The Mycobiome of the Human Eye

Abstract Fungi are an important component of the ocular surface microbiota and are known to colonize the ocular surface of the normal human eye and under diseased conditions. Mycobiome approach using next-generation sequencing of the ITS2 region demonstrated a greater diversity and abundance of fungi on the conjunctiva of normal eye and the conjunctiva and cornea of eyes with ocular diseases. Further, the ocular surface mycobiome of individuals with keratitis, endophthalmitis and post-fever retinitis is different compared to the mycobiome of the healthy eye. There is a need to extend mycobiome studies to other ocular diseases such as conjunctivitis, blepharitis, orbital cellulitis, scleritis, dry eye disease, retinitis, uveitis, meibomian gland dysfunction etc. so that their involvement in ocular diseases is established and the data could help in diagnosis and treatment especially in cases where the fungi are not cultivable.

Keywords Mycobiome · Ocular surface · Ocular disease · NGS · Amplicon sequencing · Metagenome

6.1 Introduction

Fungi are ubiquitous in nature, and apart from being present in various environmental habitats, they are also known to colonize and live on (the skin) and in (oral cavity, gastrointestinal tract, lungs etc.) the human body. As early as the twentieth century, fungi were discovered on the ocular surface (Köllner 1906). But the ability to culture fungi from the conjunctival swabs ranged from 3 to 32% (Wang et al. 2020; Segal et al. 1977; Ando and Takatori 1982; Williamson et al. 1968; Gopinathan et al. 2009; Wu et al. 2003; Sehgal et al. 1981; Behrens-Baumann and Rüchel 1999), thus making the identification of fungi from the ocular surface a challenge (Segal et al. 1977; Ozkan et al. 2017; Wu et al. 2003). Further, the low success rate in cultivating fungi from ocular samples is also influenced by several confounding factors such as the age of the subject (Williamson et al. 1968; Behrens-Baumann and Rüchel 1999) and the geographical region (Wilson et al. 1969). Longitudinal studies indicated that the fungi are probably acquired from the surrounding (Williamson et al. 1968; Wilson et al. 1969), the conjunctivae (5%) and the lid margins (80%) and are thus

S. Shivaji, *Human Ocular Microbiome*,
https://doi.org/10.1007/978-981-19-1754-7_6

not permanent residents of the eye (Wilson et al. 1969; Olson 1969). These aspects on the cultivable fungi of the eye are highlighted in Chap. 5. In the present chapter, the results of next-generation sequencing or the mycobiome approach are highlighted with respect to the total fungal diversity of the eye, the confounding factors that influence the mycobiome and the dynamics in diversity in the diseased eye including the mycobiome in idiopathic ocular surface disorders.

6.2 What Is a Mycobiome?

The word 'mycobiome' was coined by Gillevet et al. (2009). It is a combination of two words ('mukēs' meaning 'fungus' and 'biome' meaning 'life'). Two approaches that have been routinely used to generate a mycobiome are the amplicon sequencing or targeted sequencing approach and the metagenome approach (see Fig. 3.1). Both these methods are very similar to that described for the bacterial microbiome approach in Chap. 3, and thus only the differences would be highlighted.

6.2.1 Amplicon Sequencing

The various steps involved in amplicon sequencing include (see Fig. 3.2):

1. DNA extraction: Extraction of genomic DNA, quality check and storage at −80 °C.
2. Amplification of the target gene: The sequence of ITS2 (the internal transcribed spacer 2 from the eukaryotic ribosomal RNA gene) is used as a phylogenetic marker for taxonomic identification of fungi. The primers that were used to amplify ITS2 region included ITS2-Fw (5′-GCATCGATGAAGAACGCAGC-3′) and ITS2-Rev (5′-TCCTCCGCTTATTGATATGC-3′) (Shivaji et al. 2019) (Fig. 6.1). PCR reagents without template DNA served as a negative control. In addition, buffer extracts of swabs, all reagents used for DNA extraction and PCR, including water, PCR primers etc., were also analysed as controls. Further to

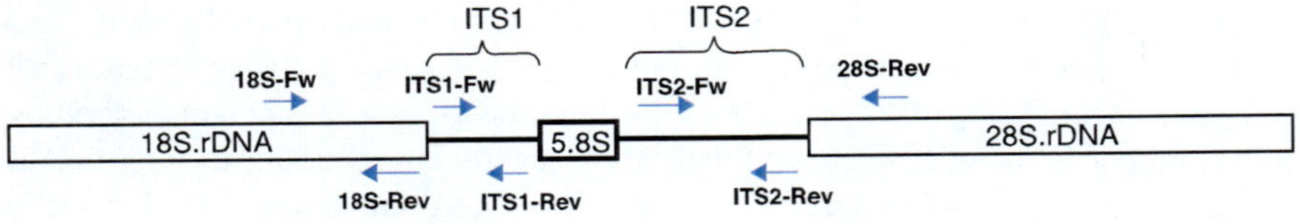

Fig. 6.1 Schematic representation of the ITS1 and ITS2 regions in the fungal RNA gene. The primers used for the PCR amplification of the ITS1 and ITS2 regions respectively are indicated. Fw and Rev refer to the forward and reverse primers respectively. The schematic figure is based on *Aspergillus* spp. (Kebeish and El-Sayed 2012)

avoid contamination from possible sources in the laboratory DNA extraction, PCR and gel electrophoresis are performed in dedicated laboratory space.

3. Library preparation: The amplicon libraries were generated as described by Dehingia et al. (2015).
4. Cluster generation: Prior to sequencing the amplicon sequences were grouped together into clusters.
5. Sequencing: Illumina MiSeq was used for sequencing the library.

6.2.2 *Metagenome Sequencing*

More recently metagenomic next-generation sequencing using NGS has become the method of choice (Fig. 6.2) for clinical samples such as blood, cerebrospinal fluid, respiratory samples, ocular fluid and gastrointestinal fluid. In this method, the total DNA in a sample, containing mixed populations of microorganisms, is sequenced and assigned to their reference genomes. Workflow for mNGS includes (Fig. 6.2) (1) extraction of genomic DNA, fragmentation and quality check, (2) attachment of adaptors to the fragments for barcoding and library sequencing, (3) simultaneous sequencing of the DNA fragments, (4) removal of human-related DNA sequence reads, (5) assembling of contigs of long DNA stretches from shorter overlapping sequences and (6) alignment of contigs to a reference database for taxonomic classification (Fig. 6.2).

6.2.3 *Data Analysis*

6.2.3.1 Taxonomic Classification of Amplicon Sequencing Reads

Paired-end reads were merged into single reads using FLASH (fast length adjustment of short reads) (Magoc and Salzberg 2011) (Fig. 6.2). Chimeric sequences and sequences with a mean Phred score <25 were eliminated prior to OTU picking which was performed using QIIME (quantitative insights into microbial ecology), and the sequences were assigned to taxonomic lineages (Kalyana Chakravarthy et al. 2019; Jayasudha et al. 2018). Sparse OTUs ($<0.001\%$ of the total number of high-quality reads) were not analysed (Fig. 6.2).

6.2.3.2 Annotation of Metagenome Sequencing Reads

The resulting Illumina raw reads devoid of adapter sequences were quality filtered to remove low-quality sequences and host DNA sequences, and the resulting high-quality reads were assembled using Megahit v1.0.2 (Li et al. 2016) and classified as

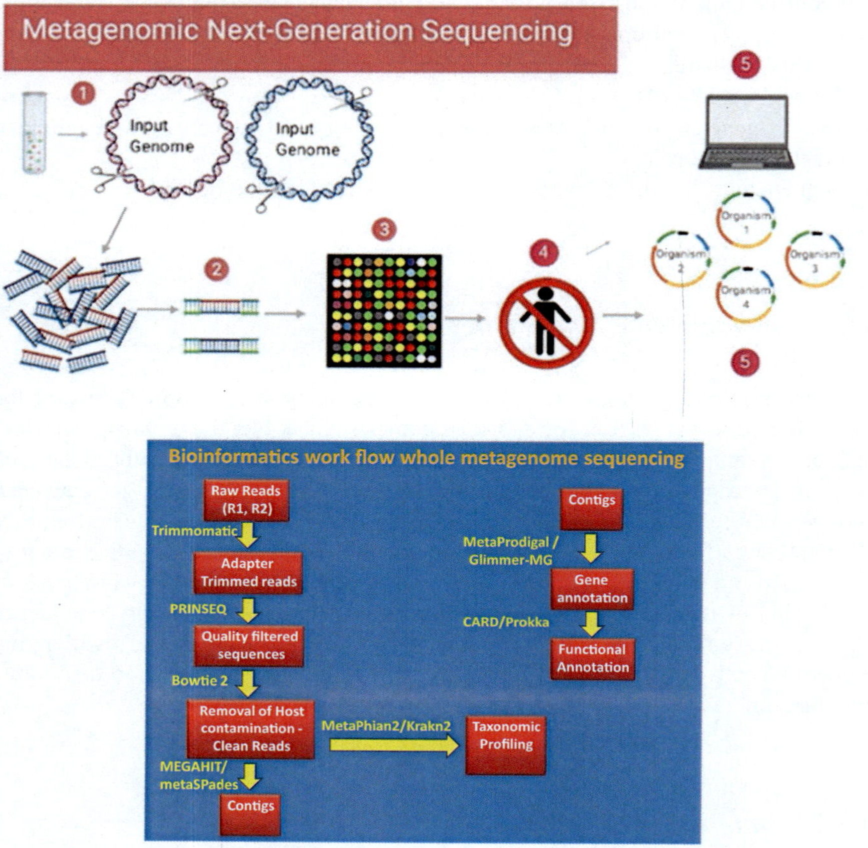

Fig. 6.2 Steps in metagenomic next-generation sequencing (**a**) and bioinformatics workflow (**b**). In a (1) extraction of genomic DNA, fragmentation, and quality check, (2) attachment of adaptors to the fragments for barcoding and library sequencing, (3) simultaneous sequencing of the DNA fragments, (4) removal of human-related DNA sequence reads, (5) assembling of contigs of long DNA stretches from shorter overlapping sequences and (6) alignment of contigs to a reference database for taxonomic classification. Modified from Lee (2019). (**b**) The various programmes used for the bioinformatics analyses are indicated in yellow adjacent to the arrow

fungi, bacteria, viruses, protozoa and archaea using the Kraken database v2.0.9 (Wood et al. 2019) (Fig. 6.2). For α-diversity analysis, Kraken results are normalized using cumulative sum scaling of the metagenome Seq package of Rv1.10.0, and the normalized counts were aggregated at the phylum to the species level (Fig. 6.2). The Shannon and the inverse Simpson indices of diversity were calculated with the diversity function of the vegan package of R (v2.5-6). For functional analysis contigs are aligned using specific databases like Resistance Gene Identifier (RGI) tool v5.1.0 (https://toolshed.g2.bx.psu.edu/view/card/rgi/f96cbb663aa9) and MEGARes 2.0 database (Doster et al. 2020), which is a collation of multiple databases (CARD,

ARG-ANNOT and ResFinder), to confirm the presence of resistance-conferring SNPs. A sequence was designated as an AMR gene when the sequence exhibited ≥25% coverage with ≥90% sequence identity to the reference gene sequence. The mycology antifungal resistance database (MARDy) (http://www.mardy.net) was used to detect antifungal drug resistance genes. Counts of reads aligned to the antimicrobial resistance genes (ARGs) were recorded and used for downstream comparative analyses (Fig. 6.2). This chapter describes the fungi associated with the eye mycobiome, to highlight the diversity of fungi in the healthy human eye and under diseased conditions.

6.3 Mycobiome of the Normal Human Eye

NGS does not depend on culturing of the fungi from a given biological sample, but DNA is extracted from the sample, and the ITS2 region located between 5.8S and 28S ribosomal DNA is amplified (Fig. 6.1) following which an amplicon library is generated which is then sequenced. Sequence variability in ITS2 sequence serves as a reliable proxy to identify fungi up to the genera/species level. Suitable controls are also subjected to the above procedure to enumerate the authenticity of the identified fungi and rule out contaminating DNA from the swabs used, the reagents used for DNA extraction or PCR, contamination from the surrounding areas of the eye etc. DNA extracted from the conjunctival swabs of healthy eyes following NGS analysis revealed 3 phyla and 65 genera (Fig. 6.3) versus just only 1 genus and 2 species identified by the conventional cultivable method (Shivaji et al. 2019). Further, all the alpha diversity indices (Shannon diversity index, Simpson index, observed number of OTUs and Chao1 index) remained unaltered between the eyes and gender of the subjects (Shivaji et al. 2019). Beta diversity analyses also indicated that the fungal microbiome was unchanged between the two eyes (left/right) and was not influenced by age. But between the sexes, the fungal community showed significant variation (Fig. 6.4) (Shivaji et al. 2019).

6.3.1 Core Ocular Mycobiome

Several bacterial genera were shared across all gut microbiomes, and such genera constituted the 'core microbiome' (Turnbaugh et al. 2007; Hamady and Knight 2009). But in many studies, the genera were shared by >80% of the microbiomes, and such genera were designated as the 'putative core microbiome'. Further changes in the environment, lifestyle and physiological differences alter the microbes in a particular niche, and such microbes constitute the 'variable microbiome' (Turnbaugh et al. 2007; Hamady and Knight 2009). In the eye of Indian subjects, 65 fungal genera were identified as associated with the healthy ocular surface out of which 4 genera each constituted the 'core' and 'putative core' mycobiomes respectively

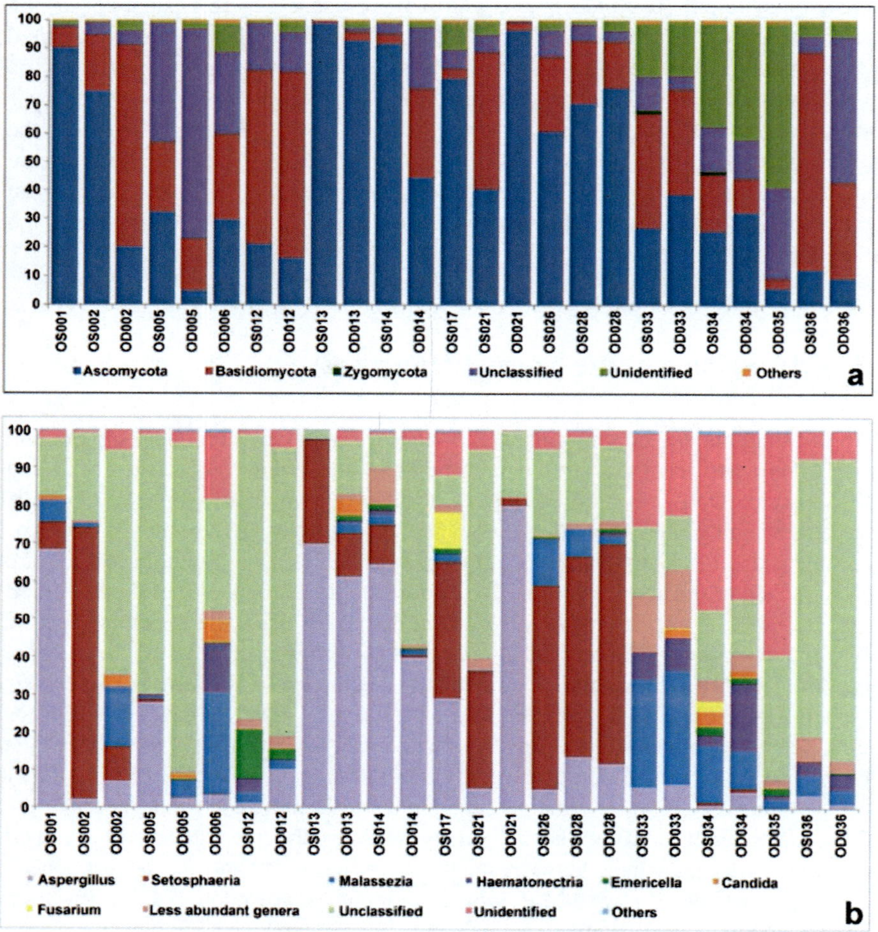

Fig. 6.3 Mycobiomes of conjunctivae represented as the abundance of the fungi at the phyla (**a**) and genera (**b**) level from healthy individuals (OD = 12 and OS = 13). Singletons, unassigned reads and sparse OTUs (with <0.001% of total high-quality reads) are represented together as others. In (**b**), only the fungal genera with >0.50% abundance (*Aspergillus* spp., *Setosphaeria* spp., *Malassezia* spp., *Haematonectria* spp., *Emericella* spp., *Candida* spp. and *Fusarium* spp.) were included. Data from Shivaji et al. (2019)

(Shivaji et al. 2019). In Chinese the ocular surface diversity was more pronounced with 94 fungal genera out of which 5 and 3 genera constituted the 'core' and the 'putative core' mycobiomes respectively (Wang et al. 2020). Further, in the ocular mycobiomes of Indians and Chinese, 11 of the 14 genera detected by Wang et al. (2020) in 50% of the conjunctival swab mycobiomes were also reported by Shivaji et al. (2019). These observed differences may be related to the different geographic locations of the individuals. Of particular importance is that many of these abundant

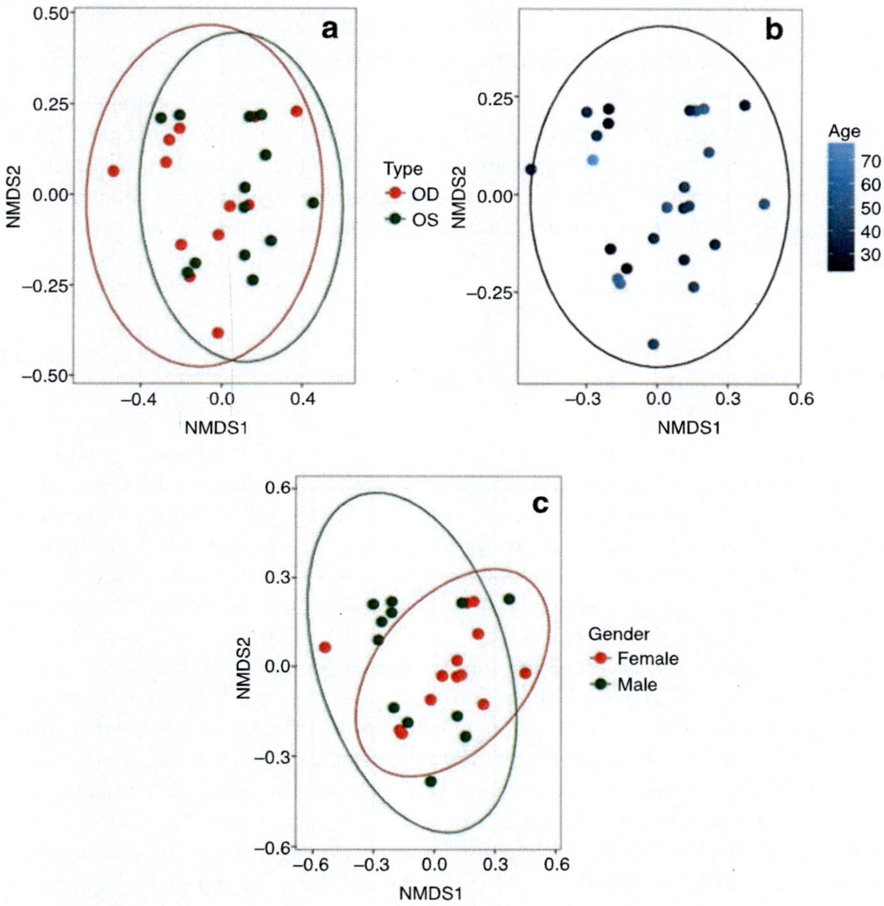

Fig. 6.4 Beta diversity analysis (NMDS plots) of the ocular surface mycobiomes of healthy controls (OD = 12 and OS = 13) indicating that the fungal community did not vary between the left and right eye (PERMANOVA, $P = 0.533$) (**a**) or the age (PERMANOVA, $P = 0.859$) of the sampled individual (**b**). Gender altered the fungal community significantly (PERMANOVA, $P = 0.009$) (**c**). Data from Shivaji et al. (2019)

genera (*Aspergillus* spp., *Emericella* spp., *Malassezia* spp., *Fusarium* spp., *Setosphaeria* spp., *Cladosporium* spp., *Haematonectria* spp., *Rhodotorula* spp., *Choiromyces* spp., *Davidiella* spp., *Meyerozyma* spp., *Alternaria* spp. and *Petromyces* spp.) are pathogenic (Wang et al. 2020; Shivaji et al. 2019) and included several rare fungal species (*Lasiodiplodia theobromae*, *Fonsecaea pedrosoi*, *Cylindrocarpon* spp., *Metarhizium anisopliae*, *Scedosporium prolificans*, *Pythium insidiosum and Paecilomyces* spp.) which cause mycotic keratitis (Maharana et al. 2016).

6.4 Mycobiome of the Diseased Human Eye: Next-Generation Sequencing Approach

Ocular mycobiomes of individuals with keratitis, endophthalmitis and post-fever retinitis are described. Mycobiomes of patients with other ocular diseases (such as conjunctivitis, scleritis, blepharitis, *dry eye disease*, retinitis, *uveitis*, orbital cellulitis, *meibomian gland dysfunction*, *SJS* and Sjogren's syndrome) are yet to be analysed.

6.4.1 Mycobiome Associated with Keratitis

Cultivation of fungi from ocular samples (conjunctival swabs or corneal scrapings) using standard fungal media lacks sensitivity, and at best fungi were detected in about 12.5% of the samples collected from individuals with keratitis. In contrast when NGS approach is used, the percentage of identified fungi could be as high as 73.5% of the conjunctival swabs (Shivaji et al. 2019). In the same study, it was demonstrated that fungi could be cultured only from 6% of the healthy control conjunctival swabs (HC-SW) (Shivaji et al. 2019). Compared to conjunctival swabs, all the corneal scrapings from fungal keratitis patients (FK-CR) were positive for fungi. In HC-SW, *Aspergillus flavus* and *A. niger* were the only fungi identified, whereas in the corneal scrapings of fungal keratitis individuals (FK-CR), *Aspergillus fumigatus*, *A. flavus*, *Fusarium* spp., *Fusarium solani*, *Colletotrichum coccodes*, *Exserohilum rostratum*, *Cladosporium* spp., *Curvularia lunata*, dematiaceous fungi and septate and aseptate fungal filaments were identified. The mycobiomes from the conjunctival swabs of healthy controls (HC-SW), from the conjunctival swabs of fungal keratitis individuals (FK-SW) and from the corneal scrapings of fungal keratitis individuals were analysed by sequencing the ITS2 region (Prashanthi et al. 2019). Fungi affiliated to the phyla *Ascomycota* increased, whereas those affiliated to *Basidiomycota* decreased in both FK-SW and FK-CR compared to HC-SW (Table 6.1). Furthermore, the four alpha diversity indices indicated that diversity, number of observed OTUs, evenness and richness were significantly decreased in FK-CR compared to HC-SW, whereas when the mycobiomes of FK-CR were compared to FK-SW, only two indices (number of observed OTUs and Chao1 index) were significantly decreased in FK-CR. Thus, dysbiosis in the mycobiomes of fungal keratitis patients (FK-SW and FK-CR) compared to HC-SW was observed.

At the genus level, 111 genera were identified in HC-SW, FK-SW and FK-CR (Fig. 6.5a, b) (Prashanthi et al. 2019). In a vast majority of the eyes (89 eyes), *Aspergillus* was the most predominant genus, and the other genera that were dominantly present in >70% of the healthy control eyes (HC-SW) included *Malassezia*, *Haematonectria*, *Emericella*, *Cladosporium*, *Fusarium*, *Choiromyces*, *Candida* and *Cochliobolus* (Prashanthi et al. 2019). Of the above genera,

Table 6.1 Mean abundance (%) of fungal phyla in the mycobiomes from conjunctival swabs of healthy controls (HC-SW, $n = 37$), conjunctival swabs of fungal keratitis patients (FK-SW, $n = 23$) and corneal scrapings of fungal keratitis patients (FK-CR, $n = 29$)

	HC-SW		FK-SW		FK-CR		*p*-value			
Phylum	Mean	Present out of 37 samples	Mean	Present out of 23 samples	Mean	Present out of 29 samples	HC-SW vs. FK-SW vs. FK-CR	HC-SW vs. FK-SW	HC-SW vs. FK-CR	FK-SW vs. FK-CR
Ascomycota	35.66	37	69.99	23	88.36	29	0.001	0.001	0.001	0.006
Basidiomycota	37.05	37	10.12	23	0.35	23	0.001	0.001	0.001	0.001
Zygomycota	0.05	2	0	1	0	0	0.461	0.845	0.215	0.278
Unclassified	24.96	37	16.07	23	7.47	29	0.001	0.012	0.001	0.006
Others[a]	2.28	37	3.82	23	3.82	29	NA	NA	NA	NA

[a]Singletons and sparse operational taxonomic units (OTUs) (with <0.001% of total high-quality reads) are together represented as others. Modified from Prashanthi et al. (2019)

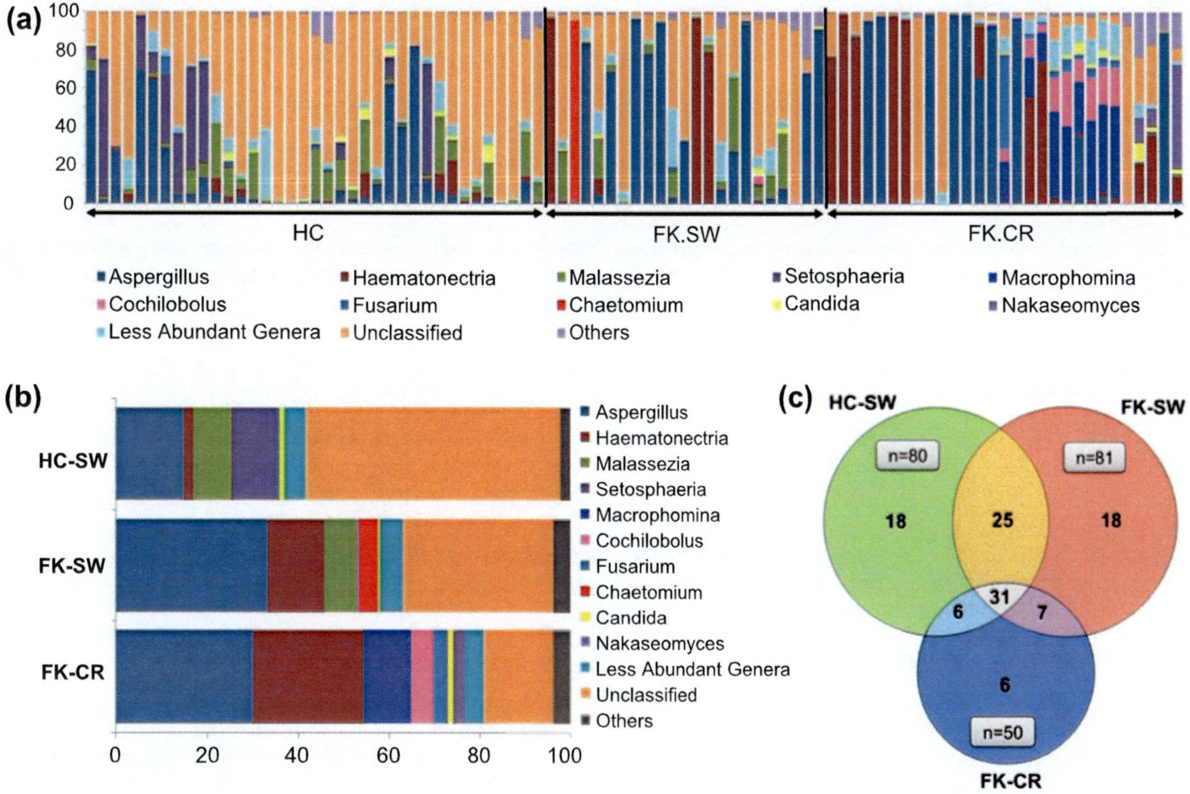

Fig. 6.5 Abundance changes (%) (**a**) and average abundance changes (%) (**b**) in fungal genera from conjunctival swabs of healthy controls (HC-SW, $n = 37$) and conjunctival swabs of fungal keratitis patients (FK-SW, $n = 23$) and in corneal scrapings of fungal keratitis patients (FK-CR, $n = 29$). (**c**) Venn diagram represents the number of fungal genera shred between the three cohorts. Modified from Prashanthi et al. (2019)

Malassezia, *Haematonectria*, *Cladosporium*, *Fusarium*, *Choiromyces*, *Candida* and *Cochliobolus* were equally abundant in FK-SW. In contrast, only three genera were present in >70% of FK-CR samples, and they included *Fusarium*, *Haematonectria* and *Cochliobolus*. Both HC-SW and FK-SW with 80 genera each shared 56 genera implying that FK-SW microbiome is related to HC-SW (Fig. 6.5c), whereas FK-CR shared only 37 genera with HC-SW. Further, at the genera level only three genera exhibited a significant differential abundance between HC-SW and FK-SW microbiomes. In contrast, in FK-CR the abundance of 33 genera differed in abundance compared to HC-SW. Heat map analysis differentiated the microbiomes of HC-SW and FK-CR (Prashanthi et al. 2019). The genera common to the three cohorts are likely to be opportunistic pathogens, whereas the seven genera shared between FK-SW and FK-CR are pathogenic (Prashanthi et al. 2019). This study demonstrated dysbiosis in the fungal microbiomes of conjunctival swabs and corneal scrapings in FK individuals compared with the healthy conjunctival swabs (Prashanthi et al. 2019). Interaction networks also demonstrated that the mycobiomes of the three cohorts are different. In HC-SW the genus *Setosphaeria*, an opportunistic human pathogen, constituted the largest hub genera and is known to cause keratitis, sinusitis, CNS vasculitis and mycoses. *Setosphaeria* in the interaction network exhibited negative interactions with several pathogens (*Pseudozyma*, *Peniophora*, *Trichomonascus*, *Talaromyces* etc.) and also interacted positively with genera like *Neosartorya*, *Bipolaris* and *Aspergillus* (Revankar and Sutton 2010) (Fig. 6.6a). In FK-SW, five hub genera were detected, and these included the pathogens *Candida*, *Cochliobolus*, *Choiromyces*, *Fusarium* and *Haematonectria*. Of these *Candida* and *Cochliobolus* exhibited negative interactions with *Botryosphaeria*, *Colletotrichum*, *Rhizoctonia*, *Cytospora*, *Ceratobasidium*, *Galactomyces* and *Epicoccum* which are plant pathogens and with *Acremonium* a human pathogen (Fig. 6.6b). In FK-CR 12 unique hub genera were detected (Fig. 6.6c) and included the plant pathogens *Macrophomina*, *Cephaliophora*, *Gibberella*, *Diaporthe*, *Lasiodiplodia*, *Retroconis*, *Colletotrichum*, *Rhizoctonia*, *Botryosphaeria*, *Bipolaris*, *Myrothecium* and *Xeromyces*. These 12 hub genera except the hub genus *Colletotrichum* exhibited both negative and positive interactions. Overall, the hub genera in HC-SW exhibit negative interactions with other pathogens but interact positively with beneficial fungi. In contradiction it was also observed that *Setosphaeria* also positively regulates the pathogens *Bipolaris*, *Neosartorya* and *Aspergillus*. So also, it is difficult to comprehend the observed positive and negative interactions exhibited by the hub genera in FK-SW and FK-CR fungal microbiomes (Fig. 6.6a–c). Spearman correlation coefficients indicated that the genera in HC-SW exhibited negative correlation with genera in FK-SW and FK-CR (Fig. 6.7) (Prashanthi et al. 2019). Positive correlation coefficients between genera are predictive of metabolic or other dependencies, whereas those genera that are negatively correlated may compete in ways that inhibit the growth of each other (Srinivasan et al. 2012).

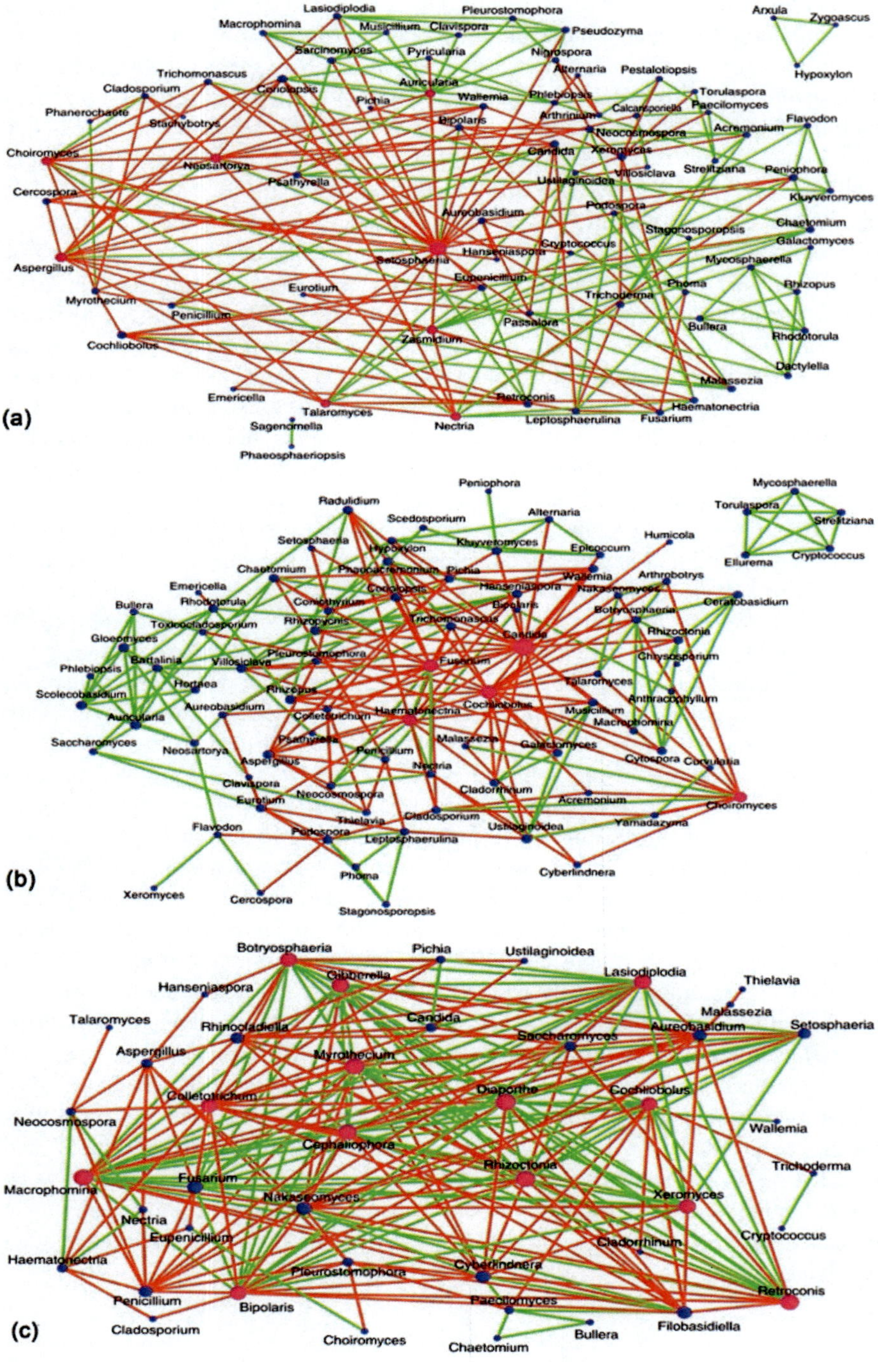

Fig. 6.6 Interactions network between the genera in the conjunctival mycobiomes of healthy controls (HC-SW) (**a**), fungal keratitis patients (FK-SW) (**b**) and the corneal mycobiome of fungal

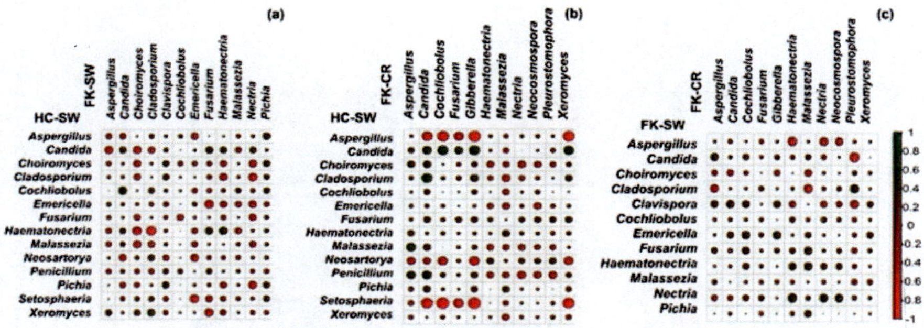

Fig. 6.7 Spearman correlation analysis of fungal genera (with median abundance >0.0001%) in the mycobiomes of conjunctiva swabs of healthy individuals (HC-SW), conjunctiva swabs of fungal keratitis patients (FK-SW) and corneal scrapings of fungal keratitis patients (FK-CR). In (**a**) HC-SW vs. FK-SW, in (**b**) HC-SW vs. FK-CR and in (**c**) FK-SW vs. FK-CR. Modified from Prashanthi et al. (2019)

6.4.2 Mycobiome Associated with Endophthalmitis

NGS analysis of ITS2 sequences in the vitreous fluid of 30 healthy controls was negative (Deshmukh et al. 2019). But the vitreous fluids from three culture-negative endophthalmitis patients indicated predominantly the presence of unidentified species indicating limitations in the existing fungal sequence databases in identifying fungi. The other identified fungi include *Candida* spp., *Malassezia* spp., *Aspergillus* spp., *Rhodotorula* spp. and *Scedosporium* spp. Based on NGS analysis, fungal pathogens were identified in 36/57 vitreous fluids from clinically infectious endophthalmitis patients, and *Fusarium* spp., *Exserohilum* spp., *Candida* spp. *and Aspergillus* spp. were the most predominant genera in the culture-negative endophthalmitis cases. Conventional culture-based methods also identified the above fungi (Gandhi et al. 2019).

6.4.3 Mycobiome Associated with Post-Fever Retinitis (PFR)

Mycobiome of the vitreous of patients with post-fever retinitis (PFR) was compared with the vitreous of individuals without ocular and systemic infection (VC) (Arunasri et al. 2020) (Fig. 6.8). At the phyla level, no significant difference was observed in

Fig. 6.6 (continued) keratitis patients (FK-CR) (**c**). Each node represents a genus, and the size of the node increases with the degree of interaction. Negative and positive interactions are depicted as green and red edges respectively. The nodes with >10 interactions are designated as the hub genera (pink). Modified from Prashanthi et al. (2019)

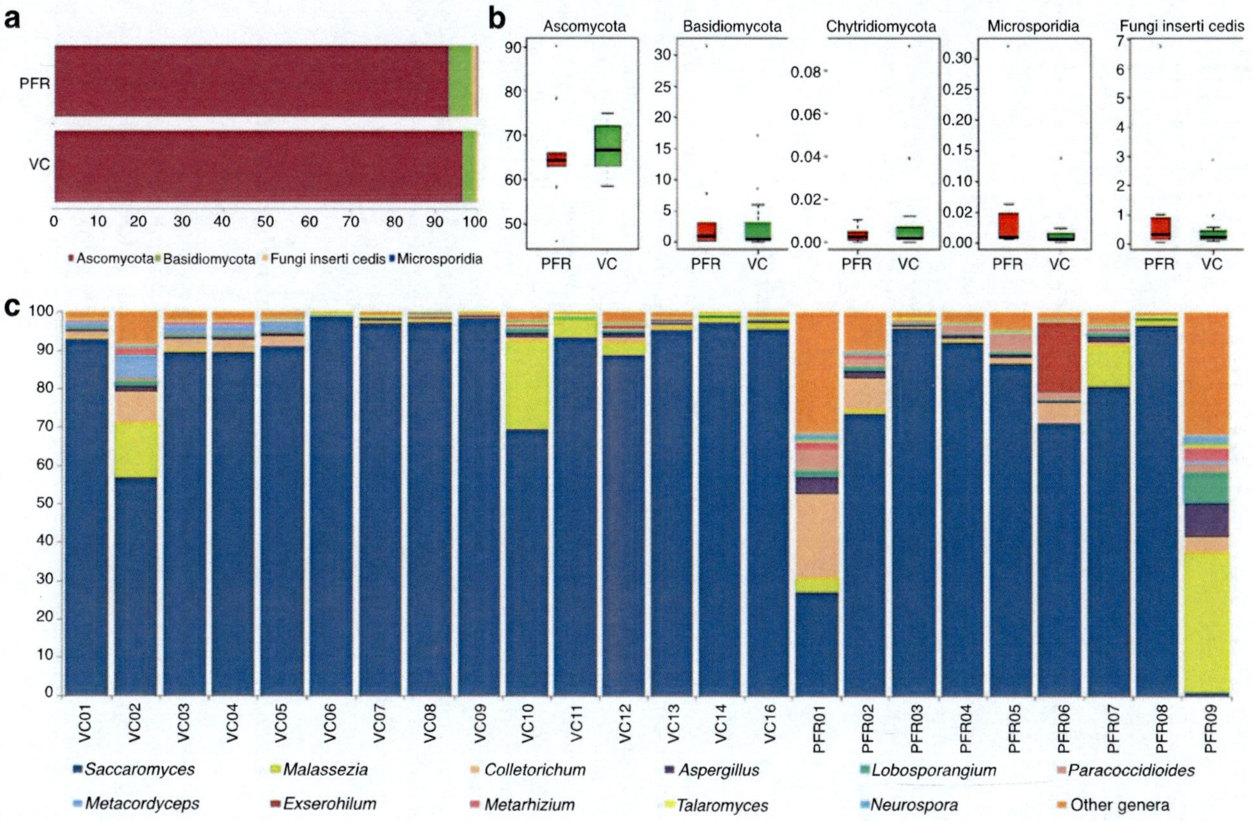

Fig. 6.8 Changes in the abundance of fungal phyla (**a**, **b**) and fungal genera (**c**) in the vitreous fluid mycobiomes of post-fever retinitis individuals (PFR, $n = 9$) and healthy controls (VC, $n = 15$). The vertical yellow line in (**c**) separates the two cohorts of VC and PFR. In the box plots of (**b**), interquartile ranges and the median values (horizontal line) have been depicted. Data from Arunasri et al. (2020)

Table 6.2 Significant abundance changes ($P < 0.05$) in the discriminative fungal genera in the mycobiomes of the vitreous fluid of controls (VC, $n = 15$) and post-fever retinitis individuals (PFR)

Genera	Mean in VC	Present in number of VC samples ($n = 15$)	Mean in PFR	Present in number of samples ($n = 69$)	*p*-value
Setosphaeria spp.	0.000522	1	1.136	8	0.0001
Arthroderma spp.	0.001621	1	0.953	8	0.0001
Isaria spp.	0.000352	1	0.011	8	0.001
Paracoccidioides spp.	0.153317	15	1.818	9	0.001
Pseudogymnoascus spp.	0.007939	14	0.058	9	0.023
Saccharomyces spp.	61.58861	15	56.8	9	0.003
Sordaria spp.	0.011009	14	0.049	9	0.002
Trichoderma spp.	0.054328	15	0.253	9	0.023
Coniosporium spp.	0.0022	12	0.448	6	0.035
Enterocytozoon spp.	0.01	15	0.033	9	0.023
Exserohilum spp.	0.063153	14	2.35	1	0.033
Kluyveromyces spp.	0.013827	15	0.041	9	0.036
Lachancea spp.	0.0075	14	0.043	8	0.008
Microsporum spp.	0.057128	15	0.17	4	0.023
Nectria spp.	0.005494	12	0	0	0.004
Nematocida spp.	0.000226	9	0	0	0.036
Paraphaeosphaeria spp.	0.03	14	0.487	9	0.018

Modified from Arunasri et al. (2020)

the vitreous mycobiome of VC and PFR individuals, and the phyla *Basidiomycota*, *Ascomycota*, *Microsporidia* and *Chytridiomycota* were detected in both VC and PFR patients, and fungi affiliated to *Ascomycota* were the most predominant in the two cohorts (Fig. 6.8a, b). Further, 134 and 133 genera were identified in healthy controls and post-fever retinitis cohorts, respectively, and the most abundant top 10 genera were *Colletotrichum* spp., *Saccharomyces* spp., *Malassezia* spp., *Lobosporangium* spp., *Aspergillus* spp., *Metacordyceps* spp., *Paracoccidioides* spp., *Talaromyces* spp., *Exserohilum* spp. and *Metarhizium* spp., and these were shared between the groups (Fig. 6.8c) (Arunasri et al. 2020). Between the two cohorts, the 17 genera showed significant difference in the abundance, and this probably has implications in the disease (Table 6.2). Pathogens affiliated to the genera *Paracoccidioides* spp., *Arthroderma* spp., *Setosphaeria* spp., *Exserohilum* spp., *Pseudogymnoascus* spp., *Trichoderma* spp., *Microsporum* spp. *and Kluyveromyces* spp. were increased in PFR group compared to VC group. Among the above pathogens, *Setosphaeria* spp., *Kluyveromyces* spp. and *Trichoderma* spp. were also increased in patients having fungal keratitis (Prashanthi et al. 2019). A few of these pathogens, namely, *Exserohilum* spp., Setosphaeria spp. and *Trichoderma* spp., were also identified in the vitreous fluid in endophthalmitis patients (Gandhi et al. 2019). In addition,

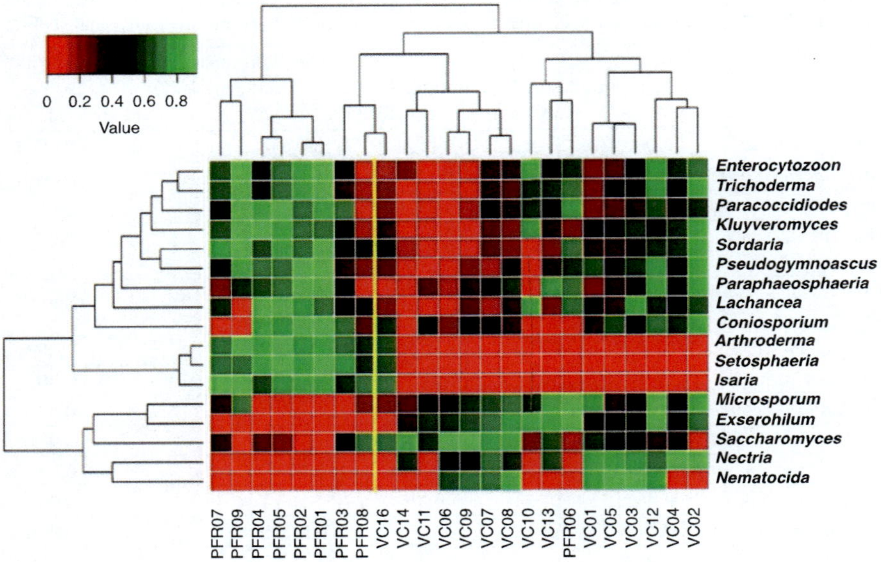

Fig. 6.9 Heat map analysis of the discriminative fungal genera in the vitreous fluid mycobiomes of post-fever retinitis individuals (PFR, $n = 9$) and healthy controls (VC, $n = 15$) showing changes in the abundance at the genera level ($P < 0.05$) *l*. In the heat map, the abundances of the individual genera were scaled from 0 to 1 and depicted by red and green colour, respectively. Data from Arunasri et al. (2020)

Candida spp. and *Fusarium* spp. were also identified in the vitreous fluid of VC and PFR cohorts, but their abundance did not exhibit any significant difference. Beta diversity analysis by heat map and principal co-ordinate analysis discriminated the VC and PFR mycobiomes into two distinct clusters (Fig. 6.9) implying dysbiotic changes in the fungal vitreous mycobiome of patients with PFR (Arunasri et al. 2020).

6.5 Conclusions

In the cultivable approach, fungi were identified as associated with several ocular diseases. Mycobiome analysis using next-generation sequencing (NGS) also confirmed the presence of diverse fungi on the ocular surface of normal healthy individuals and also in patients with endophthalmitis, keratitis and post-fever retinitis. Mycobiome analysis of the other ocular diseases is not described either because fungi are not known to be associated with the disease (e.g. ocular graft-vs.-host disease and lax eyelid syndrome) or because mycobiome studies have not yet been attempted (e.g. scleritis, blepharitis, retinitis, conjunctivitis, meibomian gland dysfunction, orbital cellulitis, uveitis, dry eye disease, SJS and Sjogren's syndrome).

Thus, there is a need to extend mycobiome studies using NGS to the above ocular diseases so that their involvement in ocular diseases is established and the data could also help in diagnosis and treatment especially in cases where the fungi are not cultivable.

References

Ando N, Takatori K (1982) Fungal flora of the conjunctival sac. Am J Ophthalmol 94:67–74. https://doi.org/10.1016/0002-9394(82)90193-3

Arunasri K, Mahesh M, Sai Prashanthi G et al (2020) Mycobiome changes in the vitreous of post fever retinitis patients. PLoS One 15:e0242138. https://doi.org/10.1371/journal.pone.0242138

Behrens-Baumann W, Rüchel R (1999) Mycosis of the eye and its adnexa, vol 32. Karger Medical and Scientific Publishers, Basel

Dehingia M, Devi KT, Talukdar NC et al (2015) Gut bacterial diversity of the tribes of India and comparison with the worldwide data. Sci Rep 5:18563. https://doi.org/10.1038/srep18563

Deshmukh D, Joseph J, Chakrabarti M et al (2019) New insights into culture negative endophthalmitis by unbiased next generation sequencing. Sci Rep 9:844. https://doi.org/10.1038/s41598-018-37502-w

Doster E, Lakin SM, Dean CJ et al (2020) MEGARes 2.0: a database for classification of antimicrobial drug, biocide and metal resistance determinants in metagenomic sequence data. Nucleic Acids Res 48:561–569. https://doi.org/10.1093/nar/gkz1010

Gandhi J, Jayasudha R, Naik P et al (2019) Targeted high-throughput sequencing identifies predominantly fungal pathogens in patients with clinically infectious, culture-negative endophthalmitis in South India. Microorganisms 7:100411. https://doi.org/10.3390/microorganisms7100411

Gillevet PM, Sikaroodi M, Torzilli AP (2009) Analyzing salt-marsh fungal diversity: comparing ARISA fingerprinting with clone sequencing and pyrosequencing. Fungal Ecol 2:160–167. https://doi.org/10.1016/j.funeco.2009.04.001

Gopinathan U, Sharma S, Garg P et al (2009) Review of epidemiological features, microbiological diagnosis and treatment outcome of microbial keratitis: experience of over a decade. Indian J Ophthalmol 57:273–279. https://doi.org/10.4103/0301-4738.53051

Hamady M, Knight R (2009) Microbial community profiling for human microbiome projects: tools, techniques, and challenges. Genome Res 19:1141–1152. https://doi.org/10.1101/gr.085464.108

Jayasudha R, Chakravarthy SK, Prashanthi GS et al (2018) Alterations in gut bacterial and fungal microbiomes are associated with bacterial Keratitis, an inflammatory disease of the human eye. J Biosci 43:835–856

Kalyana Chakravarthy S, Jayasudha R, Ranjith K et al (2019) Correction: alterations in the gut bacterial microbiome in fungal Keratitis patients. PLoS One 14:e0211757. https://doi.org/10.1371/journal.pone.0211757

Kebeish RM, El-Sayed AS (2012) Morphological and molecular characterization of L-methioninase producing Aspergillus species. Afr J Biotechnol 11:15280–15290. https://doi.org/10.5897/AJB12.845

Köllner H (1906) Schimmelpilzerkrankung der Sklera. Ophthalmologica 16:441–447. https://doi.org/10.1159/000291135

Lee R (2019) Metagenomic next generation sequencing: how does it work and is it coming to your clinical microbiology. American Society of Microbiology, Washington

Li D, Luo R, Liu CM et al (2016) MEGAHIT v1.0: a fast and scalable metagenome assembler driven by advanced methodologies and community practices. Methods 102:3–11. https://doi.org/10.1016/j.ymeth.2016.02.020

Magoc T, Salzberg SL (2011) FLASH: fast length adjustment of short reads to improve genome assemblies. Bioinformatics 27:2957–2963. https://doi.org/10.1093/bioinformatics/btr507

Maharana PK, Sharma N, Nagpal R et al (2016) Recent advances in diagnosis and management of mycotic keratitis. Indian J Ophthalmol 64:346–357. https://doi.org/10.4103/0301-4738.185592

Olson CL (1969) Fungal contamination of conjunctiva and lid margin. Arch Ophthalmol 81:351–355. https://doi.org/10.1001/archopht.1969.00990010353009

Ozkan J, Nielsen S, Diez-Vives C et al (2017) Temporal stability and composition of the ocular surface microbiome. Sci Rep 7:9880. https://doi.org/10.1038/s41598-017-10494-9

Prashanthi GS, Jayasudha R, Chakravarthy SK et al (2019) Alterations in the ocular surface fungal microbiome in fungal keratitis patients. Microorganisms 7:90309. https://doi.org/10.3390/microorganisms7090309

Revankar SG, Sutton DA (2010) Melanized fungi in human disease. Clin Microbiol Rev 23:884–928. https://doi.org/10.1128/CMR.00019-10

Segal E, Romano A, Eylan E et al (1977) Fungal flora of the normal conjunctival sac. Mykosen 20: 9–14. https://doi.org/10.1111/j.1439-0507.1977.tb01475.x

Sehgal SC, Dhawan S, Chhiber S et al (1981) Frequency and significance of fungal isolations from conjunctival sac and their role in ocular infections. Mycopathologia 73:17–19. https://doi.org/10.1007/BF00443007

Shivaji S, Jayasudha R, Sai Prashanthi G et al (2019) The human ocular surface fungal microbiome. Invest Ophthalmol Vis Sci 60:451–459. https://doi.org/10.1167/iovs.18-26076

Srinivasan S, Hoffman NG, Morgan MT et al (2012) Bacterial communities in women with bacterial vaginosis: high resolution phylogenetic analyses reveal relationships of microbiota to clinical criteria. PLoS One 7:e37818. https://doi.org/10.1371/journal.pone.0037818

Turnbaugh PJ, Ley RE, Hamady M et al (2007) The human microbiome project. Nature 449:804–810. https://doi.org/10.1038/nature06244

Wang Y, Chen H, Xia T et al (2020) Characterization of fungal microbiota on normal ocular surface of humans. Clin Microbiol Infect 26:123. https://doi.org/10.1016/j.cmi.2019.05.011

Williamson J, Gordon AM, Wood R et al (1968) Fungal flora of the conjunctival sac in health and disease. Influence of topical and systemic steroids. Br J Ophthalmol 52:127–137. https://doi.org/10.1136/bjo.52.2.127

Wilson LA, Ahearn DG, Jones DB et al (1969) Fungi from the normal outer eye. Am J Ophthalmol 67:52–56. https://doi.org/10.1016/0002-9394(69)90007-5

Wood DE, Lu J, Langmead B (2019) Improved metagenomic analysis with Kraken 2. Genome Biol 20:257. https://doi.org/10.1186/s13059-019-1891-0

Wu T, Mitchell B, Carothers T et al (2003) Molecular analysis of the pediatric ocular surface for fungi. Curr Eye Res 26:33–36. https://doi.org/10.1076/ceyr.26.1.33.14253

Chapter 7
Viruses of the Human Eye

Abstract The methods used for the identification of the viruses of the eye are dealt with briefly, and the challenges involved are highlighted. Various ocular viral diseases like conjunctivitis, keratitis, interstitial keratitis, blepharitis, uveitis etc. are described. Ocular involvement of specific viruses like varicella zoster virus, cytomegalovirus, herpes simplex virus, Epstein-Barr virus, molluscum contagiosum virus and rubeola virus is highlighted with reference to their clinical presentation, diagnosis, incidence, epidemiological features, treatment etc.

Keywords Ocular virus · Viral disease · Varicella zoster virus · Cytomegalovirus · Herpes simplex virus · Epstein-Barr virus · Molluscum contagiosum virus · Rubeola virus

7.1 Introduction

Bacteria, archaea, eukaryotes and viruses together constitute the human gut microbiome. Among the viruses, >90% present in the human gut are phages which are largely unexplored, with respect to their diversity and abundance (Mirzaei and Maurice 2017; Gregory et al. 2019). A rough estimate indicates that bacteriophages are 100 times more abundant than the eukaryotic cells in our body (Weinbauer 2004), and our genomes are imprinted with an enormous number of endogenous viral fragments, which constitute approximately 8% of our genome (Wallace et al. 2018). The endogenous viral proteins have been implicated in physiological functions, such as the morphogenesis of the placenta (Chuong 2018).

7.2 Enumeration of Viruses

The method used to characterize viruses essentially involves viral isolation from cells and subsequently screening for cytopathic effects on cell lines or to study the effects of intracerebral inoculation in suckling mice (Adriaenssens et al. 2020; Mirzaei et al. 2021). Subsequently, immunological methods were used to detect

S. Shivaji, *Human Ocular Microbiome*,
https://doi.org/10.1007/978-981-19-1754-7_7

viral antigens; with time, the polymerase chain reaction (PCR) method became the technique of choice for viral detection (Cassedy et al. 2021). With the advent of next-generation sequencing techniques, viral metagenomics allowed the discovery of new viruses (Munang'andu et al. 2017; Reyes et al. 2012; Breitbart et al. 2004). Compared to bacteria and fungi, which have well-characterized phylogenetic markers like 16S rRNA and ITS2 respectively, it has been extremely challenging to enumerate viruses in the absence of a universal viral marker. The standard procedure in enumerating viromes has been based on sequencing the viruses followed by sequence homology searches (BLAST or Kraken) (Wood and Salzberg 2014) with various sequence databases such as viral RefSeq (https://www.ncbi.nlm.nih.gov/genome/viruses/), Integrated Microbial Genome/Virus (IMG/VR) (Paez-Espino et al. 2019), the human Gut Virome Database (GVD) (Bolduc et al. 2017), Markov model databases (e.g. Prokaryotic Virus Orthologous Groups) etc. (Grazziotin et al. 2017). Several virus identification tools including MARVEL (Amgarten et al. 2018), DeepVirFinder (Ren et al. 2020), VirSorter (Roux et al. 2015) and VIBRANT (Kieft et al. 2019) facilitate the identification process. Virus-specific databases, such as VOGDB (Thannesberger et al. 2017), HVPC (Elbehery et al. 2018), pVOGs (Grazziotin et al. 2017), GLUVAB (Coutinho et al. 2019), IMG/VR (Paez-Espino et al. 2019), Virus-Host DB (Mihara et al. 2016), MVP (Gao et al. 2018), Pfam (Finn et al. 2014), eggNOG (Jensen et al. 2008; Huerta-Cepas et al. 2019), COGNIZER (Bose et al. 2015) or KEGG (Kanehisa and Goto 2000), are commonly used for functional annotation. Despite the above generalized workflow for virome studies, results on frequency of virus detection between studies are highly variable (Mirzaei and Maurice 2017) and influenced by differences in sample processing (Shkoporov et al. 2018), inadequate reference databases (Roux et al. 2019; Wang 2020), absence of appropriate culturable host gut microbes (Wang 2020) and inter-individual variation (Shkoporov et al. 2018).

7.3 Challenges in Studying Human Virome

Mirzaei et al. (2021) have recently highlighted the challenges of studying the human virome. Phages constituted a majority of the estimated 10^{31} viruses on earth and includes both double-stranded DNA (dsDNA) tailed phages and ssRNA phages. The dominance of these phages seems to vary depending on whether the viruses were characterized by electron microscopy and culture-based methods (Brum et al. 2013; Hopkins et al. 2014; Sutton and Hill 2019). Further, the multi-omics approach revealed a greater degree of diversity (Roux et al. 2019), and uncultivated viruses constituted the majority ($\geq$95%) of the viral diversity in public databases (Roux et al. 2019). Alternative methods to detect phages were based on clustering viral contigs by comparison with both viral and nonviral sequences (Garretto et al. 2019) using PHAST and PHASTER for the annotation of prophages within bacterial genomes (Arndt et al. 2016). Machine-learning methods like VirFinder (Ren et al. 2020), Virsorter (Roux et al. 2015) and MARVEL have also been used to detect dsDNA phages following metagenome sequencing (Amgarten et al. 2018). All these approaches have their own challenges and limitations related to sample and downstream processing, sequencing annotation and in silico predictions (Shkoporov and Hill 2019; Reyes et al. 2012; Carding et al. 2017). A few of the limitations include:

1. Use of 0.2–0.45 μm pore size filters, which would eliminate larger viruses.
2. Amplification of viral nucleic acids by random amplified shotgun library (RASL) is restricted to dsDNA.
3. Requirement of a high template concentration for the linker-amplified shotgun library (LASL) approach (Hindiyeh et al. 2019).
4. Overamplifcation of circular single-stranded DNA (ssDNA) by multiple displacement amplification (MDA).
5. A high background noise that confounds metagenomic data used for viral taxonomic characterization (Ajami et al. 2018).
6. Viral databases are limited and inadequate and rely on de novo assembly to recover viral genomes from metagenomes (Sutton and Hill 2019).
7. Viral metagenomes are highly mosaic and show high metagenomic complexity and strain-level diversity (Sutton and Hill 2019).

All these limitations are responsible for the underestimation of viral richness and diversity (Fig. 7.1).

7.4 Viral Infections of the Eye

Viral infections of the eye can affect any part of the eye and the surrounding regions including the eyelids to the retina and the optic nerve (Hoffman and Foster 2020). Eye infections occur due to direct contact of the eye with viruses from infected secretions in the birth canal (human papillomavirus and herpes simplex virus), from airborne particles (rhinovirus) and from fomites (adenovirus). Viruses are also acquired during viraemia (measles virus and human cytomegalovirus). Other mechanisms by which viruses get access to the eye include extension from contiguous adnexal disease (herpes simplex virus), from transplacental passage of infectious virus (rubella virus) and from the upper respiratory tract (rhinovirus). Rarely, ocular infection may disseminate to other parts of the body (enterovirus 70) (Chodosh et al. 2008). Ocular viral infections, such as viral conjunctivitis caused due to adenovirus or influenza virus, are short lived, with limited ocular complications. However, other viral infections can cause serious complications, such as stromal keratitis due to herpes simplex virus (HSV) or retinal detachment resulting from cytomegalovirus (CMV) retinitis. Some of the ocular viral diseases are discussed below.

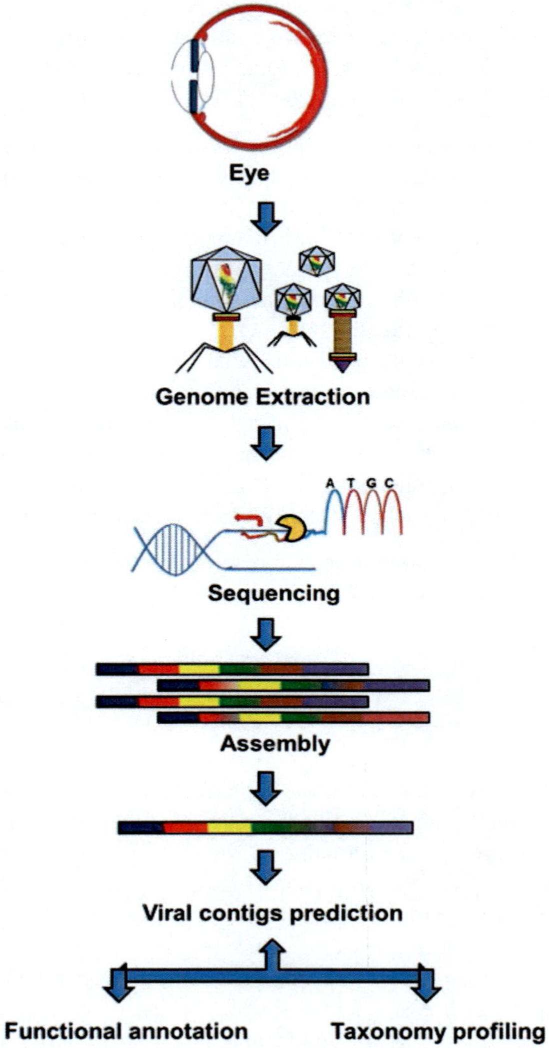

Fig. 7.1 Characterization of the free phage fraction of the human eye. (Modified from Mirzaei et al. 2021)

7.4.1 Viral Conjunctivitis

Clinical Presentation and Diagnosis Conjunctivitis, normally referred to as 'pink eye', is an infectious inflammation of the conjunctiva with dilation of the conjunctival blood vessels (Fig. 7.2a). Inflammation of the blood vessels causes the white of the eyes to appear reddish or pink. Bacteria and viruses are the common infectious agents causing conjunctivitis which is known to be highly contagious (Watson et al. 2018). Viral conjunctivitis results in vasodilatation, serous discharge, hyperplasia of the lymphoid follicles of the conjunctiva and enlargement of the corresponding

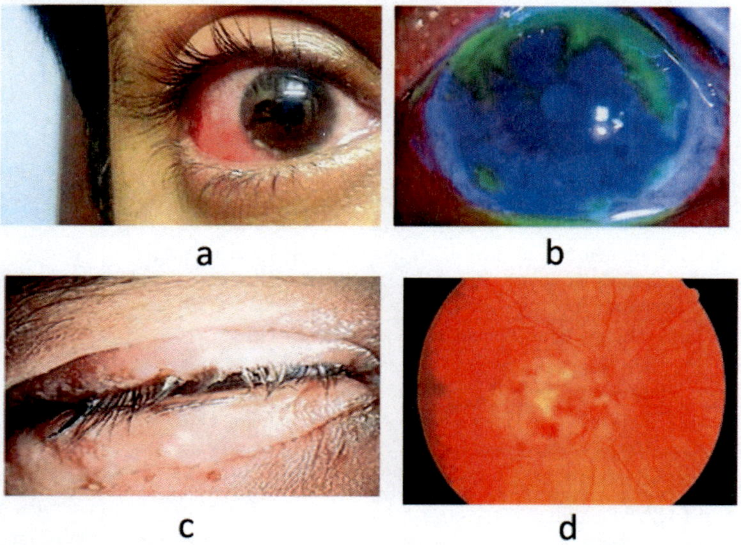

Fig. 7.2 Viral diseases of the eye. (**a**) Viral conjunctivitis (Creative Commons Attribution-ShareAlike 4.0); (**b**) herpes simplex virus keratitis (Creative Commons Attribution 2.5); (**c**) herpetic blepharitis (Creative Commons Attribution-ShareAlike 3.0); (**d**) fundus, photograph-CMV retinitis (National Eye Institute, National Institutes of Health. Ref#: EDA07Public domain)

draining lymph nodes. Patients with severe conjunctival infection present with permanent scarring of the globe and turning in of the eyelashes against the eye (Chodosh et al. 2008). Conjunctivitis is more common in adults than in children. Several RNA viruses (Table 7.1) like rhinovirus, influenza virus, respiratory syncytial virus and parainfluenza virus are known to cause conjunctivitis. But conjunctivitis caused by these viruses is mild and self-limiting, and patients rarely seek medical advice (Chodosh et al. 2008). Newcastle disease virus, an RNA virus, is known to cause follicular conjunctivitis and gives rise to fatal epidemics in poultry and also infects the poultry handlers.

DNA viruses (Table 7.2) like adenovirus types 3, 4 and 7 are responsible for follicular conjunctivitis and is accompanied by pharyngoconjunctival fever. Further, adenovirus types 8, 19 and 37 cause epidemic keratoconjunctivitis which is associated with inflammatory conjunctival membranes which can permanently scar the eyelids, whereas acute haemorrhagic conjunctivitis (AHC) may be caused by adenovirus type 11 which is also known to cause acute haemorrhagic cystitis (Chodosh et al. 2008). Among the DNA viruses, seven of the eight known human herpes viruses cause ocular disorders. Occasionally, when herpes simplex or zoster viruses are the causative agents, antivirals such as acyclovir ointment or ganciclovir gel are prescribed.

Treatment Viral conjunctivitis is generally self-limiting, and cold compresses, artificial tears or topical antihistamines are recommended (Azari and Barney

Table 7.1 Human RNA viruses and ocular infections

Virus	Ocular target
Rift Valley fever virus, influenza virus, measles (rubeola) virus, mumps virus, Newcastle disease virus, parainfluenza virus, respiratory syncytial virus, poliovirus, Coxsackie virus, echovirus, enterovirus, rhinovirus, human coronavirus, dengue virus, yellow fever virus, Colorado tick fever virus	Conjunctiva
Human coronavirus, hepatitis C virus, West Nile virus, influenza virus, measles (rubeola) virus, human immunodeficiency virus, rubella virus	Retina
Hepatitis C virus, influenza virus, measles (rubeola) virus, mumps virus, Newcastle disease virus, poliovirus, Coxsackie virus, echovirus, enterovirus, human T cell lymphotropic virus-1, rabies virus, rubella virus	Cornea
Hepatitis C virus, mumps virus, human immunodeficiency virus	Lacrimal glands
West Nile virus, influenza virus, measles (rubeola) virus, mumps virus, human T cell lymphotropic virus-1, rubella virus	Uvea
West Nile virus, influenza virus, measles (rubeola) virus, mumps virus	Optic nerve
West Nile virus, influenza virus, measles (rubeola) virus, mumps virus, poliovirus, Coxsackie virus, echovirus, enterovirus	Cranial nerves
Influenza virus, mumps virus	Lacrimal gland
Influenza virus	Episclera
Mumps virus	Sclera
Mumps virus, rubella virus	Trabecular meshwork
Rubella virus	Globe
Rubella virus	Lens

Modified from Chodosh et al. (2008)

2013). Compared to viruses, conjunctivitis due to bacteria is less frequent, and the common bacteria are *Streptococcus pneumoniae*, *Haemophilus influenza* and *Staphylococcus aureus* (Sheikh and Hurwitz 2006). *Gonococcal conjunctivitis* by *Neisseria gonorrhoeae* is uncommon and occurs in neonates and sexually active young adults (Morrow and Abbott 1998). In addition, *Chlamydial conjunctivitis* is unilateral and has concurrent genital infection (Morrow and Abbott 1998). Oral azithromycin or doxycycline is used for effective treatment (Katusic et al. 2003).

7.4.2 Viral Keratitis

Clinical Presentation and Diagnosis Infection of the cornea by a virus is referred to as viral keratitis, and it causes visual impairment which could also lead to corneal blindness (Allan and Dart 1995) in working-age adults (Whitcher et al. 2001; Sharma and Taniguchi 2017). The predisposing factors for infectious keratitis include contact lens use, physical and chemical trauma, corneal abrasions, refractive surgery, diabetes, use of topical steroids and immunosuppressive disease in the patients (Sharma and Taniguchi 2017; Green et al. 2008) and are also known to be influenced by climate and geographical region. Bacteria (*S. aureus*,

Table 7.2 Human DNA viruses and ocular infections

Virus	Ocular target
Adenovirus, herpes simplex virus (type 1), herpes simplex virus (type 2), varicella zoster virus (HHV3), Epstein-Barr virus (HHV4), human herpes virus 8 (HHV8), human papillomavirus, molluscum contagiosum virus, smallpox (variola) virus, vaccinia virus	Conjunctiva
Adenovirus, herpes simplex virus (type 1), herpes simplex virus (type 2), varicella zoster virus (HHV3), Epstein-Barr virus (HHV4), human papillomavirus, molluscum contagiosum virus, smallpox (variola) virus, vaccinia virus	Cornea
Herpes simplex virus (type 1), herpes simplex virus (type 2), varicella zoster virus (HHV3), human papillomavirus, molluscum contagiosum virus, orf virus, smallpox (variola) virus, vaccinia virus	Eyelid
Herpes simplex virus (type 1), herpes simplex virus (type 2), varicella zoster virus (HHV3)	Trabecular meshwork
Herpes simplex virus (type 1), herpes simplex virus (type 2), varicella zoster virus (HHV3), Epstein-Barr virus (HHV4), human cytomegalovirus (HHV5), human herpes virus 6 (HHV6)	Retina
Herpes simplex virus (type 1), herpes simplex virus (type 2), varicella zoster virus (HHV3), Epstein-Barr virus (HHV4), smallpox (variola) virus	Uvea
Varicella zoster virus (HHV3), Epstein-Barr virus (HHV4), human cytomegalovirus (HHV5), smallpox (variola) virus	Optic nerve
Epstein-Barr virus (HHV4)	Lacrimal gland

Modified from Chodosh et al. (2008)

coagulase-negative staphylococci, *S. pneumoniae* and *Pseudomonas aeruginosa*) (Sharma and Taniguchi 2017; Green et al. 2008), fungi, acanthamoebae and viruses are the infectious agents (Stapleton et al. 1995). Viral infection of the corneal epithelium induces punctate epithelial keratitis when the patients present with swollen epithelial cells or punctate epithelial erosions when a loss of individual epithelial cells is observed (Chodosh et al. 2008).

7.4.2.1 Herpes Simplex Virus Keratitis

Miyagawa and Yamanishi (1999) reported eight human herpes simplex viruses (HSV) which include HSV-1, HSV-2, cytomegalovirus, varicella zoster virus, Epstein-Barr virus, HSV associated with Kaposi sarcoma and lymphomas (Stewart et al. 1995) and HSV associated with roseola infantum and febrile convulsions (Safrin 1992). Herpes simplex virus keratitis in many developed countries is an important cause of infectious blindness (Liesegang et al. 1989) (Fig. 7.2b). Recurrences of viral keratitis are triggered by several other factors such as fever, ultraviolet exposure, hormonal changes, psychological stress, trigeminal nerve manipulation and ocular trauma (Wilhelmus 1998). The global incidence of herpes simplex keratitis per 100,000 people per year ranges from a minimum of 4.1 in Croatia (Ribaric 1976) to a maximum of 20.7 in Rochester (Liesegang et al. 1989; Norn

1970), and the incidence is 6 times higher in patients (120 per 100,000 people per year) who have undergone corneal transplantation (for non-herpetic corneal disease) (Remeijer et al. 1997; Ficker et al. 1989). HSV-1 and HSV-2 are acquired in utero by ascending or transplacental infection, by postnatal exposure to relatives or attendants and by exposure to genital lesions during delivery (Jeffries 1991). Corneal infections of herpes simplex keratitis are classified depending on the layer of the cornea that is involved as epithelial, stromal, endothelial or mixed. Epithelial herpes simplex keratitis affecting the epithelium manifests as an ulcer (Azher et al. 2017) and appears like a branching tree (dendritic) with terminal bulbs. The dendritic herpes simplex keratitis heals within 5–10 days but could also progress, despite treatment with topical steroids, to form a geographical ulcer. Both, the dendritic and geographical ulcer, could resolve without residual scarring. But in some patients, secondary stromal scarring may persist after the previous epithelial disease as grey white subepithelial stromal opacities to a denser opaque cicatrix. The most serious consequence of epithelial infection is recurrent stromal disease. Wilhelmus et al. (1981a, b) found that 25% of patients develop stromal inflammation (Liesegang 1989; Whitcher et al. 1976) and the patients with stromal herpes simplex keratitis present with opacity or haze of the stroma, with or without scarring, ulceration or vascularization. Endothelial keratitis is characterized by precipitates on the endothelium and with corneal oedema (Guess et al. 2007).

Treatment Management of stromal and endothelial keratitis involves oral antivirals (acyclovir or valaciclovir) and topical steroids (White and Chodosh 2014). Trifluridine, originally developed as an anticancer agent (Kaufman and Heidelberger 1964), is used for the treatment of HSV epithelial keratitis. Physical debridement of the ulcer followed by antiviral drop is practised. Vidarabine ointment 3% may be useful in paediatric cases. For HSV immune stromal keratitis, topical steroids, in conjunction with topical antiviral coverage, are the mainstay of therapy (Wilhelmus et al. 1994). Topical corticosteroids have also proved effective in treating patients with endotheliitis, iridocyclitis and trabeculitis. Oral acyclovir has been employed successfully in the treatment of primary HSV infection (Collum et al. 1985; Schwab 1988) in immunocompromised patients (Young et al. 1989), in infants and in patients with iridocyclitis not responding to topical corticosteroids (Goodman 1994).

7.4.2.1.1 Epidemiological Features of HSV Keratitis

A few epidemiological features of ocular HSV from across the world are summarized below:

1. Typically, the clinical presentations of primary HSV include acute follicular conjunctivitis, keratoconjunctivitis, periocular and eyelid skin vesicles and keratitis (Darougar et al. 1978, 1985; Wishart et al. 1984; Liesegang et al. 1989; Goodman 1994).
2. HSV disease is either unilateral (81%) or bilateral (19%) (Liesegang 2001), and recurrence (32%) was more frequent in young adults under the age of 20 years

(Wishart et al. 1987). In the United States, the recurrence of epithelial herpes in individuals with at least two prior episodes was 24.5% within 12 months and 32.9% within 24 months (Shuster et al. 1981).

3. The seasonal occurrence of ocular HSV disease was conflicting with no seasonal trend in a few studies as opposed to several studies that reported a more frequent occurrence during the winter months (Gamus et al. 1995) as in the United States (Liesegang 1989; Liesegang et al. 1989) and Japan (Uchio et al. 1994).
4. Ocular HSV is less common than other herpetic diseases (Liesegang 2001).
5. In children, HSV keratitis is more severe, and the individuals present with reduction of vision, increase in the incidence of astigmatism and recurrences as compared to adults (Beigi et al. 1994). Epithelial dendritic keratitis and interstitial keratitis were the most frequent forms of HSV keratitis in the paediatric population.
6. Incidence of ocular HSV is also influenced by prior acyclovir ointment, and a decrease in HSV serology was observed among adults in Japan (Ukkonen et al. 1984). In South India, the mean age of the patients with ocular HSV was 29 years (range, 9 months to 65 years), and the incidence was not changed with the season (Pramod et al. 1999).
7. Recent studies suggest an older age of onset and perhaps overall more severe ocular disease (Liesegang 2001). It was also observed that acyclovir-resistant herpes keratitis was increasingly common among patients with HIV (Castelo-Soccio et al. 2010).
8. Recurrences were more frequent in men both in Japan and South India, consistent with other studies (Laibson and Leopold 1964; Liesegang 1989; Dawson and Togni 1976). The predominant ocular disease was epithelial (up to 70%) (dendritic keratitis, geographic keratitis and punctate epithelial keratitis), and the remaining exhibited a stromal disease (disciform keratitis and necrotizing keratitis) (Ukkonen et al. 1984; Pramod et al. 1999).
9. Patients treated with 400 mg acyclovir exhibited beneficial effect of acyclovir, and there was no rebound after acyclovir was withdrawn (Sundmacher 1978; Shimomura et al. 1987). It was also reported that treatment with topical corticosteroids was significantly better compared to the placebo in reducing progression or persistence and in reducing the duration of HSV stromal keratitis (Wilhelmus et al. 1994). Further, HSV epithelial keratitis recurrence was not significantly different between the placebo and the drug (topical steroid treated with acyclovir) though Porter et al. (1990) had demonstrated a protective effect of either oral acyclovir or topical antiviral (Porter et al. 1990).
10. The prognostic factors for recurrence of HSV were average age of 49 years, male gender, prior dendritic ulcer (78%), stromal keratitis and 5% developed ocular hypertension (Wilhelmus et al. 1981a, b). It was also observed that sunlight, trauma, a febrile illness, emotional stress, menstruation and infectious diseases have been found to activate a recurrence of herpetic disease. The severity and frequency of disease are modulated by emotional, physical and immunological stress as well as the viral genome, its virulence and host immunoregulation (Pavan-Langston 1994).

Ocular HSV is not a major problem in HIV-positive patients (Rosenwasser and Greene 1992; Young et al. 1989) though one study reported that HIV patients exhibit more than a threefold increased risk of developing ocular HSV keratitis (Sobol et al. 2016). HSV was also detected in a case of severe bilateral necrotizing herpes simplex keratitis in an immunocompetent patient (Yoshida et al. 2019).

7.4.2.2 Congenital and Neonatal HSV Keratitis

Clinical Presentation HSV keratitis in neonates could be traced to genital infection in 30% of mothers who were infected with HSV around the time of delivery (Margolis and Atherton 1996). In neonates, the mean age of onset of disease was day 7, but the onset was delayed to 14 days when the mothers had recurrent herpes disease (HSV-1 and HSV-2). Neonatal HSV occurs in 20–25% of neonates (Nahmias et al. 1976; Corey et al. 1988), and the individuals present with retinitis, corneal ulceration, conjunctivitis, keratitis, anterior uveitis, dermatitis, vitreal inflammation, cataract formation, optic atrophy and chorioretinitis (Corey et al. 1988; Reynolds et al. 1986; Nahmias et al. 1976). Retinal involvement is apparent 1 month or later (Whitley 1990), and a higher prevalence of ocular infection (60%) occurs when HSV encephalitis is present (Nahmias et al. 1976). Neonatal HSV infection is not influenced by gender (Margolis and Atherton 1996), it is bilateral (Whitley et al. 1980) and up to 50% infants are born prematurely. Skin involvement is observed in 80% of the neonates (Whitley et al. 1980; Margolis and Atherton 1996).

Epidemiology The incidence of neonatal HSV varied between countries. In Denmark, it varied from 2.36/100,000 to 4.56/100,000 live births from 1977 to 1991; in California the incidence of neonatal HSV, between 1985 and 1995, was approximately 11.4/100,000 live births (Gutierrez et al. 1999); and in Britain (Tookey and Peckham 1996), it was 1.65/100,000 live births (76 infants). The majority of the infants was infected with either HSV-1, HSV-2 or an unknown virus. In a Centre for Diseases Control and Prevention study of neonatal herpes in 1989 (Stone et al. 1989), a significant proportion of mothers (22%) had a history of genital infection, and 9% had genital lesions at the time of delivery. Caesarean delivery failed to prevent infection. This study indicated that not all mothers of infected neonates have a history of genital HSV and are usually asymptomatic at delivery (Brown et al. 1991). In contrast, a study in Sweden on 39 children with neonatal HSV indicated 13 mothers had primary infection (6 with HSV-1 and 7 with HSV-2) (Malm et al. 1995). These findings put limitations on current preventive strategies.

7.4.2.3 Epidemics of HSV

Epidemics due to viral infections have been reported in immunocompromised individuals, health-care workers, in intensive care units, in neonatal nurseries, in dental practices, nurseries and day-care centres when exposed to infected persons, due to herpetic whitlows, nosocomial outbreaks etc. (Callen 1983). HSV-1 skin infections have been transmitted in rugby players, wrestlers (herpes gladiatorum) (Skinner et al. 1996), burn patients (Bourdarias et al. 1996) and prostitutes. Homosexual men with HSV-2 infection can easily spread the disease and cause an epidemic. But most epidemics of HSV have been reported in wrestlers and do not involve the eye (Holland et al. 1992) with follicular conjunctivitis, blepharitis and phlyctenular disease.

7.4.2.4 Necrotizing Stromal Keratitis

In patients with recurrent attacks of herpetic keratitis, necrotizing stromal keratitis (NSK) is a rare manifestation of HSV in which direct viral invasion of the corneal stroma is obvious (Lobo et al. 2019). It is characterized by ulceration, necrosis and dense infiltration of the corneal stroma. Uveitis is nearly always present and can be severe. This combination of replicating HSV along with a severe host inflammatory response leads to rapid corneal thinning and subsequently perforation. NSK is refractory to treatment with anti-inflammatory and antiviral medications. The specific contributions of the viral infection and the host immune response to the severe corneal destruction are unknown.

7.4.2.5 Immune Stromal Keratitis or Interstitial Keratitis

Immune stromal keratitis (ISK) or interstitial keratitis is a common chronic recurrent manifestation of HSV infection and occurs in 20% of patients with ocular ISK (Liesegang 1989; Wilhelmus et al. 1994). Viral antigen retained within the stroma triggers intrastromal inflammation which may be localized or diffused (Wilhelmus 1987). Stromal inflammation is often accompanied by ciliary flush, iridocyclitis and pain. If ISK is untreated or under-treated, inflammation can lead to stromal thinning, scarring, lipid deposition, neovascularization and severe vision loss. Some patients with ISK develop corneal stromal oedema without stromal infiltration and may exhibit permanent loss of endothelial cells, often termed endotheliitis. Clinical findings include keratic precipitates, iridocyclitis and stromal oedema (Knox and Holsclaw 1998). A secondary trabeculitis may also occur causing an acute intraocular pressure rise and secondary glaucoma. The mechanism by which HSV causes iridocyclitis is not fully understood though live viruses appear to play a role (Reijo et al. 1983). Cytological examination of specimens stained with Giemsa or Wright stains reveals multinucleated giant cells and intranuclear inclusions typical of herpes virus infection.

7.4.2.6 Other Diseases Caused by HSV

7.4.2.6.1 Acute Retinal Necrosis

Acute retinal necrosis (ARN) is caused by HSV-1 in younger patients and varicella zoster virus (VZV) in older patients (Thompson et al. 1994). It is a form of necrotizing retinitis (Fig. 7.2c), and analyses of the aqueous or vitreous samples of 13 ARN patients indicated VZV in all the 13 patients; in addition HSV-1 was detected in 7 patients, HSV-2 in 6 patients and cytomegalovirus in 1 patient (Ganatra et al. 2000). It also indicated that VZV or HSV-1 causes ARN in older patients (>25 years) whereas HSV-2 in the younger patients (<25 years). In Japan, HSV-2 and VZV were associated with ARN, and none were associated with HSV-1 (Itoh et al. 2000). The differences between the patients in the United States and Japan could be due to genetic and biological differences or also host sensitivity differences. ARN normally begins with retinal vasculitis and then diffuse uveitis. Patients frequently develop retinal detachments, and intravenous acyclovir hastens the resolution of the retinitis. The diseases could be either uniocular or unilateral, and treatment decreases the risk of developing the disease in the second eye (Schlingemann et al. 1996). The presentation and course of ARN in immunocompromised patients may be more severe than in immunocompetent ones (Ritterband and Friedberg 1998).

7.4.2.6.2 HSV Involvement in Various Syndromes

HSV has been implicated in various syndromes like patients with the Posner-Schlossman syndrome (Yamamoto et al. 1995), Fuchs' uveitis syndrome (Mitchell et al. 1996) and iridocorneal endothelial syndrome (Alvarado et al. 1994) and also in patients with early corneal graft failure (Cleator et al. 1994; Cockerham et al. 1997), keratitis linearis migrans, herpetic corneal endotheliopathy, autoimmune corneal endotheliopathy and idiopathic corneal endotheliopathy (Olsen et al. 1994; Robin et al. 1985). Several of these latter patients disclosed a positive antibody to HSV antigen (Robin et al. 1985). HSV is also known to exhibit viral latency and thus remains dormant within the host cell (Liesegang 1992), thus confirming that the cornea other than nervous tissue could harbour latent virus. Antiviral drugs do not eliminate latent virus.

7.4.2.6.3 HSV in Blepharitis

Herpetic blepharitis is characterized by vesicular lesions of the eyelid with surrounding erythema and subsequently progresses to ulceration and crusting and heals without scarring (Fig. 7.2d). Herpetic ulcerative blepharitis may reappear but generally involves only the lids and rarely progresses to keratitis and iridocyclitis. HSV may also cause an acute or chronic follicular conjunctivitis (Ritterband and Friedberg 1998).

7.4.3 Ocular Complications of AIDS

HIV infection is also known to be associated with ocular complications and manifests as (1) HIV retinopathy, (2) ocular adnexal neoplasms, (3) opportunistic ocular infections, (4) drug-induced manifestations and (5) neuro-ophthalmic lesions (Tay-Kearney and Jabs 1996). In AIDS patients, HIV retinopathy appears as multifocal infarcts of the retinal nerve fibre layer or cotton wool patches. Less often intraretinal haemorrhages occur. Several ocular infections are caused by opportunistic ocular infections including CMV retinitis (Fig. 7.2c), microsporidial keratoconjunctivitis and choroiditis exclusively in AIDS patients. Other infections, such as ocular syphilis, toxoplasmosis retinochoroiditis, molluscum contagiosum of the eyelids and herpes zoster ophthalmicus, are also seen in HIV-infected patients. HSV-1, HSV-2 or varicella zoster virus which cause acute retinal necrosis occur more commonly in HIV-infected patients than in normal patients, and the infection is refractory to antiviral therapy and can result in unilateral or bilateral blindness. Kaposi sarcoma of the eyelids or conjunctiva, associated with HSV-8 infection and non-Hodgkin's lymphomas of the orbit, also occurs frequently in AIDS patients. Recently, squamous cell carcinoma of the ocular surface (conjunctiva and cornea) has been suggested as a marker for AIDS. Neuro-ophthalmic lesions are commonly caused by cryptococcal meningitis or other opportunistic infections in AIDS patients but may occur directly due to HIV infection of the central nervous system. Uveitis and retinitis due to anti-HIV medications can be confused with opportunistic intraocular infections (Chodosh et al. 2008). Farrell and co-workers (1988) describe a patient with uveitis in whom HIV was cultured from the aqueous and inflammation responded to treatment with zidovudine (Lucca et al. 1990).

7.4.4 Ocular Diseases Caused by Varicella Zoster Virus

Clinical Presentation Varicella is an acute, highly contagious, viral disease, and the etiological agent is varicella zoster virus (VZV) which causes skin eruption. It spreads either by contact with blisters or by droplet infection from an infected person. The disease is common in children, and in the United States 90% of cases occurs in children less than 3 years of age (Ritterband and Friedberg 1998). The disease is rarely severe and is accompanied by severe symptoms in both infants and adults. The average incubation period is 10–16 days. Ocular involvement in varicella frequently occurs on the lids and manifests as papillary conjunctivitis, dendritic keratitis or delayed immune stromal keratitis (Nesburn et al. 1974). Several other complications include corneal melting, gangrene of the eyelids, chronic uveitis, optic neuritis, extraocular muscle palsies, cataract, central retinal artery occlusion and choroidal lesions (Friedberg and Micale 1994; Deegan and Duker 1994). In the immunocompromised and the elderly patients, ocular involvement could be due to reactivation of the latent varicella zoster (Pepose 1997) as in patients with herpes zoster ophthalmicus (HZO), a clinical syndrome involving the ophthalmic branch of the trigeminal nerve (Hutchinson 1866). Ocular complications occur in 50–72% of HZO patients and may include conjunctivitis, scleritis, episcleritis, keratitis, secondary glaucoma, uveitis, cranial nerve palsies, cataract, retinitis, post-herpetic neuralgia and contracture of the eyelids (Harding et al. 1987) that may necessitate surgical repair (Marsh and Cooper 1993). About two thirds of HZO patients with ocular disease develop corneal disease and may develop secondary complications like serpiginous ulceration and sclerokeratitis (Mondino et al. 1978). Kerato-uveitis is common in HZO patients with corneal stromal oedema, Descemet folds, keratic precipitates and uveitis. Secondary glaucoma due to viral infection of the trabecular meshwork or ischaemic vasculitis of the ciliary body is also common (Harding et al. 1987). Corneal sensitivity is depressed in almost all patients with corneal involvement from HZO. Signs of corneal neuropathy may begin insidiously with decreased corneal lustre and mild punctate epitheliopathy. Over time a grey haze develops within the epithelium, the tear break time is shortened and epithelial defects with boggy stromal ulcers develop in the lower cornea. Superinfection, corneal thinning, fibrovascular scarring and progression to perforation are all possible sequelae. Diagnosis of HZO is usually made from the characteristic dermatomal rash using several staining techniques to confirm the presence of multinucleated giant cells. Immunofluorescent, EIA and PCR methods are more sensitive in detecting the virus (Liesegang 1991; Nahass et al. 1992).

Treatment Intravenous acyclovir is recommended for VZV infections (O'Brien and Campoli-Richards 1989; Beutner et al. 1995) and antiviral therapy for HZO (Pepose 1997). Topical corticosteroids are beneficial in alleviating the inflammatory and immune reactions and the vasculitis in some of the corneal and uveal manifestations (disciform keratitis, sclerokeratitis, interstitial keratitis, kerato-uveitis and trabeculitis). Corneal transplantation can be successful when judiciously planned (Marsh and Cooper 1989).

7.4.4.1 Retinal Involvement of HZV

HZV can cause an infectious retinitis, ARN syndrome and progressive outer retinal.

necrosis (PORN) (Engstrom et al. 1994). In PORN inflammation, retinal vasculitis and optic nerve involvement are decreased. PORN could be unilateral or bilateral and can also be seen in patients with other retinal or choroidal infections. Treatment includes acyclovir with foscarnet or ganciclovir (Spaide et al. 1996).

7.4.5 Ocular Diseases Caused by Cytomegalovirus

Presentation and Diagnosis Cytomegalovirus (CMV) can cause retinitis (Fig. 7.2c) in congenitally infected neonates with an incidence of 1% among live births in the United States; it is usually subclinical, and 10–15% of infants have symptoms which include mental retardation, developmental abnormalities and hearing loss. Retinitis develops in 5–30% of infants (Alford et al. 1990). In the past two decades, CMV retinitis has increased in the adult population (25–40%) and more so in patients following withdrawal of immunosuppression drugs and in patients who are profoundly immunocompromised (Kuppermann et al. 1993a, b). Retinitis due to CMV disease in AIDS patients is extremely high (85%), and patients may complain of flashes of light in their vision, blurred vision, floaters or loss of visual field (Baldassano et al. 1995). Clinically CMV retinitis produces a haemorrhagic necrotic retinitis with multiple punctate lesions which is common in the posterior portion of the fundus. Untreated CMV lesions expand, and eventually the entire retina is destroyed. Lesions frequently reactivate and progress, particularly with systemic therapy (Holland and Shuler 1992). CMV disease could extend into the macula and cause visual loss in patients and retinal detachment (Freeman et al. 1993). CMV can also develop in the fellow eye as well as elsewhere in the body even though retinitis in the involved eye is well controlled (Martin et al. 1994).

Diagnosis CMV diagnosis is best done by PCR (Shinkai et al. 1997; Dodt et al. 1997) using intraocular samples from both aqueous and vitreous (Short et al. 1997). Viral culture has also been unrewarding in these fluids because of the small quantities available (McCann et al. 1995).

Treatment The antiviral drugs foscarnet ganciclovir and cidofovir are used for the treatment of CMV diseases (Drew et al. 1995). All these drugs inhibit CMV DNA polymerase (Chou et al. 1995a, b) and also affect the polymerase gene (Sullivan et al. 1993). The continued progression of retinitis despite systemic therapy is attributed to viral resistance (Dunn et al. 1995) or due to delivery of subtherapeutic levels of medication to the eye (Kuppermann et al. 1993a, b). This problem has been overcome by the use of implantation devices which release drug at a steady rate into the vitreous (Martin et al. 1994; Musch et al. 1997). Regression of CMV lesions has been reported in AIDS patients treated with highly active antiretroviral therapy (HAART) (Whitcup et al. 1997).

7.4.6 Ocular Diseases Caused by Adenovirus

Presentation and Diagnosis Adenoviruses are endemic in the population and often associated with respiratory and gastrointestinal infections (Fox et al. 1977). Certain serotypes predominate in ocular diseases (Ritterband and Friedberg 1998) and cause significant ocular morbidity. Transmission occurs from contamination of fingers, ophthalmic instruments, swimming pools, sexual contacts and eye clinics (Pettit and Holland 1979). Handwashing and cleansing of ophthalmic instruments are the mainstay of control. Adenoviruses cause conjunctivitis which could be either epidemic keratoconjunctivitis (EKC) or pharyngoconjunctival fever (PCF). EKC patients present with corneal lesions and PCF with concurrent pharyngitis and fever. Symptoms of EKC include rapid onset of watery, red, sore, gritty-feeling eyes and could be unilateral or bilateral. Petechial haemorrhages, pseudomembranes and membranes are present in the most severe cases. Lid oedema and preauricular adenopathy are associated findings. Normally conjunctivitis resolves in 10–14 days, but secondary corneal involvement may last much longer. The corneal involvement typically begins with a diffuse fine punctate epithelial keratitis which gradually develops into a focal epithelial keratitis with irregular grey-white dots. Subsequently, subepithelial infiltrates with symptoms of increased photophobia and decreased vision are observed. These infiltrates normally get cleared over weeks or months, but at times they could persist longer and at times take years before they fade away (Pettit and Holland 1979).

Treatment Treatment of ocular disease by adenovirus is palliative, and warm or cold compresses, topical vasoconstrictors, lubrication and cycloplegic drops are suggested. Prophylactic antibiotic treatment is unnecessary because the risk of bacterial superinfection is very low. Treatment with topical steroids is effective in relieving symptoms of subepithelial infiltrates, but this treatment is limited to patients with marked visual changes or for those who are unable to perform normal activities. Cidofovir, as an antiviral drug, is being screened in clinical trials for the treatment of adenovirus conjunctivitis (Safrin et al. 1997).

7.4.7 Ocular Diseases Caused by Molluscum Contagiosum Virus

Presentation and Diagnosis Molluscum contagiosum virus produces raised umbilicated, dome-shaped papules in different parts of the body including the face, abdomen and genital region (Jordan 1994). Transmission of the virus in children is by direct contact, whereas in adults it is by sexual contact and has also been reported in immunocompromised patients as in AIDS (Kohn 1987). Eyelid lesions are caused by the shedding of virus particles into the tear film, and the lesions may be inconspicuous and hidden by the lashes. At times the lesions are found on the conjunctiva but rarely on the cornea (Charles and Friedberg 1992). Punctate keratopathy, a corneal pannus resembling trachoma and subepithelial infiltrates may also be seen (Jordan 1994). Diagnosis is made clinically or from light microscopy of the lesions.

Treatment The disease is self-limiting, and it is best managed by surgical treatments which include resection, curettage, cryotherapy and carbon dioxide laser. The eye findings clear rapidly after eradication of the lesions (Jordan 1994).

7.4.8 Ocular Diseases Caused by Epstein-Barr Virus

Presentation and Diagnosis EBV is a herpes virus implicated in Sjogren's syndrome, nasopharyngeal carcinoma, Burkitt's lymphoma and oral hairy leucoplakia, and the virus is transmitted through oral secretions (Matoba et al. 1986; Niederman 1994). Ocular findings include neuro-ophthalmologic and anterior segment diseases (follicular conjunctivitis, dendritic epithelial keratitis and stromal keratitis) (Matoba et al. 1986; Wilhelmus 1981). Stromal keratitis is characterized by granular or ring-shaped anterior stromal opacities and affect the cornea and can be associated with variable degrees of superficial and deep corneal vascularization (Matoba et al. 1986). Retinal involvement is rare in EBV, but a variety of presentations including macular oedema, retinal haemorrhages, chorioretinitis and papilledema have been described (Tanner 1954). Diagnosis depends on associated systemic conditions and laboratory evaluation. Chronic EBV disease can result in anterior uveitis and vitritis as well as macular oedema and optic nerve swelling (Wong et al. 1987).

7.4.9 Ocular Diseases Caused by Rubeola Virus

Presentation and Diagnosis Rubeola (measles), a paramyxovirus, causes epidemics in children and young adults. Ocular manifestations are limited to conjunctivitis which is mild and self-limiting. Koplik spots (small red-coloured spots with a bluish-white centre) may be seen on the conjunctiva, semilunar fold and caruncle (Baxby 1997). Rubeola also causes keratitis which is punctate, generally occurs centrally and typically resolves within several weeks. Patients complain of tearing, photophobia, foreign body sensation and decreased vision. Treatment is symptomatic and permanent sequelae rare (Deckard and Bergstrom 1981). Retinal involvement is uncommon but may manifest as neuro-retinitis or disc and macular oedema. A secondary pigmentary degeneration may occur (Scheie and Morse 1972). Patients with subacute sclerosing panencephalitis, caused by a variant of rubeola, may exhibit focal choroiditis and macular and optic disc oedema as well as optic atrophy (Robb and Watters 1970).

7.4.10 Conclusions

Viral eye diseases may be self-limiting but could pose several challenges related to diagnosis and therapy. Recurrence of the viral ocular infections is yet another issue that is not easily addressed. Laboratory testing for viral infections as part of diagnosis is very crucial and should shift from traditional culture methods to PCR- and NGS-based technologies. Continuous and sustaining research into discovering antiviral compounds is a must for the treatment of viral diseases of the eye.

References

Adriaenssens EM, Sullivan MB, Knezevic P et al (2020) Taxonomy of prokaryotic viruses: 2018-2019 update from the ICTV Bacterial and Archaeal Viruses Subcommittee. Arch Virol 165: 1253–1260. https://doi.org/10.1007/s00705-020-04577-8

Ajami NJ, Wong MC, Ross MC et al (2018) Maximal viral information recovery from sequence data using VirMAP. Nat Commun 9:3205. https://doi.org/10.1038/s41467-018-05658-8

Alford CA, Stagno S, Pass RF et al (1990) Congenital and perinatal cytomegalovirus infections. Rev Infect Dis 12(Suppl 7):S745–S753. https://doi.org/10.1093/clinids/12.supplement_7.s745

Allan BD, Dart JK (1995) Strategies for the management of microbial keratitis. Br J Ophthalmol 79: 777–786. https://doi.org/10.1136/bjo.79.8.777

Alvarado JA, Underwood JL, Green WR et al (1994) Detection of herpes simplex viral DNA in the iridocorneal endothelial syndrome. Arch Ophthalmol 112:1601–1609. https://doi.org/10.1001/archopht.1994.01090240107034

Amgarten D, Braga LPP, da Silva AM et al (2018) MARVEL, a tool for prediction of bacteriophage sequences in metagenomic bins. Front Genet 9:304. https://doi.org/10.3389/fgene.2018.00304

Arndt D, Grant JR, Marcu A et al (2016) PHASTER: a better, faster version of the PHAST phage search tool. Nucleic Acids Res 44:W16–W21. https://doi.org/10.1093/nar/gkw387

Azari AA, Barney NP (2013) Conjunctivitis: a systematic review of diagnosis and treatment. JAMA 310:1721–1729. https://doi.org/10.1001/jama.2013.280318

Azher TN, Yin XT, Tajfirouz D et al (2017) Herpes simplex keratitis: challenges in diagnosis and clinical management. Clin Ophthalmol 11:185–191. https://doi.org/10.2147/OPTH.S80475

Baldassano V, Dunn JP, Feinberg J et al (1995) Cytomegalovirus retinitis and low CD4+ T-lymphocyte counts. N Engl J Med 333:670. https://doi.org/10.1056/NEJM199509073331019

Baxby D (1997) The diagnosis of the invasion of measles from a study of the exanthema as it appears on the buccal mucous membrane. Rev Med Virol 7:71–74. https://doi.org/10.1002/(sici)1099-1654(199707)7:2<71::Aid-rmv185>3.0.Co;2-s

Beigi B, Algawi K, Foley-Nolan A et al (1994) Herpes simplex keratitis in children. Br J Ophthalmol 78:458–460. https://doi.org/10.1136/bjo.78.6.458

Beutner KR, Friedman DJ, Forszpaniak C et al (1995) Valaciclovir compared with acyclovir for improved therapy for herpes zoster in immunocompetent adults. Antimicrob Agents Chemother 39:1546–1553. https://doi.org/10.1128/AAC.39.7.1546

Bolduc B, Youens-Clark K, Roux S et al (2017) iVirus: facilitating new insights in viral ecology with software and community data sets imbedded in a cyberinfrastructure. ISME J 11:7–14. https://doi.org/10.1038/ismej.2016.89

Bose T, Haque MM, Reddy C et al (2015) COGNIZER: a framework for functional annotation of metagenomic datasets. PLoS One 10:e0142102. https://doi.org/10.1371/journal.pone.0142102

Bourdarias B, Perro G, Cutillas M et al (1996) Herpes simplex virus infection in burned patients: epidemiology of 11 cases. Burns 22:287–290. https://doi.org/10.1016/0305-4179(95)00146-8

Breitbart M, Felts B, Kelley S et al (2004) Diversity and population structure of a near-shore marine-sediment viral community. Proc Biol Sci 271:565–574. https://doi.org/10.1098/rspb.2003.2628

Brown ZA, Benedetti J, Ashley R et al (1991) Neonatal herpes simplex virus infection in relation to asymptomatic maternal infection at the time of labor. N Engl J Med 324:1247–1252. https://doi.org/10.1056/NEJM199105023241804

Brum JR, Schenck RO, Sullivan MB (2013) Global morphological analysis of marine viruses shows minimal regional variation and dominance of non-tailed viruses. ISME J 7:1738–1751. https://doi.org/10.1038/ismej.2013.67

Callen JP (1983) Epidemic herpes simplex virus infection. Am J Dis Child 137:182–184. https://doi.org/10.1001/archpedi.1983.02140280074022

Carding SR, Davis N, Hoyles L (2017) Review article: the human intestinal virome in health and disease. Aliment Pharmacol Ther 46:800–815. https://doi.org/10.1111/apt.14280

Cassedy A, Parle-McDermott A, O'Kennedy R (2021) Virus detection: a review of the current and emerging molecular and immunological methods. Front Mol Biosci 8:637559. https://doi.org/10.3389/fmolb.2021.637559

Castelo-Soccio L, Bernardin R, Stern J et al (2010) Successful treatment of acyclovir-resistant herpes simplex virus with intralesional cidofovir. Arch Dermatol 146:124–126. https://doi.org/10.1001/archdermatol.2009.363

Charles NC, Friedberg DN (1992) Epibulbar molluscum contagiosum in acquired immune deficiency syndrome. Case report and review of the literature. Ophthalmology 99:1123–1126. https://doi.org/10.1016/s0161-6420(92)31842-1

Chodosh J, Chintakuntlawar AV, Robinson CM (2008) Human eye infections. In: Mahy BWJ, Van Regenmortel MHV (eds) Encyclopedia of virology. Academic, Oxford, pp 491–497. https://doi.org/10.1016/b978-012374410-4.00396-4

Chou S, Erice A, Jordan MC et al (1995a) Analysis of the UL97 phosphotransferase coding sequence in clinical cytomegalovirus isolates and identification of mutations conferring ganciclovir resistance. J Infect Dis 171:576–583. https://doi.org/10.1093/infdis/171.3.576

Chou S, Guentzel S, Michels KR et al (1995b) Frequency of UL97 phosphotransferase mutations related to ganciclovir resistance in clinical cytomegalovirus isolates. J Infect Dis 172:239–242. https://doi.org/10.1093/infdis/172.1.239

Chuong EB (2018) The placenta goes viral: retroviruses control gene expression in pregnancy. PLoS Biol 16:e3000028. https://doi.org/10.1371/journal.pbio.3000028

Cleator GM, Klapper PE, Dennett C et al (1994) Corneal donor infection by herpes simplex virus: herpes simplex virus DNA in donor corneas. Cornea 13:294–304. https://doi.org/10.1097/00003226-199407000-00003

Cockerham GC, Krafft AE, McLean IW (1997) Herpes simplex virus in primary graft failure. Arch Ophthalmol 115:586–589. https://doi.org/10.1001/archopht.1997.01100150588001

Collum LM, Akhtar J, McGettrick P (1985) Oral acyclovir in herpetic keratitis. Trans Ophthalmol Soc U K 104(Pt 6):629–632

Corey L, Stone EF, Whitley R et al (1988) Difference between herpes simplex virus Type I and Type 2 neonatal encephalitis in neurological outcome. Lancet 331:1–4. https://doi.org/10.1016/s0140-6736(88)90997-x

Coutinho FH, Edwards RA, Rodriguez-Valera F (2019) Charting the diversity of uncultured viruses of Archaea and Bacteria. BMC Biol 17:109. https://doi.org/10.1186/s12915-019-0723-8

Darougar S, Hunter PA, Viswalingam M et al (1978) Acute follicular conjunctivitis and keratoconjunctivitis due to herpes simplex virus in London. Br J Ophthalmol 62:843–849. https://doi.org/10.1136/bjo.62.12.843

Darougar S, Wishart MS, Viswalingam ND (1985) Epidemiological and clinical features of primary herpes simplex virus ocular infection. Br J Ophthalmol 69:2–6. https://doi.org/10.1136/bjo.69.1.2

Dawson CR, Togni B (1976) Herpes simplex eye infections: clinical manifestations, pathogenesis and management. Surv Ophthalmol 21:121–135. https://doi.org/10.1016/0039-6257(76)90090-4

Deckard PS, Bergstrom TJ (1981) Rubeola keratitis. Ophthalmology 88:810–813. https://doi.org/10.1016/s0161-6420(81)34944-6

Deegan WF 3rd, Duker JS (1994) Unifocal choroiditis in primary varicella zoster (chickenpox). Arch Ophthalmol 112:735–736. https://doi.org/10.1001/archopht.1994.01090180033011

Dodt KK, Jacobsen PH, Hofmann B et al (1997) Development of cytomegalovirus (CMV) disease may be predicted in HIV-infected patients by CMV polymerase chain reaction and the antigenemia test. AIDS 11:F21–F28. https://doi.org/10.1097/00002030-199703110-00001

Drew WL, Ives D, Lalezari JP et al (1995) Oral ganciclovir as maintenance treatment for cytomegalovirus retinitis in patients with AIDS. Syntex Cooperative Oral Ganciclovir Study Group. N Engl J Med 333:615–620. https://doi.org/10.1056/NEJM199509073331002

Dunn JP, MacCumber MW, Forman MS et al (1995) Viral sensitivity testing in patients with cytomegalovirus retinitis clinically resistant to foscarnet or ganciclovir. Am J Ophthalmol 119:587–596. https://doi.org/10.1016/s0002-9394(14)70217-x

Elbehery AHA, Feichtmayer J, Singh D et al (2018) The Human Virome Protein Cluster Database (HVPC): A Human Viral Metagenomic Database for Diversity and Function Annotation. Front Microbiol 9:1110. https://doi.org/10.3389/fmicb.2018.01110

Engstrom RE Jr, Holland GN, Margolis TP et al (1994) The progressive outer retinal necrosis syndrome. A variant of necrotizing herpetic retinopathy in patients with AIDS. Ophthalmology 101:1488–1502. https://doi.org/10.1016/s0161-6420(94)31142-0

Farrell PL, Heinemann MH, Roberts CW et al (1988) Response of human immunodeficiency virus-associated uveitis to zidovudine. Am J Ophthalmol 106:7–10. https://doi.org/10.1016/s0002-9394(14)76379-2

Ficker LA, Kirkness CM, Rice NS et al (1989) The changing management and improved prognosis for corneal grafting in herpes simplex keratitis. Ophthalmology 96:1587–1596. https://doi.org/10.1016/s0161-6420(89)32668-6

Finn RD, Bateman A, Clements J et al (2014) Pfam: the protein families database. Nucleic Acids Res 42:D222–D230. https://doi.org/10.1093/nar/gkt1223

Fox JP, Hall CE, Cooney MK (1977) The Seattle virus watch. VII Observations of adenovirus infections. Am J Epidemiol 105:362–386. https://doi.org/10.1093/oxfordjournals.aje.a112394

Freeman WR, Friedberg DN, Berry C et al (1993) Risk factors for development of rhegmatogenous retinal detachment in patients with cytomegalovirus retinitis. Am J Ophthalmol 116:713–720. https://doi.org/10.1016/s0002-9394(14)73471-3

Friedberg MA, Micale AJ (1994) Monocular blindness from central retinal artery occlusion associated with chickenpox. Am J Ophthalmol 117:117–118. https://doi.org/10.1016/s0002-9394(14)73029-6

Gamus D, Romano A, Sucher E et al (1995) Herpetic eye attacks: variability of circannual rhythms. Br J Ophthalmol 79:50–53. https://doi.org/10.1136/bjo.79.1.50

Ganatra JB, Chandler D, Santos C et al (2000) Viral causes of the acute retinal necrosis syndrome. Am J Ophthalmol 129:166–172. https://doi.org/10.1016/s0002-9394(99)00316-5

Gao NL, Zhang C, Zhang Z et al (2018) MVP: a microbe-phage interaction database. Nucleic Acids Res 46:D700–D707. https://doi.org/10.1093/nar/gkx1124

Garretto A, Hatzopoulos T, Putonti C (2019) virMine: automated detection of viral sequences from complex metagenomic samples. PeerJ 7:e6695. https://doi.org/10.7717/peerj.6695

Goodman JL (1994) Infections caused by herpes simplex viruses. Infectious diseases: a treatise of infectious processes, 5th edn. Lippincott, Philadelphia

Grazziotin AL, Koonin EV, Kristensen DM (2017) Prokaryotic virus orthologous groups (pVOGs): a resource for comparative genomics and protein family annotation. Nucleic Acids Res 45:D491–D498. https://doi.org/10.1093/nar/gkw975

Green M, Apel A, Stapleton F (2008) Risk factors and causative organisms in microbial keratitis. Cornea 27:22–27. https://doi.org/10.1097/ICO.0b013e318156caf2

Gregory AC, Zablocki O, Howell A et al. (2019) The human gut virome database. bioRxiv 655910. https://doi.org/10.1101/655910

Guess S, Stone DU, Chodosh J (2007) Evidence-based treatment of herpes simplex virus keratitis: a systematic review. Ocul Surf 5:240–250. https://doi.org/10.1016/s1542-0124(12)70614-6

Gutierrez KM, Falkovitz Halpern MS, Maldonado Y et al (1999) The epidemiology of neonatal herpes simplex virus infections in California from 1985 to 1995. J Infect Dis 180:199–202. https://doi.org/10.1086/314848

Harding SP, Lipton JR, Wells JC (1987) Natural history of herpes zoster ophthalmicus: predictors of postherpetic neuralgia and ocular involvement. Br J Ophthalmol 71:353–358. https://doi.org/10.1136/bjo.71.5.353

Hindiyeh M, Mor O, Pando R et al (2019) Comparison of the new fully automated extraction platform eMAG to the MagNA PURE 96 and the well-established easyMAG for detection of common human respiratory viruses. PLoS One 14:e0211079. https://doi.org/10.1371/journal.pone.0211079

Hoffman J, Foster A (2020) Viral diseases of the eye. Community Eye Health 33:65–67

Holland GN, Shuler JD (1992) Progression rates of cytomegalovirus retinopathy in ganciclovir-treated and untreated patients. Arch Ophthalmol 110:1435–1442. https://doi.org/10.1001/archopht.1992.01080220097029

Holland EJ, Mahanti RL, Belongia EA et al (1992) Ocular involvement in an outbreak of herpes gladiatorum. Am J Ophthalmol 114:680–684. https://doi.org/10.1016/s0002-9394(14)74044-9

Hopkins M, Kailasan S, Cohen A et al (2014) Diversity of environmental single-stranded DNA phages revealed by PCR amplification of the partial major capsid protein. ISME J 8:2093–2103. https://doi.org/10.1038/ismej.2014.43

Huerta-Cepas J, Szklarczyk D, Heller D et al (2019) eggNOG 5.0: a hierarchical, functionally and phylogenetically annotated orthology resource based on 5090 organisms and 2502 viruses. Nucleic Acids Res 47:D309–D314. https://doi.org/10.1093/nar/gky1085

Hutchinson J (1866) A clinical report in herpes zoster frontails seu ophthalmicus (shingles affecting the forehead and nose). R London Ophthalmic Hospital Rep 5:191–196

Itoh N, Matsumura N, Ogi A et al (2000) High prevalence of herpes simplex virus type 2 in acute retinal necrosis syndrome associated with herpes simplex virus in Japan. Am J Ophthalmol 129: 404–405. https://doi.org/10.1016/s0002-9394(99)00391-8

Jeffries DJ (1991) Intra-uterine and neonatal herpes simplex virus infection. Scand J Infect Dis Suppl 80:21–26

Jensen LJ, Julien P, Kuhn M et al (2008) eggNOG: automated construction and annotation of orthologous groups of genes. Nucleic Acids Res 36:D250–D254. https://doi.org/10.1093/nar/gkm796

Jordan M (1994) Molluscum contagiosum. In: Infectious diseases, 5th edn. Lippincott, Philadelphia

Kanehisa M, Goto S (2000) KEGG: Kyoto encyclopedia of genes and genomes. Nucleic Acids Res 28:27–30. https://doi.org/10.1093/nar/28.1.27

Katusic D, Petricek I, Mandic Z et al (2003) Azithromycin vs doxycycline in the treatment of inclusion conjunctivitis. Am J Ophthalmol 135:447–451. https://doi.org/10.1016/s0002-9394(02)02094-9

Kaufman HE, Heidelberger C (1964) Therapeutic antiviral action of 5-Trifluoromethyl-2-′-Deoxyuridine in herpes simplex keratitis. Science 145:585–586. https://doi.org/10.1126/science.145.3632.585

Kieft K, Zhou Z, Anantharaman K (2019) VIBRANT: automated recovery, annotation and curation of microbial viruses, and evaluation of virome function from genomic sequences. bioRxiv 855387. https://doi.org/10.1101/855387

Knox CM, Holsclaw DS (1998) Interstitial keratitis. Int Ophthalmol Clin 38:183–195. https://doi.org/10.1097/00004397-199803840-00017

Kohn SR (1987) Molluscum contagiosum in patients with acquired immunodeficiency syndrome. Arch Ophthalmol 105:458. https://doi.org/10.1001/archopht.1987.01060040028015

Kuppermann BD, Petty JG, Richman DD et al (1993a) Correlation between CD4+ counts and prevalence of cytomegalovirus retinitis and human immunodeficiency virus-related noninfectious retinal vasculopathy in patients with acquired immunodeficiency syndrome. Am J Ophthalmol 115:575–582. https://doi.org/10.1016/s0002-9394(14)71453-9

Kuppermann BD, Quiceno JI, Flores-Aguilar M et al (1993b) Intravitreal ganciclovir concentration after intravenous administration in AIDS patients with cytomegalovirus retinitis: implications for therapy. J Infect Dis 168:1506–1509. https://doi.org/10.1093/infdis/168.6.1506

Laibson PR, Leopold IH (1964) An evaluation of double-blind IDU therapy in 100 cases of herpetic keratitis. Trans Am Acad Ophthalmol Otolaryngol 68:22–34

Liesegang TJ (1989) Epidemiology of ocular herpes simplex. Natural history in Rochester, Minn, 1950 through 1982. Arch Ophthalmol 107:1160–1165. https://doi.org/10.1001/archopht.1989.01070020226030

Liesegang TJ (1991) Diagnosis and therapy of herpes zoster ophthalmicus. Ophthalmology 98: 1216–1229. https://doi.org/10.1016/s0161-6420(91)32163-8

Liesegang TJ (1992) Biology and molecular aspects of herpes simplex and varicell-zoster virus infections. Ophthalmology 99:781–799. https://doi.org/10.1016/s0161-6420(92)31921-9

Liesegang TJ (2001) Herpes simplex virus epidemiology and ocular importance. Cornea 20:1–13. https://doi.org/10.1097/00003226-200101000-00001

Liesegang TJ, Melton LJ 3rd, Daly PJ et al (1989) Epidemiology of ocular herpes simplex. Incidence in Rochester, Minn, 1950 through 1982. Arch Ophthalmol 107:1155–1159. https://doi.org/10.1001/archopht.1989.01070020221029

Lobo AM, Agelidis AM, Shukla D (2019) Pathogenesis of herpes simplex keratitis: the host cell response and ocular surface sequelae to infection and inflammation. Ocul Surf 2019(17):40–49. https://doi.org/10.1016/j.jtos.2018.10.002

Lucca JA, Farris RL, Bielory L et al (1990) Keratoconjunctivitis sicca in male patients infected with human immunodeficiency virus type 1. Ophthalmology 97:1008–1010. https://doi.org/10.1016/s0161-6420(90)32472-7

Malm G, Berg U, Forsgren M (1995) Neonatal herpes simplex: clinical findings and outcome in relation to type of maternal infection. Acta Paediatr 84:256–260. https://doi.org/10.1111/j.1651-2227.1995.tb13625.x

Margolis TP, Atherton SS (1996) Herpes simplex virus diseases: posterior segment of the eye. Ocular Infection and Immunity. Mosby, St. Louis

Marsh RJ, Cooper M (1989) Ocular surgery in ophthalmic zoster. Eye (Lond) 3(Pt 3):313–317. https://doi.org/10.1038/eye.1989.45

Marsh RJ, Cooper M (1993) Ophthalmic herpes zoster. Eye (Lond) 7(Pt 3):350–370. https://doi.org/10.1038/eye.1993.74

Martin DF, Parks DJ, Mellow SD et al (1994) Treatment of cytomegalovirus retinitis with an intraocular sustained-release ganciclovir implant. A randomized controlled clinical trial. Arch Ophthalmol 112:1531–1539. https://doi.org/10.1001/archopht.1994.01090240037023

Matoba AY, Wilhelmus KR, Jones DB (1986) Epstein-Barr viral stromal keratitis. Ophthalmology 93:746–751. https://doi.org/10.1016/s0161-6420(86)33668-6

McCann JD, Margolis TP, Wong MG et al (1995) A sensitive and specific polymerase chain reaction-based assay for the diagnosis of cytomegalovirus retinitis. Am J Ophthalmol 120:219–226. https://doi.org/10.1016/s0002-9394(14)72610-8

Mihara T, Nishimura Y, Shimizu Y et al (2016) Linking virus genomes with host taxonomy. Viruses 8:66. https://doi.org/10.3390/v8030066

Mirzaei MK, Maurice CF (2017) Menage a trois in the human gut: interactions between host, bacteria and phages. Nat Rev Microbiol 15:397–408. https://doi.org/10.1038/nrmicro.2017.30

Mirzaei KM, Xue J, Costa R et al (2021) Challenges of studying the human virome—relevant emerging technologies. Trends Microbiol 29:171–181. https://doi.org/10.1016/j.tim.2020.05.021

Mitchell SM, Phylactou L, Fox JD et al (1996) The detection of herpesviral DNA in aqueous fluid samples from patients with Fuchs' heterochromic cyclitis. Ocul Immunol Inflamm 4:33–38. https://doi.org/10.3109/09273949609069125

Miyagawa H, Yamanishi K (1999) The epidemiology and pathogenesis of infections caused by the high numbered human herpesviruses in children: HHV-6, HHV-7 and HHV-8. Curr Opin Infect Dis 12:251–255. https://doi.org/10.1097/00001432-199906000-00016

Mondino BJ, Brown SI, Mondzelewski JP (1978) Peripheral corneal ulcers with herpes zoster ophthalmicus. Am J Ophthalmol 86:611–614. https://doi.org/10.1016/0002-9394(78)90176-9

Morrow GL, Abbott RL (1998) Conjunctivitis. Am Fam Physician 57:735–746

Munang'andu HM, Mugimba KK, Byarugaba DK et al (2017) Current advances on virus discovery and diagnostic role of viral metagenomics in aquatic organisms. Front Microbiol 8:406. https://doi.org/10.3389/fmicb.2017.00406

Musch DC, Martin DF, Gordon JF et al (1997) Treatment of cytomegalovirus retinitis with a sustained-release ganciclovir implant. The Ganciclovir Implant Study Group. N Engl J Med 337:83–90. https://doi.org/10.1056/NEJM199707103370203

Nahass GT, Goldstein BA, Zhu WY et al (1992) Comparison of Tzanck smear, viral culture, and DNA diagnostic methods in detection of herpes simplex and varicella-zoster infection. JAMA 268:2541–2544

Nahmias AJ, Visintine AM, Caldwell DR et al (1976) Eye infections with herpes simplex viruses in neonates. Surv Ophthalmol 21:100–105. https://doi.org/10.1016/0039-6257(76)90086-2

Nesburn AB, Borit A, Pentelei-Molnar J et al (1974) Varicella dendritic keratitis. Invest Ophthalmol 13:764–770

Niederman J (1994) Infectious mononucleosis. In: Infectious diseases, 5th edn. Lippincott, Philadelphia

Norn MS (1970) Dendritic (herpetic) keratitis. I. Incidence—seasonal variations—recurrence rate—visual impairment—therapy. Acta Ophthalmol 48:91–107. https://doi.org/10.1111/j.1755-3768.1970.tb06577.x

O'Brien JJ, Campoli-Richards DM (1989) Acyclovir. An updated review of its antiviral activity, pharmacokinetic properties and therapeutic efficacy. Drugs 37:233–309. https://doi.org/10.2165/00003495-198937030-00002

Olsen TW, Hardten DR, Meiusi RS et al (1994) Linear endotheliitis. Am J Ophthalmol 117:468–474. https://doi.org/10.1016/s0002-9394(14)70006-6

Paez-Espino D, Roux S, Chen IA et al (2019) IMG/VR v.2.0: an integrated data management and analysis system for cultivated and environmental viral genomes. Nucleic Acids Res 47:D678–D686. https://doi.org/10.1093/nar/gky1127

Pavan-Langston D (1994) Viral disease of the cornea and external eye. In: System of ophthalmology. WB Saunders, Philadelphia

Pepose JS (1997) The potential impact of the varicella vaccine and new antivirals on ocular disease related to varicella-zoster virus. Am J Ophthalmol 123:243–251. https://doi.org/10.1016/s0002-9394(14)71042-6

Pettit TH, Holland GN (1979) Chronic keratoconjunctivitis associated with ocular adenovirus infection. Am J Ophthalmol 88:748–751. https://doi.org/10.1016/0002-9394(79)90677-9

Porter SM, Patterson A, Kho P (1990) A comparison of local and systemic acyclovir in the management of herpetic disciform keratitis. Br J Ophthalmol 74:283–285. https://doi.org/10.1136/bjo.74.5.283

Pramod NP, Rajendran P, Kannan KA et al (1999) Herpes simplex keratitis in South India: clinico-virological correlation. Jpn J Ophthalmol 43:303–307. https://doi.org/10.1016/s0021-5155(99)00026-x

Reijo A, Antti V, Jukka M (1983) Endothelial cell loss in herpes zoster keratouveitis. Br J Ophthalmol 67:751–754. https://doi.org/10.1136/bjo.67.11.751

Remeijer L, Doornenbal P, Geerards AJ et al (1997) Newly acquired herpes simplex virus keratitis after penetrating keratoplasty. Ophthalmology 104:648–652. https://doi.org/10.1016/s0161-6420(97)30257-7

Ren J, Song K, Deng C et al (2020) Identifying viruses from metagenomic data using deep learning. Quant Biol 8:64–77. https://doi.org/10.1007/s40484-019-0187-4

Reyes A, Semenkovich NP, Whiteson K et al (2012) Going viral: next-generation sequencing applied to phage populations in the human gut. Nat Rev Microbiol 10:607–617. https://doi.org/10.1038/nrmicro2853

Reynolds JD, Griebel M, Mallory S et al (1986) Congenital herpes simplex retinitis. Am J Ophthalmol 102:33–36. https://doi.org/10.1016/0002-9394(86)90205-9

Ribaric V (1976) The incidence of herpetic keratitis among population. Ophthalmologica 173:19–22. https://doi.org/10.1159/000307814

Ritterband DC, Friedberg DN (1998) Virus infections of the eye. Rev Med Virol 8:187–201. https://doi.org/10.1002/(sici)1099-1654(199810)8:4<187::aid-rmv221>3.0.co;2-s

Robb RM, Watters GV (1970) Ophthalmic manifestations of subacute sclerosing panencephalitis. Arch Ophthalmol 83:426–435. https://doi.org/10.1001/archopht.1970.00990030426007

Robin JB, Steigner JB, Kaufman HE (1985) Progressive herpetic corneal endotheliitis. Am J Ophthalmol 100:336–337. https://doi.org/10.1016/0002-9394(85)90811-6

Rosenwasser GO, Greene WH (1992) Simultaneous herpes simplex types 1 and 2 keratitis in acquired immunodeficiency syndrome. Am J Ophthalmol 113:102–103. https://doi.org/10.1016/s0002-9394(14)75764-2

Roux S, Enault F, Hurwitz BL et al (2015) VirSorter: mining viral signal from microbial genomic data. PeerJ 3:e985. https://doi.org/10.7717/peerj.985

Roux S, Adriaenssens EM, Dutilh BE et al (2019) Minimum information about an uncultivated virus genome (MIUViG). Nat Biotechnol 37:29–37. https://doi.org/10.1038/nbt.4306

Safrin S (1992) Treatment of acyclovir-resistant herpes simplex virus infections in patients with AIDS. J Acquir Immune Defic Syndr (1988) 5(Suppl 1):S29–S32

Safrin S, Cherrington J, Jaffe HS (1997) Clinical uses of cidofovir. Rev Med Virol 7:145–156. https://doi.org/10.1002/(sici)1099-1654(199709)7:3<145::aid-rmv196>3.0.co;2-0

Scheie HG, Morse PH (1972) Rubeola retinopathy. Arch Ophthalmol 88:341–344. https://doi.org/10.1001/archopht.1972.01000030343025

Schlingemann RO, Bruinenberg M, Wertheim-van Dillen P et al (1996) Twenty years' delay of fellow eye involvement in herpes simplex virus type 2-associated bilateral acute retinal necrosis syndrome. Am J Ophthalmol 122:891–892. https://doi.org/10.1016/s0002-9394(14)70390-3

Schwab IR (1988) Oral acyclovir in the management of herpes simplex ocular infections. Ophthalmology 95:423–430. https://doi.org/10.1016/s0161-6420(88)33156-8

Sharma A, Taniguchi J (2017) Review: emerging strategies for antimicrobial drug delivery to the ocular surface: implications for infectious keratitis. Ocul Surf 15:670–679. https://doi.org/10.1016/j.jtos.2017.06.001

Sheikh A, Hurwitz B (2006) Antibiotics versus placebo for acute bacterial conjunctivitis. Cochrane Database Syst Rev:CD001211. https://doi.org/10.1002/14651858.CD001211.pub2

Shimomura Y, Ohashi Y, Maeda N et al (1987) Herpetic keratitis therapy to reduce recurrence. Curr Eye Res 6:105–110. https://doi.org/10.3109/02713688709020076

Shinkai M, Bozzette SA, Powderly W et al (1997) Utility of urine and leukocyte cultures and plasma DNA polymerase chain reaction for identification of AIDS patients at risk for developing human cytomegalovirus disease. J Infect Dis 175:302–308. https://doi.org/10.1093/infdis/175.2.302

Shkoporov AN, Hill C (2019) Bacteriophages of the human gut: the "known unknown" of the microbiome. Cell Host Microbe 25:195–209. https://doi.org/10.1016/j.chom.2019.01.017

Shkoporov AN, Ryan FJ, Draper LA et al (2018) Reproducible protocols for metagenomic analysis of human faecal phageomes. Microbiome 6:68. https://doi.org/10.1186/s40168-018-0446-z

Short GA, Margolis TP, Kuppermann BD et al (1997) A polymerase chain reaction-based assay for diagnosing varicella-zoster virus retinitis in patients with acquired immunodeficiency syndrome. Am J Ophthalmol 123:157–164. https://doi.org/10.1016/s0002-9394(14)71031-1

Shuster JJ, Kaufman HE, Nesburn AB (1981) Statistical analysis of the rate of recurrence of herpesvirus ocular epithelial disease. Am J Ophthalmol 91:328–331. https://doi.org/10.1016/0002-9394(81)90285-3

Skinner GR, Davies J, Ahmad A et al (1996) An outbreak of herpes rugbiorum managed by vaccination of players and sociosexual contacts. J Infect 33:163–167. https://doi.org/10.1016/s0163-4453(96)92129-5

Sobol EK, Fargione RA, Atiya M et al (2016) Case-control study of herpes simplex eye disease: Bronx epidemiology of human immunodeficiency virus eye studies. Cornea 35:801–806. https://doi.org/10.1097/ICO.0000000000000814

Spaide RF, Martin DF, Teich SA et al (1996) Successful treatment of progressive outer retinal necrosis syndrome. Retina 16:479–487. https://doi.org/10.1097/00006982-199616060-00003

Stapleton F, Dart JK, Seal DV et al (1995) Epidemiology of Pseudomonas aeruginosa keratitis in contact lens wearers. Epidemiol Infect 114:395–402. https://doi.org/10.1017/s0950268800052109

Stewart JA, Reef SE, Pellett PE et al (1995) Herpesvirus infections in persons infected with human immunodeficiency virus. Clin Infect Dis 21(Suppl 1):S114–S120. https://doi.org/10.1093/clinids/21.supplement_1.s114

Stone KM, Brooks CA, Guinan ME et al (1989) National surveillance for neonatal herpes simplex virus infections. Sex Transm Dis 16:152–156. https://doi.org/10.1097/00007435-198907000-00007

Sullivan V, Biron KK, Talarico C et al (1993) A point mutation in the human cytomegalovirus DNA polymerase gene confers resistance to ganciclovir and phosphonylmethoxyalkyl derivatives. Antimicrob Agents Chemother 37:19–25. https://doi.org/10.1128/AAC.37.1.19

Sundmacher R (1978) Trifluorothymidin-prophylaxis of dendritic keratitis in steroid-treated herpetic keratouveitis (author's transl). Klin Monbl Augenheilkd 173:516–519

Sutton TDS, Hill C (2019) Gut bacteriophage: current understanding and challenges. Front Endocrinol (Lausanne) 10:784. https://doi.org/10.3389/fendo.2019.00784

Tanner OR (1954) Ocular manifestations of infectious mononucleosis. AMA Arch Ophthalmol 51: 229–241. https://doi.org/10.1001/archopht.1954.00920040231010

Tay-Kearney ML, Jabs DA (1996) Treatment of CMV retinopathy in patients with AIDS. Ocul Immunol Inflamm 4:113–118. https://doi.org/10.3109/09273949609079641

Thannesberger J, Hellinger HJ, Klymiuk I et al (2017) Viruses comprise an extensive pool of mobile genetic elements in eukaryote cell cultures and human clinical samples. FASEB J 31: 1987–2000. https://doi.org/10.1096/fj.201601168R

Thompson WS, Culbertson WW, Smiddy WE et al (1994) Acute retinal necrosis caused by reactivation of herpes simplex virus type 2. Am J Ophthalmol 118:205–211. https://doi.org/10.1016/s0002-9394(14)72900-9

Tookey P, Peckham CS (1996) Neonatal herpes simplex virus infection in the British Isles. Paediatr Perinat Epidemiol 10:432–442. https://doi.org/10.1111/j.1365-3016.1996.tb00070.x

Uchio E, Hatano H, Mitsui K et al (1994) A retrospective study of herpes simplex keratitis over the last 30 years. Jpn J Ophthalmol 38:196–201

Ukkonen P, Hovi T, von Bonsdorff CH et al (1984) Age-specific prevalence of complement-fixing antibodies to sixteen viral antigens: a computer analysis of 58,500 patients covering a period of eight years. J Med Virol 13:131–148. https://doi.org/10.1002/jmv.1890130204

Wallace AD, Wendt GA, Barcellos LF et al (2018) To ERV is human: a phenotype-wide scan linking polymorphic human endogenous retrovirus-K insertions to complex phenotypes. Front Genet 9:298. https://doi.org/10.3389/fgene.2018.00298

Wang D (2020) 5 challenges in understanding the role of the virome in health and disease. PLoS Pathog 16:e1008318. https://doi.org/10.1371/journal.ppat.1008318

Watson S, Cabrera-Aguas M, Khoo P (2018) Common eye infections. Aust Prescr 41:67–72. https://doi.org/10.18773/austprescr.2018.016

Weinbauer MG (2004) Ecology of prokaryotic viruses. FEMS Microbiol Rev 28:127–181. https://doi.org/10.1016/j.femsre.2003.08.001

Whitcher JP, Dawson CR, Hoshiwara I et al (1976) Herpes simplex keratitis in a developing country. Natural history and treatment of epithelial ulcers in Tunisia. Arch Ophthalmol 94: 587–592. https://doi.org/10.1001/archopht.1976.03910030279006

Whitcher JP, Srinivasan M, Upadhyay MP (2001) Corneal blindness: a global perspective. Bull World Health Organ 79:214–221

Whitcup SM, Fortin E, Nussenblatt RB et al (1997) Therapeutic effect of combination antiretroviral therapy on cytomegalovirus retinitis. JAMA 277:1519–1520. https://doi.org/10.1001/jama.1997.03540430031027

White ML, Chodosh J (2014) Herpes simplex virus keratitis: a treatment guideline. Hoskins Center for Quality Eye Care and American Academy of Ophthalmology

Whitley RJ (1990) Herpes simplex virus. Infectious diseases of the fetus and newborn. W.B. Saunders, Philadelphia

Whitley RJ, Nahmias AJ, Visintine AM et al (1980) The natural history of herpes simplex virus infection of mother and newborn. Pediatrics 66:489–494

Wilhelmus KR (1981) Ocular involvement in infectious mononucleosis. Am J Ophthalmol 91:117–118. https://doi.org/10.1016/0002-9394(81)90358-5

Wilhelmus KR (1987) Diagnosis and management of herpes simplex stromal keratitis. Cornea 6: 286–291. https://doi.org/10.1097/00003226-198706040-00011

Wilhelmus KR (1998) Epidemiology of ocular infections vol 2. Duane's foundations of clinical ophthalmology. Lippincott Williams & Wilkins, Philadelphia

Wilhelmus KR, Coster DJ, Donovan HC et al (1981a) Prognostic indicators of herpetic keratitis. Analysis of a five-year observation period after corneal ulceration. Arch Ophthalmol 99:1578–1582. https://doi.org/10.1001/archopht.1981.03930020452009

Wilhelmus KR, Falcon MG, Jones BR (1981b) Bilateral herpetic keratitis. Br J Ophthalmol 65: 385–387. https://doi.org/10.1136/bjo.65.6.385

Wilhelmus KR, Gee L, Hauck WW et al (1994) Herpetic eye disease study. A controlled trial of topical corticosteroids for herpes simplex stromal keratitis. Ophthalmology 101:1883–1895.; discussion 1895-1886. https://doi.org/10.1016/s0161-6420(94)31087-6

Wishart PK, James C, Wishart MS et al (1984) Prevalence of acute conjunctivitis caused by chlamydia, adenovirus, and herpes simplex virus in an ophthalmic casualty department. Br J Ophthalmol 68:653–655. https://doi.org/10.1136/bjo.68.9.653

Wishart MS, Darougar S, Viswalingam ND (1987) Recurrent herpes simplex virus ocular infection: epidemiological and clinical features. Br J Ophthalmol 71:669–672. https://doi.org/10.1136/bjo.71.9.669

Wong KW, D'Amico DJ, Hedges TR 3rd et al (1987) Ocular involvement associated with chronic Epstein-Barr virus disease. Arch Ophthalmol 105:788–792. https://doi.org/10.1001/archopht.1987.01060060074036

Wood DE, Salzberg SL (2014) Kraken: ultrafast metagenomic sequence classification using exact alignments. Genome Biol 15:R46. https://doi.org/10.1186/gb-2014-15-3-r46

Yamamoto S, Pavan-Langston D, Tada R et al (1995) Possible role of herpes simplex virus in the origin of Posner-Schlossman syndrome. Am J Ophthalmol 119:796–798. https://doi.org/10.1016/s0002-9394(14)72788-6

Yoshida M, Hosogai M, Yokokura S et al (2019) Bilateral necrotizing herpes simplex keratitis in an immunocompetent patient with genetic analysis of herpes simplex virus 1. Cornea 38:1185–1188. https://doi.org/10.1097/ICO.0000000000002026

Young TL, Robin JB, Holland GN et al (1989) Herpes simplex keratitis in patients with acquired immune deficiency syndrome. Ophthalmology 96:1476–1479. https://doi.org/10.1016/s0161-6420(89)32706-0

Chapter 8
Virome of the Healthy Human Eye

Abstract Diversity of the virome at different niches of the human body like the gastrointestinal tract, urogenital tract, respiratory tract, etc. is discussed with respect to their identity and abundance and the development of the virome with age. This served as a background to the eye virome which highlights the conjunctival and ocular fluid viromes and the factors that may alter their virome.

Keywords Virome · Normal eye · Age · Conjunctiva · Ocular fluid

8.1 Introduction

The human body harbours bacteria, fungi and viruses which together are termed the microbiome. The least represented are the viruses collectively termed the 'virome'. It is estimated that approximately 10^{13} virus particles are associated with a human being. The virome varies in number and heterogeneity depending on the particular niche (different body sites) the virome occupies. Human virome comprises both the eukaryotic and prokaryotic viruses which impact human health: the former viruses cause acute or chronic infections in humans and could have serious or fatal consequences whereas the later affect the diversity and function of the bacterial communities that make up the human microbiome and thus indirectly affect human health (Raghavendra and Pullaiah 2018; Minot et al. 2012, 2013). Despite the advances in identifying viruses using sequence data, a vast majority of the viral sequences remain unidentified in available databases and are referred to as the 'dark matter' (Liang and Bushman 2021; Aggarwala et al. 2017; Shkoporov and Hill 2019). Viruses are commonly colonized in the human gastrointestinal tract, and the virome is highly variable in healthy individuals. Several factors such as anatomical location, diet, age, geographic location and disease status of the individual sampled influence the virome (Liang and Bushman 2021). This chapter would focus briefly on the virome of the healthy individuals at different sites of the body and highlight the virome of the healthy human eye.

S. Shivaji, *Human Ocular Microbiome*,
https://doi.org/10.1007/978-981-19-1754-7_8

8.2 Diversity of the Human Virome

The human virome has a preponderance of tailed phages represented by *Caudovirales* and the icosahedral non-tailed phages (*Microviridae*), and these phages could be lytic phages infecting host cells, producing viral particles and lysing the host cells and eventually the viral particles are dispersed. Unlike the lytic phages, the lysogenic phages integrate their genomes with the host genome cells and get integrated into the bacterial chromosome, and this prophage is quiescent initially, but under specific conditions the phage genome gets excised and exhibits lytic growth (Jacob et al. 1962). In pseudo-lysogeny, the phage injects its genome into the bacterial cell, but unlike in lysogeny it does not exhibit lytic growth. Some phages like the filamentous phages affiliated to the family *Inoviridae* infect host cells and undergo budding to produce new phage particles but preserve the infected cell host. A few of these phages could cause acute infections, whereas others could engage in benign colonization and associate as commensals. These commensals have been studied in healthy individuals, and some new lineages of viral commensals affiliated to the family *Anelloviridae* (such as torque teno virus, torque teno midi virus and torque teno mini virus) (Nishizawa et al. 1997; Takahashi et al. 2000; Ninomiya et al. 2007) and *Redondoviridae*, a common resident of the respiratory tract, were discovered (Abbas et al. 2019; Spezia et al. 2020; Lázaro-Perona et al. 2020). These viruses in the gut exhibit high inter-individual variation but ~80 to 90% of viral contigs were relatively stable over time (1–2.5 years) in an individual (Minot et al. 2011; Shkoporov and Hill 2019). The oral virome also revealed similar stability (Abeles et al. 2014, 2015).

8.3 Virome of Different Body Sites

The ubiquitous distribution of viruses on the human body led to several studies which enumerated the viral abundance and diversity at specific sites of the body. The variation in diversity and abundance of phages at different anatomical sites of the human body was attributed to host bacteria and eukaryotic cells present at the different sites. Some notable site-specific features of the virome of the gastrointestinal tract, oral cavity, respiratory tract, skin, blood, semen, urinogenital system, nervous system, etc. are distinct and summarized below in Fig. 8.1 and in several reviews (Popgeorgiev et al. 2013; Liang and Bushman 2021). The diverse viruses of the human body may have a role in shaping the microbial diversity in the human beings.

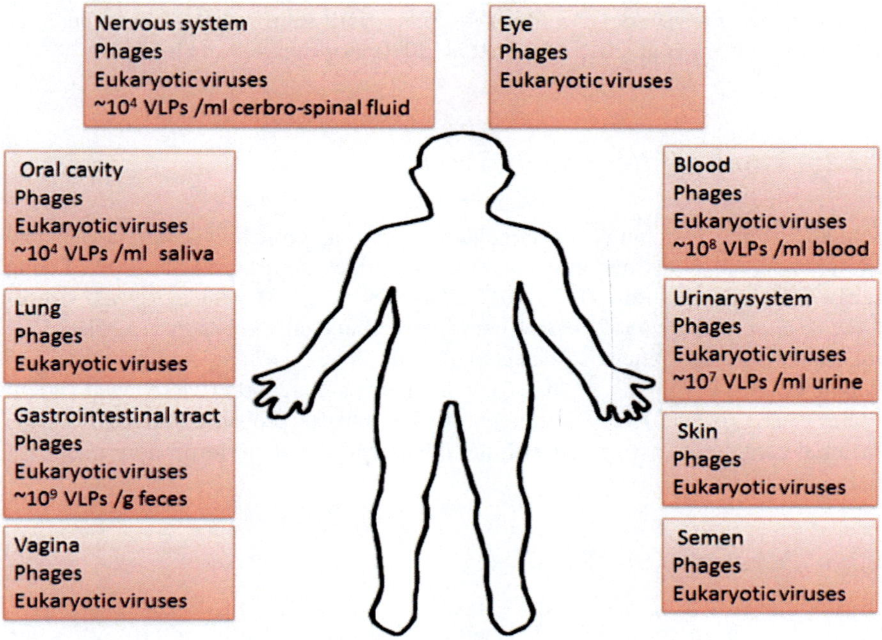

Fig. 8.1 Phages and eukaryotic viruses at different body sites. (Modified from Liang and Bushman 2021; Zhang et al. 2006; Aggarwala et al. 2017; Shkoporov and Hill 2019; Young et al. 2015; Abbas et al. 2019; Spezia et al. 2020; Lázaro-Perona et al. 2020; Mirzaei and Maurice 2017; Abeles et al. 2014, 2015; Dutilh et al. 2014; Yutin et al. 2018; Shkoporov et al. 2018; Sutton and Hill 2019; Robles-Sikisaka et al. 2013; Abbas et al. 2017; Naidu et al. 2014; Baker et al. 2017; Pride et al. 2012; Breitbart and Rohwer 2005; Moustafa et al. 2017; Castillo et al. 2019; Foulongne et al. 2012; Hannigan; et al. 2015; Tirosh et al. 2018; Santiago-Rodriguez et al. 2015; Garretto et al. 2019; Jakobsen et al. 2020; Ghose et al. 2019; Lamade and Strank 2012; McGavern and Kang 2011 and Popgeorgiev et al. 2013)

8.3.1 Virome of Gastrointestinal Tract

The virome of the gastrointestinal tract (GI) is the most colonized site (~10^9VLPs per gram of intestinal contents) with phages belonging to the order *Caudovirales* being the most predominant, and the next abundant virus included the spherical *Microviridae* (Aggarwala et al. 2017; Shkoporov and Hill 2019; Mirzaei and Maurice 2017; Virgin 2014). Cross-assembly phages (crAssphage) affiliated to the *Caudovirales* are global in their distribution and predominantly prevalent in >50% of human gut samples (Dutilh et al. 2014; Guerin et al. 2018; Edwards et al. 2019). crAssphages replicate via pseudo-lysogeny. In the healthy human gut, eukaryotic viruses are relatively low in abundance and include viruses affiliated to *Anelloviridae*, *Herpesviridae*, *Geminiviridae*, *Papillomaviridae*, *Nanoviridae*, *Parvoviridae*, *Adenoviridae*, *Polyomaviridae* and *Circoviridae* (Rascovan et al. 2016; Popgeorgiev et al. 2013). The common RNA viruses in the GI tract are

Caliciviridae, *Reoviridae* and *Picornaviridae* and some plant viruses such as *Virgaviridae* (Lim et al. 2015; Liang et al. 2020; Popgeorgiev et al. 2013).

8.3.2 Virome of Respiratory Tract

Samples of the respiratory tract including sputum, bronchoalveolar lavage and nasopharyngeal swabs harboured viral communities (Willner et al. 2009; Young et al. 2015; Abbas et al. 2017, 2019; Clarke et al. 2018) with the DNA viruses *Anelloviridae* and *Redondoviridae* being very abundant (Abbas et al. 2019). The eukaryotic viruses frequently detected include *Herpesviridae*, *Adenoviridae* and *Papillomaviridae* (Young et al. 2015; Willner et al. 2009; Clarke et al. 2018; Abbas et al. 2017, 2019). Phages are common and are probably derived from the bacterial community associated with the mouth and the upper respiratory tract.

8.3.3 Virome of the Blood

Viruses have been consistently detected in good numbers in blood (10^8 VLPs per millilitre of blood) (Breitbart and Rohwer 2005; Moustafa et al. 2017; Liu et al. 2018). Sequencing of the blood DNA from 8240 healthy individuals free from infectious diseases detected viral sequences that mapped to 94 different viruses, which included DNA viruses, proviruses and RNA viruses (herpesviruses, papillomaviruses, anelloviruses, adenoviruses, three polyomaviruses, hepatitis B, hepatitis C, HIV, HTLV, influenza virus and parvovirus B19) in 42% of the study participants (Moustafa et al. 2017). Among the *Anelloviridae*, the torque teno viruses (TTVs) were common (Popgeorgiev et al. 2013). Parvoviruses were also detected in the plasma of healthy blood donors (Fryer et al. 2007). Several insect viruses affiliated to dengue virus have also been identified in the blood of apparently healthy individuals (Dias et al. 2012). Viruses affiliated to *Marseilleviridae*, *Mimiviridae* and *Phycodnaviridae* in blood may represent environmental contamination (Castillo et al. 2019; Zárate et al. 2017) and probably enter systemic circulation by crossing the gut epithelial cell layers (Nguyen et al. 2017).

8.3.4 Virome of the Skin

Several eukaryotic viruses affiliated to *Circoviridae*, *Papillomaviridae* and *Polyomaviridae* have been reported based on metagenomic analysis of skin swabs (Foulongne et al. 2012). Further, a DNA virome study indicated that majority of the viral sequences (~95%) could not be identified to a known viral genome, and some of the sequences were affiliated with *Caudovirales* (Hannigan et al. 2015). The skin also harboured *Circoviridae*, *Papillomaviridae*, *Adenoviridae*, *Herpesviridae*, Anelloviridae and *Polyomaviridae* which are eukaryotic viruses (Hannigan et al. 2015). These eukaryotic viruses were also observed on the skin of both healthy individuals and individuals with primary immunodeficiencies (Tirosh et al. 2018).

8.3.5 Virome of the Urogenital System

Urine samples from healthy humans contain viruses (10^7 VLPs per millilitre of urine) (Santiago-Rodriguez et al. 2015). In some cohorts of individuals, the human papillomaviruses were the most predominant and were identified in >90% of subjects (Garretto et al. 2019; Santiago-Rodriguez et al. 2015). In the healthy human vaginal samples, majority of the viral sequences were affiliated to double-stranded DNA phages whereas only 4% of total sequenced reads were identified as eukaryotic viruses (Jakobsen et al. 2020). In the seminal plasma, *Anelloviridae*, *Herpesviridae* and several genotypes of *Papillomaviridae* were detected (Li et al. 2020).

8.3.6 Virome of the Nervous System

In the nervous system of healthy humans, phages affiliated to *Myoviridae*, *Siphoviridae*, *Podoviridae* and *Herpesviridae* (Ghose et al. 2019) are the most predominant. The VLP number was ~10^4 per millilitre of cerebrospinal fluid (Ghose et al. 2019). Human cytomegalovirus, varicella zoster virus and herpes simplex viruses could remain latent in the central nervous system but get activated subsequently and produce viral particles (Lamade and Strank 2012; McGavern and Kang 2011).

8.3.7 Viromes of Other Tissues

In a comprehensive study using RNA-Seq, data was generated from 51 somatic tissues of 547 individuals and analysed (Kumata et al. 2020). Consistent with several earlier studies, human herpesviruses, including EBV; herpes simplex virus type 1; cytomegalovirus; human herpesvirus 6A, 6B and 7; varicella zoster virus; HCV and torque teno virus, were detected. Some of these viruses (e.g. EBV and human papilloma virus were detected in a broad range of tissues, whereas HCV (spleen) and GBV (breast and blood) were detected in a tissue-specific manner. In addition, other non-human viruses like a plant virus (e.g. tomato spotted wilt virus) and insect viruses (e.g. deformed wing virus) are associated as contaminants. DNA viruses were dominant, and it was suggested that the DNA viruses modulate the healthy human virome at the transcriptome level (Kumata et al. 2020).

8.3.8 The Human Megavirome

Viruses with large dsDNA genomes are known as giant viruses (Colson et al. 2012; Popgeorgiev et al. 2013). Giant viruses like *Senegalvirus* and *Mimivirus* have been detected in the human gut (Breitbart et al. 2008; Reyes et al. 2010; Lagier et al. 2012; Colson et al. 2013) and stool sample (Lagier et al. 2012) and the giant blood *marseillevirus* (GBM) in human blood. Although these viruses are found in non-pathological conditions, there is a need to understand the consequences of long-term persistence of the giant viruses.

8.4 Development of the Human Virome

The stepwise assembly of the human virome from birth into adulthood was surveyed, and it was established that viruses are absent in the gut of healthy neonates. By 17 h after birth, the gut of the neonates is colonised by VLPs (Liang et al. 2020; Liang and Bushman 2021), and the bacteria are the source of prophages in the gut. By month 4, the baby has more viruses in the gut, and these colonizers were phages affiliated to *Myoviridae*, *Podoviridae*, *Microviridae* and *Siphoviridae* (Liang et al. 2020; Lim et al. 2015). It has also been observed that breast feeding alters the types of phages coinciding with the alteration in the bacteria in the infant gut through maternal immune cells, maternal antibodies, lactoferrin, milk oligosaccharides, gangliosides and mucin (Drew et al. 1981; Liang et al. 2020; Turin and Ochoa 2014; Berlutti et al. 2011; Wakabayashi et al. 2014; Pou et al. 2019; Albrecht and Arck 2020; Wiciński et al. 2020). Based on a human Gut Virome Database (GVD) of 1986 individuals representing 16 countries, Gregory et al. (2020) revealed age-dependent patterns in the viromes across healthy Westerners. The GVD identified 33,242 unique viruses, and meta-analyses showed highly personalized viromes and documented viral diversity changes from infancy into senescence (from 0 to 65 years old). The eukaryotic virus (mostly human *Annelloviruses*) richness is high at infancy and decreases with age. But the phages in the gut and specifically the bacteriophages *Siphoviridae* increased with age (Mirzaei and Maurice 2017). It was also observed that viral richness changed in proportion to bacterial richness trends and is age-dependent (Gregory et al. 2020).

8.5 Virome of the Eye

The human gut contains the most abundant number of viruses (~10^9 VLPs per gram of intestinal contents) compared to other body sites. These viruses may be acquired from the environment or from another region of the body as in the upper respiratory phage population contributing to the lower respiratory tract sites (Liang and

Bushman 2021). In fact, small circular DNA viruses (*Anelloviridae* and *Redondoviridae*) are commonly detected in respiratory samples, faecal samples and in the blood. It is possible that these small circular DNA viruses replicate in the GI tract and following swallowing of saliva may find their way into the gut. Viruses are also detected in sites remote from normal microbial colonization sites as in the cerebrospinal fluid (CSF). The presence of these viruses in CSF could be attributed to local viral replication, to systemic circulation of viruses or due to reagent contamination (Liang and Bushman 2021). It is in this context that it is very difficult to establish whether the eye has its own virome or is acquired from the skin or the environment. Metagenomic sequencing of the ocular surface indicated that the fungi and the viruses constituted approximately 1% of the abundance each and the vast majority (approximately 98%) of the sequencing reads were of bacterial origin (Wu et al. 2003; Wen et al. 2017). One study found that the healthy conjunctiva of 65% of subjects sampled had a positive signal for torque teno virus (Doan et al. 2016), a virus that is known to be associated with other ocular diseases like seasonal hyperacute panuveitis and culture-negative endophthalmitis (Lee et al. 2015). Studies have indicated that bacteriophages adhere to mucosal surfaces (Barr et al. 2013) and colonise all niches of the human body (Barr 2017); but the role of these viruses in eye health and disease is not clear.

8.5.1 Viruses on the Ocular Surface

BRiSK (biome representational in silico karyotyping) analysis revealed viruses on the ocular surface of healthy individuals. In this technique, all the DNA in a sample is restricted using a Type IIB restriction enzyme (BsaXI) to generate 27 bp double-stranded restriction fragments. These small DNA fragments are sequenced and used for the identification of viruses, fungi, bacteria and parasites present in a clinical sample (Doan et al. 2016). BRiSK sequencing analysis of the left lower eyelid from 80 healthy subjects revealed the presence of several bacterial genera and multiple viruses in the conjunctiva. The three viruses that could be identified included multiple sclerosis-associated retrovirus (MSRV), torque teno virus (TTV) and human endogenous retroviruses (HERV-K). But both MSRV and HERV-K are human endogenous retroviruses originating from the host DNA, and hence their clinical significance is unclear (Dewannieux and Heidmann 2013). TTV in healthy conjunctiva samples was also confirmed by quantitative PCR, and it was estimated that at least one TTV copy was present per 100 epithelial cells; gender and age had no influence. In addition to TTV, Abelson murine leukaemia virus, human papillomavirus (HPV) and Merkel cell polyomavirus (MCV) were detected in a few samples (Spurgeon et al. 2015; Palazzi et al. 2000). Thus the healthy conjunctiva harbours a resident DNA viral community in most subjects (Doan et al. 2016).

8.5.2 Ocular Surface Virome of Children

A recent study analysed ocular surface microbiome of 150 children of either gender in the age range 1–59 months. The analysis revealed several viruses (including *Allium cepa amalgavirus*, *Alphapapillomavirus*, *Avihepatovirus*, *Gammapapillomavirus*, *Betaherpesvirus*, *Gammaherpesvirus*, *Retrovirus K*, *Mastadenovirus*, *Kadipiro virus*, *Lactococcus* phage 936, *Lactococcus* phage 286, *measles virus*, *Moraxella phage*, *Ribgrass mosaic virus*, *Streptococcus* phage, *tomato brown rugose fruit virus*) representing eukaryotic viruses and bacteriophages (Doan et al. 2020).

8.5.3 Viruses in the Anophthalmic Conjunctiva

Siegal et al. (2021) reported that, in addition to TTV, MCV was frequently detected in healthy (5 of 20) and anophthalmic conjunctiva (19 of 20). Further, the copy number of MCV was significantly increased (average 891 copies/ng) in anophthalmic conjunctiva compared with the unaffected fellow eye (193 copies/ng) ($p < 0.001$) suggesting that the ocular surface resident virome was affected by enucleation and prosthesis placement. The copy number of TTV was also higher in the anophthalmic conjunctiva (41 copies/ng) compared to the contralateral conjunctiva (8 copies/ng), but difference was not statistically significant. Two other viruses that were detected include papilloma family viruses and SEN virus (a small DNA virus closely related to TTV) in the anophthalmic socket than in the contralateral socket (Siegal et al. 2021).

8.5.4 Virome of the Ocular Fluids

In the vitreous fluid of the posterior chamber of the eye, the top ten viral families detected in individuals without ocular infections included viruses of the family *Myoviridae*, *Phycodnaviridae*, *Siphoviridae*, *Herpesviridae*, *Iridoviridae*, *Poxviridae*, *Podoviridae*, *Baculoviridae*, *Retroviridae and* Flaviviridae (Arunasri et al. 2021). A total of 30 eukaryotic genera (Table 8.1) and 24 phages were identified (Table 8.2). The median abundance of the eukaryotic viruses ranged from 0.006 (*Oryzavirus*) to 2.177 (*Ranavirus*), and among the phages the median abundance ranged from 0.009 (unclassified *Inoviridae*) to 2.075 (unclassified *Podoviridae*). It is interesting that the viral families of vitreous fluids were also shared with CSF and plasma and the common families included *Phycodnaviridae*, *Myoviridae*, *Siphoviridae*, *Microviridae*, *Podoviridae*, *Herpesviridae*, *Poxviridae*, *Inoviridae* and *Papillomaviridae* (Ghose et al. 2019). In the present study *Anelloviruses*, a single-stranded DNA virus was detected in the vitreous fluids of

Table 8.1 Viral genera identified in the vitreous fluid of individuals without ocular infections[a]

Sl No.	Viral genera	Mean abundance
1	*Lymphocryptovirus*	0.488078
2	*Chloriridovirus*	0.737222
3	*Iridovirus*	2.013551
4	*Ranavirus*	2.177961
5	*Betapapillomavirus*	0.012454
6	*Deltapapillomavirus*	0.029408
7	*Rhopapillomavirus*	0.028845
8	*Ichnovirus*	0.010591
9	*Polyomavirus*	0.055518
10	*Aquabirnavirus*	0.040861
11	*Coltivirus*	0.017641
12	*Orthoreovirus*	0.048721
13	*Oryzavirus*	0.00604
14	*Circovirus*	0.025268
15	*Nanovirus*	0.013747
16	*Copiparvovirus*	0.018266
17	*Novirhabdovirus*	0.019379
18	*Nucleorhabdovirus*	0.030137
19	*Mamastrovirus*	0.096279
20	*Cilevirus*	0.015397
21	*Hepacivirus*	0.242584
22	*Enterovirus*	0.07924
23	*Kobuvirus*	0.089026
24	*Comovirus*	0.031522
25	*Allexivirus*	0.027912
26	*Potexvirus*	0.124534
27	*Carlavirus*	0.143193
28	*Pecluvirus*	0.011913
29	*Pomovirus*	0.031057
30	Unclassified *Reoviridae*	0.006582

[a]Modified from Arunasri et al. (2021)

healthy individuals (Arunasri et al. 2021). This virus has been reported as present in the human body (Focosi et al. 2016).

8.6 Factors that Shape the Ocular Virome

Factors that influence the human microbiome such as diet, age, host genetics and immunity, ethnicity, medication, cohabitation, geography and stochastics of colonization also often influence the virome (Liang and Bushman 2021) (Fig. 8.2). Viromes are also modulated by actively interacting with other microorganisms and

Table 8.2 Bacteriophage genera identified in the vitreous fluid of individuals without ocular infections[a]

Sl No.	Bacteriophage genera	Mean abundance
1	*Bcep78-like virus*	0.067
2	*Felixouna-like virus*	0.180
3	*Hpuna-like virus*	0.118
4	*T7-like virus*	0.167
5	*Phi29-like virus*	0.042
6	*3a-like virus*	0.148
7	*77-like virus*	0.169
8	*Bignuz-like virus*	0.106
9	*Che8-like virus*	0.696
10	*Cjwuna-like virus*	0.467
11	*D3112-like virus*	0.135
12	*Lambda-like virus*	0.445
13	*Tuna-like virus*	0.112
14	*Betalipothrix virus*	0.405
15	*Ino virus*	0.105
16	*Chlamydiamicrovirus*	0.013
17	Unassigned *Spounavirinae*	0.634
18	Unclassified *Autographivirinae*	0.152
19	Unclassified *Picovirinae*	0.096
20	Unassigned *Podoviridae*	0.153
21	Unclassified *Podoviridae*	2.075
22	Unclassified *Caudovirales*	1.452
23	Unclassified *Inoviridae*	0.009
24	Unclassified phages	1.832

[a]Modified from Arunasri et al. (2021)

disease states of the host. It is known that ocular surface viromes change with age but the possibility that some of the above factors also modulate the virome needs to be investigated.

8.7 Conclusions

Ocular surface viromes need to be characterized in the diseased eye, and the confounding factors that influence the abundance and diversity of the eye by host factors such as ethnicity, age, gender, diet, systemic health condition, etc. need to be identified.

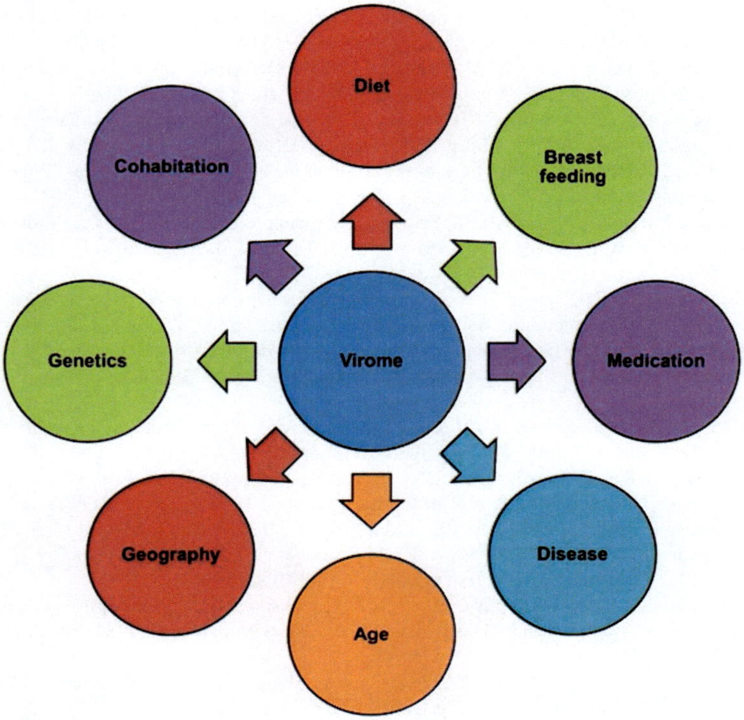

Fig. 8.2 The schematic diagram represents the major factors that shape the human virome which included breast feeding (Liang et al. 2020), diet (Minot et al. 2011), medications (Zuo et al. 2020; Liang et al. 2020), host genetics (Lim et al. 2015; Reyes et al. 2010, 2015; Clarke et al. 2018; Moreno-Gallego et al. 2019), cohabitation (Robles-Sikisaka et al. 2013), geography (Liu et al. 2018; Liang et al. 2020; Zuo et al. 2020; Rampelli et al. 2017; Gregory et al. 2020) disease and ageing (Gregory et al. 2020). (Modified from Liang and Bushman 2021)

References

Abbas AA, Diamond JM, Chehoud C et al (2017) The perioperative lung transplant virome: torque teno viruses are elevated in donor lungs and show divergent dynamics in primary graft dysfunction. Am J Transplant 17:1313–1324. https://doi.org/10.1111/ajt.14076

Abbas AA, Taylor LJ, Dothard MI et al (2019) Redondoviridae, a family of small, circular DNA viruses of the human oro-respiratory tract associated with periodontitis and critical illness. Cell Host Microbe 25(719–729):e714. https://doi.org/10.1016/j.chom.2019.04.001

Abeles SR, Robles-Sikisaka R, Ly M et al (2014) Human oral viruses are personal, persistent and gender-consistent. ISME J 8:1753–1767. https://doi.org/10.1038/ismej.2014.31

Abeles SR, Ly M, Santiago-Rodriguez TM et al (2015) Effects of long term antibiotic therapy on human oral and fecal viromes. PLoS One 10:e0134941. https://doi.org/10.1371/journal.pone.0134941

Aggarwala V, Liang G, Bushman FD (2017) Viral communities of the human gut: metagenomic analysis of composition and dynamics. Mob DNA 8:12. https://doi.org/10.1186/s13100-017-0095-y

Albrecht M, Arck PC (2020) Vertically transferred immunity in neonates: mothers, mechanisms and mediators. Front Immunol 11:555. https://doi.org/10.3389/fimmu.2020.00555

Arunasri K, Sai Prashanthi G, Tyagi M et al (2021) Intraocular viral communities associated with post-fever retinitis. Front Med 8. https://doi.org/10.3389/fmed.2021.724195

Baker JL, Bor B, Agnello M et al (2017) Ecology of the oral microbiome: beyond bacteria. Trends Microbiol 25:362–374. https://doi.org/10.1016/j.tim.2016.12.012

Barr JJ (2017) A bacteriophages journey through the human body. Immunol Rev 279:106–122. https://doi.org/10.1111/imr.12565

Barr JJ, Auro R, Furlan M et al (2013) Bacteriophage adhering to mucus provide a non-host-derived immunity. Proc Natl Acad Sci U S A 110:10771–10776. https://doi.org/10.1073/pnas.1305923110

Berlutti F, Pantanella F, Natalizi T et al (2011) Antiviral properties of lactoferrin—a natural immunity molecule. Molecules 16:6992–7018. https://doi.org/10.3390/molecules16086992

Breitbart M, Rohwer F (2005) Method for discovering novel DNA viruses in blood using viral particle selection and shotgun sequencing. BioTechniques 39:729–736. https://doi.org/10.2144/000112019

Breitbart M, Haynes M, Kelley S et al (2008) Viral diversity and dynamics in an infant gut. Res Microbiol 159:367–373. https://doi.org/10.1016/j.resmic.2008.04.006

Castillo DJ, Rifkin RF, Cowan DA et al (2019) The healthy human blood microbiome: fact or fiction? Front Cell Infect Microbiol 9:148. https://doi.org/10.3389/fcimb.2019.00148

Clarke EL, Lauder AP, Hofstaedter CE et al (2018) Microbial lineages in sarcoidosis. A metagenomic analysis tailored for low-microbial content samples. Am J Respir Crit Care Med 197:225–234. https://doi.org/10.1164/rccm.201705-0891OC

Colson P, de Lamballerie X, Fournous G et al (2012) Reclassification of giant viruses composing a fourth domain of life in the new order Megavirales. Intervirology 55:321–332. https://doi.org/10.1159/000336562

Colson P, Fancello L, Gimenez G et al (2013) Evidence of the megavirome in humans. J Clin Virol 57:191–200. https://doi.org/10.1016/j.jcv.2013.03.018

Dewannieux M, Heidmann T (2013) Endogenous retroviruses: acquisition, amplification and taming of genome invaders. Curr Opin Virol 3:646–656. https://doi.org/10.1016/j.coviro.2013.08.005

Dias LL, Amarilla AA, Poloni TR et al (2012) Detection of dengue virus in sera of Brazilian blood donors. Transfusion 52:1667–1671. https://doi.org/10.1111/j.1537-2995.2012.03729.x

Doan T, Akileswaran L, Andersen D et al (2016) Paucibacterial microbiome and resident DNA virome of the healthy conjunctiva. Invest Ophthalmol Vis Sci 57:5116–5126. https://doi.org/10.1167/iovs.16-19803

Doan T, Hinterwirth A, Worden L et al (2020) Post-antibiotic ocular surface microbiome in children: a cluster-randomized trial. Ophthalmology 127:1127–1130. https://doi.org/10.1016/j.ophtha.2020.02.014

Drew HR, Wing RM, Takano T et al (1981) Structure of a B-DNA dodecamer: conformation and dynamics. Proc Natl Acad Sci U S A 78:2179–2183. https://doi.org/10.1073/pnas.78.4.2179

Dutilh BE, Cassman N, McNair K et al (2014) A highly abundant bacteriophage discovered in the unknown sequences of human faecal metagenomes. Nat Commun 5:4498. https://doi.org/10.1038/ncomms5498

Edwards RA, Vega AA, Norman HM et al (2019) Global phylogeography and ancient evolution of the widespread human gut virus crAssphage. Nat Microbiol 4:1727–1736. https://doi.org/10.1038/s41564-019-0494-6

Focosi D, Antonelli G, Pistello M et al (2016) Torquetenovirus: the human virome from bench to bedside. Clin Microbiol Infect 22:589–593. https://doi.org/10.1016/j.cmi.2016.04.007

Foulongne V, Sauvage V, Hebert C et al (2012) Human skin microbiota: high diversity of DNA viruses identified on the human skin by high throughput sequencing. PLoS One 7:e38499. https://doi.org/10.1371/journal.pone.0038499

Fryer JF, Delwart E, Hecht FM et al (2007) Frequent detection of the parvoviruses, PARV4 and PARV5, in plasma from blood donors and symptomatic individuals. Transfusion 47:1054–1061. https://doi.org/10.1111/j.1537-2995.2007.01235.x

Garretto A, Miller-Ensminger T, Wolfe AJ et al (2019) Bacteriophages of the lower urinary tract. Nat Rev Urol 16:422–432. https://doi.org/10.1038/s41585-019-0192-4

Ghose C, Ly M, Schwanemann LK et al (2019) The virome of cerebrospinal fluid: viruses where we once thought there were none. Front Microbiol 10:2061. https://doi.org/10.3389/fmicb.2019.02061

Gregory AC, Zablocki O, Zayed AA et al (2020) The gut Virome database reveals age-dependent patterns of virome diversity in the human gut. Cell Host Microbe 28(724–740):e728. https://doi.org/10.1016/j.chom.2020.08.003

Guerin E, Shkoporov A, Stockdale SR et al (2018) Biology and taxonomy of crAss-like bacteriophages, the most abundant virus in the human gut. Cell Host Microbe 24(653–664):e656. https://doi.org/10.1016/j.chom.2018.10.002

Hannigan GD, Meisel JS, Tyldsley AS et al (2015) The human skin double-stranded DNA virome: topographical and temporal diversity, genetic enrichment, and dynamic associations with the host microbiome. mBio 6:e01578-01515. https://doi.org/10.1128/mBio.01578-15

Jacob F, Sussman R, Monod J (1962) On the nature of the repressor ensuring the immunity of lysogenic bacteria. C R Hebd Seances Acad Sci 254:4214–4216

Jakobsen RR, Haahr T, Humaidan P et al (2020) Characterization of the vaginal DNA virome in health and dysbiosis. Viruses 12(10):1143. https://doi.org/10.3390/v12101143

Kumata R, Ito J, Takahashi K et al (2020) A tissue level atlas of the healthy human virome. BMC Biol 18:55. https://doi.org/10.1186/s12915-020-00785-5

Lagier JC, Armougom F, Million M et al (2012) Microbial culturomics: paradigm shift in the human gut microbiome study. Clin Microbiol Infect 18:1185–1193. https://doi.org/10.1111/1469-0691.12023

Lamade MU, Strank C (2012) Herpesvirus infections of the central nervous system in immunocompromised patients. Ther Adv Neurol Disord 5:279–296. https://doi.org/10.1177/1756285612456234

Lázaro-Perona F, Dahdouh E, Roman-Soto S et al (2020) Metagenomic detection of two vientoviruses in a human sputum sample. Viruses 12(3):327. https://doi.org/10.3390/v12030327

Lee AY, Akileswaran L, Tibbetts MD et al (2015) Identification of torque teno virus in culture-negative endophthalmitis by representational deep DNA sequencing. Ophthalmology 122:524–530. https://doi.org/10.1016/j.ophtha.2014.09.001

Li Y, Altan E, Pilcher C et al (2020) Semen virome of men with HIV on or off antiretroviral treatment. AIDS 34:827–832. https://doi.org/10.1097/QAD.0000000000002497

Liang G, Bushman FD (2021) The human virome: assembly, composition and host interactions. Nat Rev Microbiol 19:514–527. https://doi.org/10.1038/s41579-021-00536-5

Liang G, Zhao C, Zhang H et al (2020) The stepwise assembly of the neonatal virome is modulated by breastfeeding. Nature 581:470–474. https://doi.org/10.1038/s41586-020-2192-1

Lim ES, Zhou Y, Zhao G et al (2015) Early life dynamics of the human gut virome and bacterial microbiome in infants. Nat Med 21:1228–1234. https://doi.org/10.1038/nm.3950

Liu S, Huang S, Chen F et al (2018) Genomic analyses from non-invasive prenatal testing reveal genetic associations, patterns of viral infections, and Chinese population history. Cell 175(347–359):e314. https://doi.org/10.1016/j.cell.2018.08.016

McGavern DB, Kang SS (2011) Illuminating viral infections in the nervous system. Nat Rev Immunol 11:318–329. https://doi.org/10.1038/nri2971

Minot S, Sinha R, Chen J et al (2011) The human gut virome: inter-individual variation and dynamic response to diet. Genome Res 21:1616–1625. https://doi.org/10.1101/gr.122705.111

Minot S, Grunberg S, Wu GD et al (2012) Hypervariable loci in the human gut virome. Proc Natl Acad Sci U S A 109:3962–3966. https://doi.org/10.1073/pnas.1119061109

Minot S, Bryson A, Chehoud C et al (2013) Rapid evolution of the human gut virome. Proc Natl Acad Sci U S A 110:12450–12455. https://doi.org/10.1073/pnas.1300833110

Mirzaei MK, Maurice CF (2017) Menage a trois in the human gut: interactions between host, bacteria and phages. Nat Rev Microbiol 15:397–408. https://doi.org/10.1038/nrmicro.2017.30

Moreno-Gallego JL, Chou SP, Di Rienzi SC et al (2019) Virome diversity correlates with intestinal microbiome diversity in adult monozygotic twins. Cell Host Microbe 25(261–272):e265. https://doi.org/10.1016/j.chom.2019.01.019

Moustafa A, Xie C, Kirkness E et al (2017) The blood DNA virome in 8,000 humans. PLoS Pathog 13:e1006292. https://doi.org/10.1371/journal.ppat.1006292

Naidu M, Robles-Sikisaka R, Abeles SR et al (2014) Characterization of bacteriophage communities and CRISPR profiles from dental plaque. BMC Microbiol 14:175. https://doi.org/10.1186/1471-2180-14-175

Nguyen S, Baker K, Padman BS et al (2017) Bacteriophage transcytosis provides a mechanism to cross epithelial cell layers. MBio 8(6):e01874-17. https://doi.org/10.1128/mBio.01874-17

Ninomiya M, Nishizawa T, Takahashi M et al (2007) Identification and genomic characterization of a novel human torque teno virus of 3.2 kb. J Gen Virol 88:1939–1944. https://doi.org/10.1099/vir.0.82895-0

Nishizawa T, Okamoto H, Konishi K et al (1997) A novel DNA virus (TTV) associated with elevated transaminase levels in posttransfusion hepatitis of unknown etiology. Biochem Biophys Res Commun 241:92–97. https://doi.org/10.1006/bbrc.1997.7765

Palazzi MA, Erwenne CM, Villa LL (2000) Detection of human papillomavirus in epithelial lesions of the conjunctiva. Sao Paulo Med J 118:125–130. https://doi.org/10.1590/s1516-31802000000500003

Popgeorgiev N, Temmam S, Raoult D et al (2013) Describing the silent human virome with an emphasis on giant viruses. Intervirology 56:395–412. https://doi.org/10.1159/000354561

Pou C, Nkulikiyimfura D, Henckel E et al (2019) The repertoire of maternal anti-viral antibodies in human newborns. Nat Med 25:591–596. https://doi.org/10.1038/s41591-019-0392-8

Pride DT, Salzman J, Haynes M et al (2012) Evidence of a robust resident bacteriophage population revealed through analysis of the human salivary virome. ISME J 6:915–926. https://doi.org/10.1038/ismej.2011.169

Raghavendra P, Pullaiah T (2018) Pathogen identification using novel sequencing methods. In: Raghavendra P, Pullaiah T (eds) Advances in cell and molecular diagnostics. Academic, Cambridge, pp 161–202. https://doi.org/10.1016/b978-0-12-813679-9.00007-5

Rampelli S, Turroni S, Schnorr SL et al (2017) Characterization of the human DNA gut virome across populations with different subsistence strategies and geographical origin. Environ Microbiol 19:4728–4735. https://doi.org/10.1111/1462-2920.13938

Rascovan N, Duraisamy R, Desnues C (2016) Metagenomics and the human virome in asymptomatic individuals. Annu Rev Microbiol 70:125–141. https://doi.org/10.1146/annurev-micro-102215-095431

Reyes A, Haynes M, Hanson N et al (2010) Viruses in the faecal microbiota of monozygotic twins and their mothers. Nature 466:334–338. https://doi.org/10.1038/nature09199

Reyes A, Blanton LV, Cao S et al (2015) Gut DNA viromes of Malawian twins discordant for severe acute malnutrition. Proc Natl Acad Sci U S A 112:11941–11946. https://doi.org/10.1073/pnas.1514285112

Robles-Sikisaka R, Ly M, Boehm T et al (2013) Association between living environment and human oral viral ecology. ISME J 7:1710–1724. https://doi.org/10.1038/ismej.2013.63

Santiago-Rodriguez TM, Ly M, Bonilla N et al (2015) The human urine virome in association with urinary tract infections. Front Microbiol 6:14. https://doi.org/10.3389/fmicb.2015.00014

Shkoporov AN, Hill C (2019) Bacteriophages of the human gut: the "known unknown" of the microbiome. Cell Host Microbe 25:195–209. https://doi.org/10.1016/j.chom.2019.01.017

Shkoporov AN, Khokhlova EV, Fitzgerald CB et al (2018) PhiCrAss001 represents the most abundant bacteriophage family in the human gut and infects Bacteroides intestinalis. Nat Commun 9:4781. https://doi.org/10.1038/s41467-018-07225-7

Siegal N, Gutowski M, Akileswaran L et al (2021) Elevated levels of Merkel cell polyoma virus in the anophthalmic conjunctiva. Sci Rep 11:15366. https://doi.org/10.1038/s41598-021-92642-w

Spezia PG, Macera L, Mazzetti P et al (2020) Redondovirus DNA in human respiratory samples. J Clin Virol 131:104586. https://doi.org/10.1016/j.jcv.2020.104586

Spurgeon ME, Cheng J, Bronson RT et al (2015) Tumorigenic activity of merkel cell polyomavirus T antigens expressed in the stratified epithelium of mice. Cancer Res 75:1068–1079. https://doi.org/10.1158/0008-5472.CAN-14-2425

Sutton TDS, Hill C (2019) Gut bacteriophage: current understanding and challenges. Front Endocrinol (Lausanne) 10:784. https://doi.org/10.3389/fendo.2019.00784

Takahashi K, Iwasa Y, Hijikata M et al (2000) Identification of a new human DNA virus (TTV-like mini virus, TLMV) intermediately related to TT virus and chicken anemia virus. Arch Virol 145:979–993. https://doi.org/10.1007/s007050050689

Tirosh O, Conlan S, Deming C et al (2018) Expanded skin virome in DOCK8-deficient patients. Nat Med 24:1815–1821. https://doi.org/10.1038/s41591-018-0211-7

Turin CG, Ochoa TJ (2014) The role of maternal breast milk in preventing infantile diarrhea in the developing world. Curr Trop Med Rep 1:97–105. https://doi.org/10.1007/s40475-014-0015-x

Virgin HW (2014) The virome in mammalian physiology and disease. Cell 157:142–150. https://doi.org/10.1016/j.cell.2014.02.032

Wakabayashi H, Oda H, Yamauchi K et al (2014) Lactoferrin for prevention of common viral infections. J Infect Chemother 20:666–671. https://doi.org/10.1016/j.jiac.2014.08.003

Wen X, Miao L, Deng Y et al (2017) The influence of age and sex on ocular surface microbiota in healthy adults. Invest Ophthalmol Vis Sci 58:6030–6037. https://doi.org/10.1167/iovs.17-22957

Wiciński M, Sawicka E, Gebalski J et al (2020) Human milk oligosaccharides: health benefits, potential applications in infant formulas, and pharmacology. Nutrients 12(1):266. https://doi.org/10.3390/nu12010266

Willner D, Furlan M, Haynes M et al (2009) Metagenomic analysis of respiratory tract DNA viral communities in cystic fibrosis and non-cystic fibrosis individuals. PLoS One 4:e7370. https://doi.org/10.1371/journal.pone.0007370

Wu T, Mitchell B, Carothers T et al (2003) Molecular analysis of the pediatric ocular surface for fungi. Curr Eye Res 26:33–36. https://doi.org/10.1076/ceyr.26.1.33.14253

Young JC, Chehoud C, Bittinger K et al (2015) Viral metagenomics reveal blooms of anelloviruses in the respiratory tract of lung transplant recipients. Am J Transplant 15:200–209. https://doi.org/10.1111/ajt.13031

Yutin N, Makarova KS, Gussow AB et al (2018) Discovery of an expansive bacteriophage family that includes the most abundant viruses from the human gut. Nat Microbiol 3:38–46. https://doi.org/10.1038/s41564-017-0053-y

Zárate S, Taboada B, Yocupicio-Monroy M et al (2017) Human Virome. Arch Med Res 48:701–716. https://doi.org/10.1016/j.arcmed.2018.01.005

Zhang T, Breitbart M, Lee WH et al (2006) RNA viral community in human feces: prevalence of plant pathogenic viruses. PLoS Biol 4:e3. https://doi.org/10.1371/journal.pbio.0040003

Zuo T, Sun Y, Wan Y et al (2020) Human-gut-DNA Virome variations across geography, ethnicity, and urbanization. Cell Host Microbe 28(741–751):e744. https://doi.org/10.1016/j.chom.2020.08.005

Chapter 9
Virome of the Diseased Human Eye

Abstract Next-generation sequencing approach has been used to describe the virome of the eye with ocular diseases only with respect to one disease in which the vitreous fluid virome of post-fever retinitis (PFR) patients was compared with the virome generated from the vitreous of individuals free of any ocular infection. Several viral families (68 families) and genera (213 genera) were identified in the vitreous fluid of both the cohorts. The abundance of families *Astroviridae*, *Polyomaviridae* and *Birnaviridae* was significantly different in the two cohorts of PFR and control group. Several eukaryotic viral genera (30) and bacteriophages (24) could be used to discriminate the viromes of the PFR individuals with control individuals without any ocular disease. That the viromes were different between the controls and the PFR cohorts was also confirmed by heatmap analysis, non-metric multidimensional scaling analysis, principal coordinate analysis and linear discriminating analysis. Functional analysis of the viruses based on known functions of the genera hints at a possible role for pathogenic viruses in PFR.

Keywords Virome · Diseased eye · Post-fever retinitis · Bacteriophages · Eukaryotic viral genera

9.1 Introduction

DNA or RNA viruses are known to infect eyes and cause different ocular diseases (Chaps. 7 and 8 for details). Using PCR and sequencing strategies, several viral pathogens have been reported to cause ocular diseases. These include rhinovirus, influenza virus, respiratory syncytial virus, Newcastle virus, parainfluenza virus, herpes simplex virus (HSV) and adenoviruses (Chodosh et al. 2008). Among the viruses, the HSV group exhibits a spectrum of ocular infections which includes epithelial or stromal keratitis (both in adults and neonates) (Liesegang et al. 1989), necrotizing stromal keratitis (NSK) (Thygeson 1987), immune stromal keratitis (ISK) (Liesegang et al. 1989; Wilhelmus et al. 1994) and acute retinal necrosis (ARN) (Thompson et al. 1994). HSV has also been involved in various syndromes affecting the eye like the Posner-Schlossman syndrome (Yamamoto et al. 1995), iridocorneal endothelial syndrome (Alvarado et al. 1994) and idiopathic corneal

S. Shivaji, *Human Ocular Microbiome*,
https://doi.org/10.1007/978-981-19-1754-7_9

endotheliopathy (Olsen et al. 1994; Robin et al. 1985) and has also been implicated in blepharitis (Ritterband and Friedberg 1998). HIV is also known to be associated with ocular complications like retinopathy, ocular adnexal neoplasms, neuro-ophthalmic lesions and drug-induced manifestations (Jabs et al. 2007). Other viruses that cause ocular diseases include varicella zoster virus which causes papillary conjunctivitis, dendritic keratitis or delayed immune stromal keratitis, keratouveitis (Nesburn et al. 1974; Harding et al. 1987), infectious retinitis (Engstrom et al. 1994); cytomegalovirus (CMV) which causes retinitis (Alford et al. 1990); adenovirus which causes significant ocular morbidity (Ritterband and Friedberg 1998); Epstein-Barr virus which causes neuro-ophthalmologic and anterior segment diseases (follicular conjunctivitis, dendritic epithelial keratitis and stromal keratitis) (Matoba et al. 1986; Wilhelmus 1981); and rubeola virus which causes conjunctivitis, keratitis, neuroretinitis, choroiditis, macular and optic disc oedema and optic atrophy (Robb and Watters 1970). Compared to these above studies which used PCR and specific amplicon sequencing strategies to identify viral pathogens, approaches like the virome strategy using NGS would help to identify all the viruses in a biological sample. In this context, viromes have been characterized from the skin surface (Hannigan et al. 2015), blood (Moustafa et al. 2017), breast milk (Pannaraj et al. 2018), gastrointestinal (GI) tract (Columpsi et al. 2016; Reyes et al. 2015), oral cavity (Abeles et al. 2014), cerebrospinal fluids (Ghose et al. 2019), respiratory tract (Wylie 2017), bladder (Santiago-Rodriguez et al. 2015) and vagina (Wylie et al. 2018). Viromes of the healthy individuals have also been characterized from the ocular surface from children and adults, in the anophthalmic conjunctiva and in the ocular fluids and additionally the factors that shape the healthy human ocular virome have been described (see Chap. 8). But the virome associated with only one ocular disease has been studied and described so far.

9.2 Viral Communities in the Vitreous Fluid of Patients with Post-Fever Retinitis

Post-fever retinitis (PFR), also referred to as retinitis post-febrile illness, normally occurs 2–4 weeks after febrile illness (Kawali et al. 2019). PFR is common in the Indian subcontinent, and often it is detected after viral (Unnikrishnan and Raju 2017; Lalitha et al. 2007), bacterial (Prabhushanker et al. 2017) or protozoan (Vishwanath et al. 2014) epidemics and following a viral systemic illness in immunocompetent individuals (Menia et al. 2019). Though clinical impression of retinitis and PFR is documented (Kawali et al. 2019), changes in the intraocular microbiota is scantly studied. Metagenome sequencing approach by NGS was used to unravel the virome in the vitreous of PFR patients and was compared to the vitreous in normal healthy individuals (Kotakonda et al. 2021). Both DNA and RNA viruses were extracted from the vitreous samples, amplified with random hexamers, amplicon libraries prepared and sequenced using paired-end sequencing. Sequences were annotated using viral databases from NCBI.

9.2.1 Viral Abundance in PFR

Several different groups of viruses were detected in the vitreous fluid of both control individuals and individuals with PFR (Fig. 9.1a). Among the various groups, the dsDNA viruses were the most predominant (88.1 and 87.1%) in both the cohorts (control (VC) and PFR groups), respectively. Further the dsRNA viruses, ssRNA viruses and ssDNA viruses were significantly different ($P < 0.05$) between the cohorts (Fig. 9.1b) with ssDNA viruses and ssRNA viruses increased in the diseased PFR group compared to VC group. But in the two groups, the ssRNA positive-strand viruses were similar in abundance (6.3 and 7.5 in the VC and PFR groups, respectively) in the two cohorts. The Simpson diversity index in the VC control and PFR groups was also observed to be significantly different (Kotakonda et al. 2021).

9.2.2 Discriminating Families and Genera in PFR

Comparison of the abundance at the family level indicated that Birnaviridae, Polyomaviridae and Astroviridae were significantly different between the PFR and VC groups. Further, 213 viral genera were identified, out of which, several eukaryotic viral genera (30 genera) (Table 9.1) and bacteriophages (24 genera) (Table 9.2) significantly increased in PFR compared to the VC group. The viruses that significantly increased in PFR group were *Lymphocryptovirus*, *Betapapillomavirus*, *Ichnovirus*, *Orthoreovirus*, *Circovirus*, *Nanovirus*, *Hepacivirus*, *Kobuvirus* and *Comovirus* (Table 9.1). In addition, several bacteriophages that increased in PFR included *Bcep78*-like virus, *Hpuna*-like virus, *3-alike* virus, *77*-like virus, *Lambda*-like *virus*, *Tuna*-like virus, *Inovirus*, *Chlamydiamicrovirus*, unassigned *Spounavirinae* and unclassified *Autographivirinae* (Table 9.2) (Kotakonda et al. 2021). Beta-diversity analysis of the bacteriophages and eukaryotic viruses following principal coordinate analysis and 3D non-metric multidimensional scaling (NMDS) indicated that the viromes in PFR formed a cluster distinct from the cluster formed by the control group (Fig. 9.2a–d). Heatmap analysis also discriminated the viromes of control and PFR group (Fig. 9.3a, b). The 29 eukaryotic viral genera (Fig. 9.4a) and 22 bacteriophages (Fig. 9.4b) also showed significant difference in abundance by linear discriminant analysis (LDA) combined with effect size measurements (LEfSe) (Kotakonda et al. 2021).

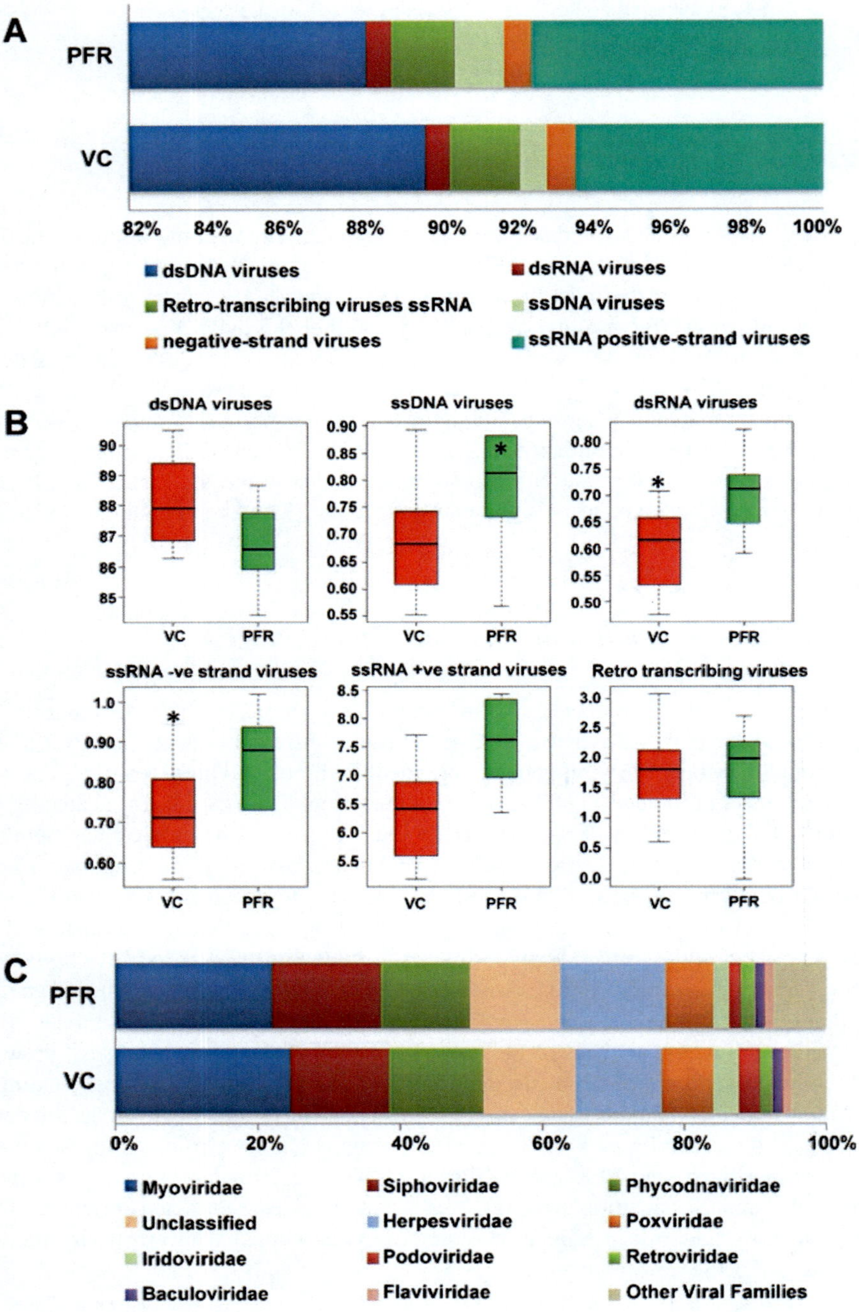

Fig. 9.1 Comparison of the vitreous fluid viromes of control (VC) and post-fever retinitis (PFR) groups showing relative abundance (%) of the different types of viruses including dsDNA viruses, ssDNA viruses, dsRNA viruses, ssRNA-negative strand viruses and ssRNA-positive strand viruses and retro transcribing viruses (**a**, **b**). (**c**) Abundance of virus families in vitreous fluid of VC and PFR groups. (Data from Kotakonda et al. 2021)

Table 9.1 Thirty viral genera that were significantly different in abundance ($p < 0.05$) in the vitreous fluid of control (VC) compared to the fluid from post-fever retinitis (PFR) individuals

S. No.	Viral genera	Mean abundance in VC	Mean abundance in PFR
1	*Lymphocryptovirus*	0.488078	0.819065
2	*Chloriridovirus*	0.737222	0.559702
3	*Iridovirus*	2.013551	1.763227
4	*Ranavirus*	2.177961	1.480304
5	*Betapapillomavirus*	0.012454	0.187051
6	*Deltapapillomavirus*	0.029408	0.017996
7	*Rhopapillomavirus*	0.028845	0.018844
8	*Ichnovirus*	0.010591	0.375047
9	*Polyomavirus*	0.055518	0.03307
10	*Aquabirnavirus*	0.040861	0.017954
11	*Coltivirus*	0.017641	0.009325
12	*Orthoreovirus*	0.048721	0.129131
13	*Oryzavirus*	0.00604	0.001589
14	*Circovirus*	0.025268	0.382411
15	*Nanovirus*	0.013747	0.098435
16	*Copiparvovirus*	0.018266	0.010248
17	*Novirhabdovirus*	0.019379	0.010984
18	*Nucleorhabdovirus*	0.030137	0.019842
19	*Mamastrovirus*	0.096279	0.036622
20	*Cilevirus*	0.015397	0.007587
21	*Hepacivirus*	0.242584	0.981777
22	*Enterovirus*	0.079240	0.042516
23	*Kobuvirus*	0.089026	0.414933
24	*Comovirus*	0.031522	0.064082
25	*Allexivirus*	0.027912	0.016201
26	*Potexvirus*	0.124534	0.103455
27	*Carlavirus*	0.143193	0.096071
28	*Pecluvirus*	0.011913	0.192004
29	*Pomovirus*	0.031057	0.017243
30	*Unclassified Reoviridae*	0.006582	0.002714

Modified from Kotakonda et al. (2021)

9.2.3 *Functional Attributes of Ocular Viruses*

Functional analysis of the viruses based on known properties of the viruses indicated that ocular pathogenic virus increased in abundance in the PFR patient samples compared to the controls. The viral genus *Lymphocryptovirus* that is increased in abundance in PFR group has been earlier reported as associated with eye diseases such as retinitis, conjunctivitis, retinal vasculitis, dry eye syndrome, oculo-glandular syndrome, episcleritis, choroiditis, dacryoadenitis, uveitis, keratitis and papillitis (Slobod et al. 2000). So also, the genera *Betapappillomavirus, Hepacivirus,*

Table 9.2 Twenty-four genera of bacteriophages that were significantly different in abundance ($p < 0.05$) in the vitreous fluid of control (VC) compared to the fluid from post-fever retinitis (PFR) individuals

Sl No.	Bacteriophage genera	Mean abundance VC	Mean abundance PFR
1	*Bcep78-like virus*	0.067	0.139
2	*Felixouna-like virus*	0.180	0.131
3	*Hpuna-like virus*	0.118	0.496
4	*T7-like virus*	0.167	0.134
5	*Phi29-like virus*	0.042	0.022
6	*3a-like virus*	0.148	0.362
7	*77-like virus*	0.169	0.366
8	*Bignuz-like virus*	0.106	0.074
9	*Che8-like virus*	0.696	0.534
10	*Cjwuna-like virus*	0.467	0.356
11	*D3112-like virus*	0.135	0.083
12	*Lambda-like virus*	0.445	0.715
13	*Tuna-like virus*	0.112	0.440
14	*Betalipothrix virus*	0.405	0.341
15	*Ino virus*	0.105	0.171
16	*Chlamydiamicrovirus*	0.013	0.184
17	Unassigned *Spounavirinae*	0.634	0.923
18	Unclassified *Autographivirinae*	0.152	0.271
19	Unclassified *Picovirinae*	0.096	0.073
20	Unassigned *Podoviridae*	0.153	0.107
21	Unclassified *Podoviridae*	2.075	1.351
22	Unclassified *Caudovirales*	1.452	1.068
23	Unclassified *Inoviridae*	0.009	0.004
24	Unclassified phages	1.832	1.404

Data from Kotakonda et al. (2021)

Kobuvirus and *Orthoreovirus* and *Circovirus* may have pathogenic effects. Several virome studies have also implicated *Papillomaviruses* in co-infection, head and neck squamous cell carcinomas (Jiang et al. 2015), oral cavity cancer, oropharyngeal and laryngeal cancer (Drop et al. 2017) and diabetes mellitus (Dworzanski et al. 2019). These results reveal that viruses co-exist under normal healthy conditions in the eye as in the controls, but in immunocompetent individuals viruses could have triggered the immune response in PFR individuals.

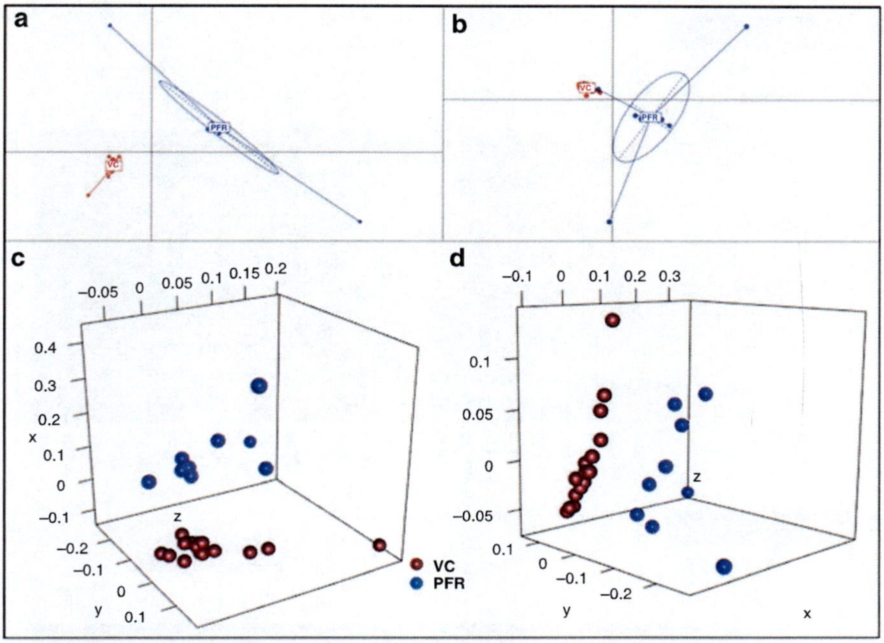

Fig. 9.2 Comparison of the vitreous viromes ((eukaryotic viral genera (**a**) and bacteriophages (**b**)) in the control (VC) and post-fever retinitis (PFR) groups by principal coordinate plots based on Jaccard distances and by non-metric multidimensional scaling analysis based on Bray-Curtis distances. (Data from Kotakonda et al. 2021)

9.3 Conclusions

The viromes of the vitreous fluid were significantly different between the normal healthy eye and the PFR and retinitis group patients, and the viruses could be discriminated. Similar studies on the virome of other ocular diseases would shed light on the involvement of viruses in ocular diseases.

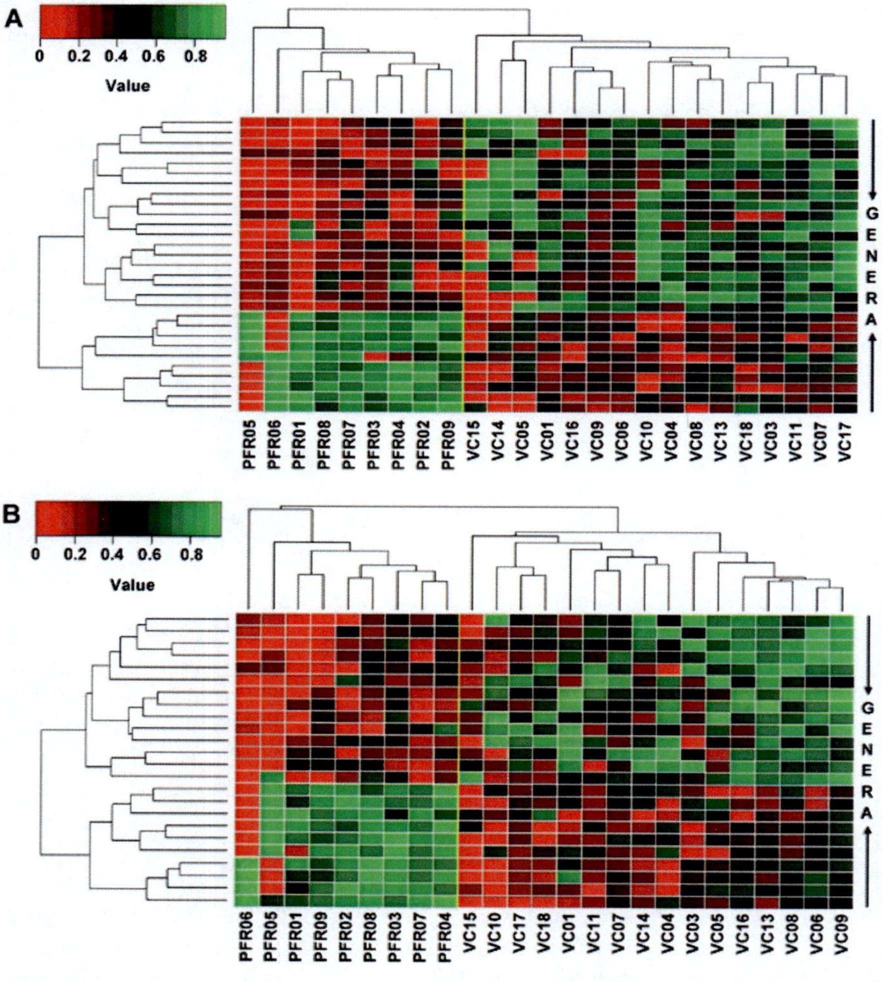

Fig. 9.3 Comparison of the vitreous viromes ((eukaryotic viruses (**a**) and bacteriophages (**b**)) in the control (VC) and post-fever retinitis (PFR) groups by heatmap analysis showing rank normalized abundances (scaled between 0 and 1). (Data from Kotakonda et al. 2021)

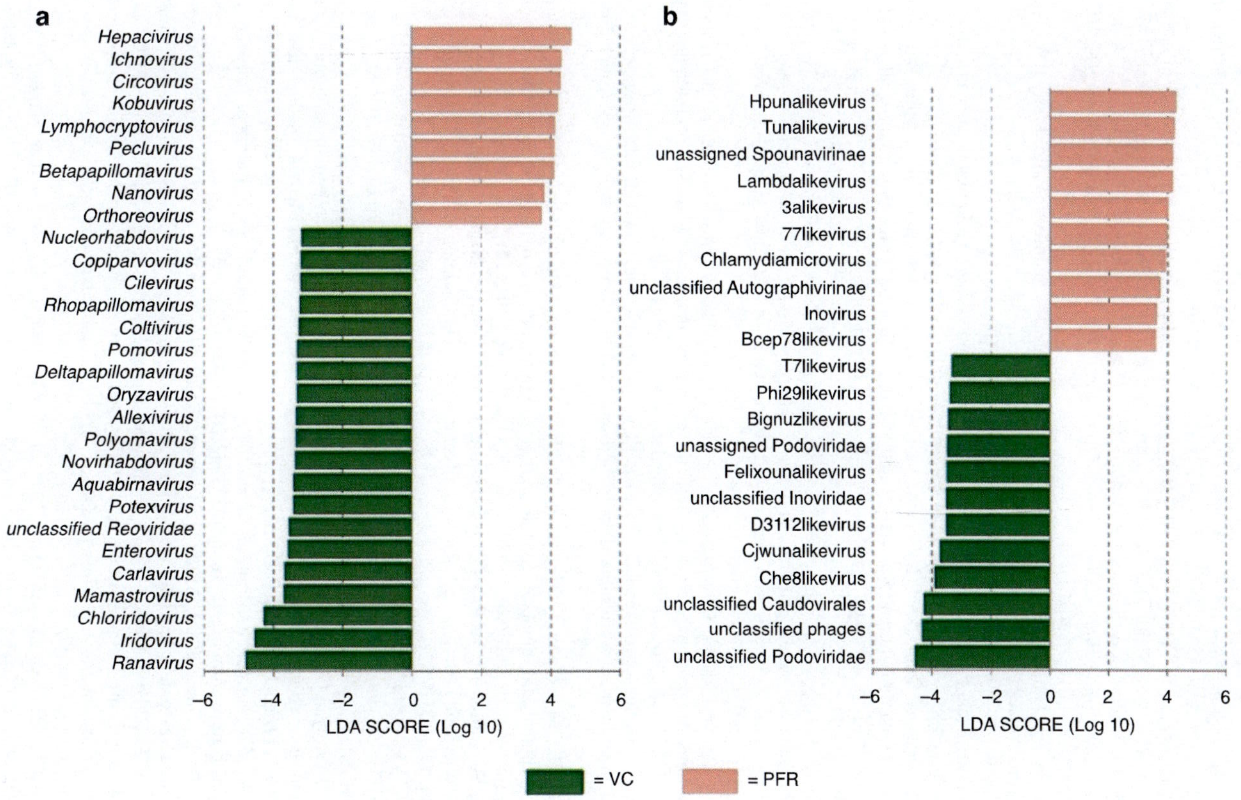

Fig. 9.4 Comparison of the vitreous viromes of eukaryotic viruses (**a**) and bacteriophages (**b**) by linear discriminant analysis (LDA) in the control (VC) and post-fever retinitis (PFR) patients. LDA combined with effect size measurements (LEfSe) was used to represent the genera that are statistically significant by *horizontal bars*. *Peach colour bars* indicate increase in the relative abundance of the genera in PFR group. Data from Kotakonda et al. (2021)

References

Abeles SR, Robles-Sikisaka R, Ly M et al (2014) Human oral viruses are personal, persistent and gender-consistent. ISME J 8:1753–1767. https://doi.org/10.1038/ismej.2014.31

Alford CA, Stagno S, Pass RF et al (1990) Congenital and perinatal cytomegalovirus infections. Rev Infect Dis 12(Suppl 7):S745–S753. https://doi.org/10.1093/clinids/12.supplement_7.s745

Alvarado JA, Underwood JL, Green WR et al (1994) Detection of herpes simplex viral DNA in the iridocorneal endothelial syndrome. Arch Ophthalmol 112:1601–1609. https://doi.org/10.1001/archopht.1994.01090240107034

Chodosh J, Chintakuntlawar AV, Robinson CM (2008) Human eye infections. In: Mahy BWJ, Van Regenmortel MHV (eds) Encyclopedia of virology. Academic, Oxford, pp 491–497. https://doi.org/10.1016/b978-012374410-4.00396-4

Columpsi P, Sacchi P, Zuccaro V et al (2016) Beyond the gut bacterial microbiota: the gut virome. J Med Virol 88:1467–1472. https://doi.org/10.1002/jmv.24508

Drop B, Strycharz-Dudziak M, Kliszczewska E et al (2017) Coinfection with Epstein-Barr virus (EBV), human papilloma virus (HPV) and Polyoma BK virus (BKPyV) in laryngeal, oropharyngeal and oral cavity cancer. Int J Mol Sci 18(12):2752. https://doi.org/10.3390/ijms18122752

Dworzanski J, Drop B, Kliszczewska E et al (2019) Prevalence of Epstein-Barr virus, human papillomavirus, cytomegalovirus and herpes simplex virus type 1 in patients with diabetes mellitus type 2 in south-eastern Poland. PLoS One 14:e0222607. https://doi.org/10.1371/journal.pone.0222607

Engstrom RE Jr, Holland GN, Margolis TP et al (1994) The progressive outer retinal necrosis syndrome. A variant of necrotizing herpetic retinopathy in patients with AIDS. Ophthalmology 101:1488–1502. https://doi.org/10.1016/s0161-6420(94)31142-0

Ghose C, Ly M, Schwanemann LK et al (2019) The virome of cerebrospinal fluid: viruses where we once thought there were none. Front Microbiol 10:2061. https://doi.org/10.3389/fmicb.2019.02061

Hannigan GD, Meisel JS, Tyldsley AS et al (2015) The human skin double-stranded DNA virome: topographical and temporal diversity, genetic enrichment, and dynamic associations with the host microbiome. mBio 6:e01578–e01515. https://doi.org/10.1128/mBio.01578-15

Harding SP, Lipton JR, Wells JC (1987) Natural history of herpes zoster ophthalmicus: predictors of postherpetic neuralgia and ocular involvement. Br J Ophthalmol 71:353–358. https://doi.org/10.1136/bjo.71.5.353

Jabs DA, Van Natta ML, Holbrook JT et al (2007) Longitudinal study of the ocular complications of AIDS: 2. Ocular examination results at enrollment. Ophthalmology 114:787–793. https://doi.org/10.1016/j.ophtha.2006.07.065

Jiang R, Ekshyyan O, Moore-Medlin T et al (2015) Association between human papilloma virus/Epstein-Barr virus coinfection and oral carcinogenesis. J Oral Pathol Med 44:28–36. https://doi.org/10.1111/jop.12221

Kawali A, Mahendradas P, Mohan A et al (2019) Epidemic retinitis. Ocul Immunol Inflamm 27:571–577. https://doi.org/10.1080/09273948.2017.1421670

Kotakonda A, Gumpili SP, Mudit T, Pappuru Rajeev R, Sisinthy S (2021) Intraocular viral communities associated with post-fever retinitis. Front Med 8:1–13. https://doi.org/10.3389/fmed.2021.724195. URL: https://www.frontiersin.org/article/10.3389/fmed.2021.724195

Lalitha P, Rathinam S, Banushree K et al (2007) Ocular involvement associated with an epidemic outbreak of chikungunya virus infection. Am J Ophthalmol 144:552–556. https://doi.org/10.1016/j.ajo.2007.06.002

Liesegang TJ, Melton LJ 3rd, Daly PJ et al (1989) Epidemiology of ocular herpes simplex. Incidence in Rochester, Minn, 1950 through 1982. Arch Ophthalmol 107:1155–1159. https://doi.org/10.1001/archopht.1989.01070020221029

Matoba AY, Wilhelmus KR, Jones DB (1986) Epstein-Barr viral stromal keratitis. Ophthalmology 93:746–751. https://doi.org/10.1016/s0161-6420(86)33668-6

Menia NK, Sharma SP, Bansal R (2019) Fungal retinitis following influenza virus type A (H1N1) infection. Indian J Ophthalmol 67:1483–1484. https://doi.org/10.4103/ijo.IJO_1691_18

Moustafa A, Xie C, Kirkness E et al (2017) The blood DNA virome in 8,000 humans. PLoS Pathog 13:e1006292. https://doi.org/10.1371/journal.ppat.1006292

Nesburn AB, Borit A, Pentelei-Molnar J et al (1974) Varicella dendritic keratitis. Investig Ophthalmol 13:764–770

Olsen TW, Hardten DR, Meiusi RS et al (1994) Linear endotheliitis. Am J Ophthalmol 117:468–474. https://doi.org/10.1016/s0002-9394(14)70006-6

Pannaraj PS, Ly M, Cerini C et al (2018) Shared and distinct features of human milk and infant stool viromes. Front Microbiol 9:1162. https://doi.org/10.3389/fmicb.2018.01162

Prabhushanker M, Topiwalla TT, Ganesan G et al (2017) Bilateral retinitis following typhoid fever. Int J Retina Vitreous 3:11. https://doi.org/10.1186/s40942-017-0065-z

Reyes A, Blanton LV, Cao S et al (2015) Gut DNA viromes of Malawian twins discordant for severe acute malnutrition. Proc Natl Acad Sci U S A 112:11941–11946. https://doi.org/10.1073/pnas.1514285112

Ritterband DC, Friedberg DN (1998) Virus infections of the eye. Rev Med Virol 8:187–201. https://doi.org/10.1002/(sici)1099-1654(1998100)8:4<187::aid-rmv221>3.0.co;2-s

Robb RM, Watters GV (1970) Ophthalmic manifestations of subacute sclerosing panencephalitis. Arch Ophthalmol 83:426–435. https://doi.org/10.1001/archopht.1970.00990030426007

Robin JB, Steigner JB, Kaufman HE (1985) Progressive herpetic corneal endotheliitis. Am J Ophthalmol 100:336–337. https://doi.org/10.1016/0002-9394(85)90811-6

Santiago-Rodriguez TM, Ly M, Bonilla N et al (2015) The human urine virome in association with urinary tract infections. Front Microbiol 6:14. https://doi.org/10.3389/fmicb.2015.00014

Slobod KS, Sandlund JT, Spiegel PH et al (2000) Molecular evidence of ocular Epstein-Barr virus infection. Clin Infect Dis 31:184–188. https://doi.org/10.1086/313932

Thompson WS, Culbertson WW, Smiddy WE et al (1994) Acute retinal necrosis caused by reactivation of herpes simplex virus type 2. Am J Ophthalmol 118:205–211. https://doi.org/10.1016/s0002-9394(14)72900-9

Thygeson P (1987) Marginal herpes simplex keratitis simulating catarrhal ulcer. Investig Ophthalmol 105:1415–1417

Unnikrishnan S, Raju K (2017) Ocular manifestations in mosquito-borne viral fevers. J Clin Ophthalmol Res 5:77–80. https://doi.org/10.4103/2320-3897.205178

Vishwanath S, Badami K, Sriprakash KS et al (2014) Post-fever retinitis: a single center experience from south India. Int Ophthalmol 34:851–857. https://doi.org/10.1007/s10792-013-9891-7

Wilhelmus KR (1981) Ocular involvement in infectious mononucleosis. Am J Ophthalmol 91:117–118. https://doi.org/10.1016/0002-9394(81)90358-5

Wilhelmus KR, Gee L, Hauck WW et al (1994) Herpetic eye disease study. A controlled trial of topical corticosteroids for herpes simplex stromal keratitis. Ophthalmology 101:1883–1895. ; discussion 1895-1886. https://doi.org/10.1016/s0161-6420(94)31087-6

Wylie KM (2017) The virome of the human respiratory tract. Clin Chest Med 38:11–19. https://doi.org/10.1016/j.ccm.2016.11.001

Wylie KM, Wylie TN, Cahill AG et al (2018) The vaginal eukaryotic DNA virome and preterm birth. Am J Obstet Gynecol 219:189 e181-189:e112. https://doi.org/10.1016/j.ajog.2018.04.048

Yamamoto S, Pavan-Langston D, Tada R et al (1995) Possible role of herpes simplex virus in the origin of Posner-Schlossman syndrome. Am J Ophthalmol 119:796–798. https://doi.org/10.1016/s0002-9394(14)72788-6

Chapter 10
Cellular and Molecular Basis of Ocular Diseases

Abstract Innate and adaptive immunity is the first line of defence on the ocular surface against ocular diseases. The cellular and molecular basis of several ocular diseases is discussed with respect to the role of accumulation of various cell types and various proteins and lipids that are either up- or downregulated in the diseased state of the eye. Nevertheless, the cellular and molecular sequence of events that trigger the pathogenesis and sustain the diseased state is yet to be clearly established. Such data would eventually help in targeted therapy of an ocular disease.

Keywords Ocular disease · Cellular · Molecular · Pathogenesis · Therapy

10.1 Introduction

The surface of the human eye harbours several microorganisms from the environment that could act as potential agents of infection. Under normal circumstances these microbes (bacteria, fungi and viruses) colonize the eyelids, conjunctiva and tear film as commensals and include the Gram-positive genera *Corynebacterium* spp., *Staphylococcus* spp., *Streptococcus* spp., *Micrococcus* spp. and *Propionibacterium* spp. (Miller and Iovieno 2009; Aoki et al. 2021). Further, the Gram-negative bacteria are less common in the eye of healthy individuals and include *Pseudomonas* spp., *Haemophilus* spp. *and* Neisseria spp. Fungi are also present on the ocular surface in healthy individuals (Wu et al. 2003). These commensal bacteria of the eye may act as a pathogen in immunocompromised patients and infections, like corneal ulcers, keratitis and conjunctivitis (Aoki et al. 2021). The corneal and conjunctival epithelial cells are in constant contact with these commensal bacteria but are not susceptible to the host immune response suggesting an innate immune response on the ocular surface that permits colonization by the commensal microbiota (Ueta 2008; Petrillo et al. 2020). A combination of anatomical, mechanical and immunological defence mechanisms associated with the eye along with the commensal bacteria constitutes the innate immune response and acts as a barrier to potential infection (Akpek and Gottsch 2003).

S. Shivaji, *Human Ocular Microbiome*,
https://doi.org/10.1007/978-981-19-1754-7_10

10.2 Innate Immunity on the Ocular Surface

Innate immunity is the first line of defence against corneal infection (Alberts et al. 2002) which includes the bony orbit, the eyelids (which guard against trauma) and the tears which keep the cornea moist, prevent drying of the cornea, eliminate foreign particles and secrete antimicrobial proteins (β-defensin, calprotectin, lactoferrin, lysozyme, lipocalin and beta-lysin) that are increased in concentration during inflammation and restrict the viability of the microbes (Garreis et al. 2010; Dartt 2011; Flanagan and Willcox 2009) and immunoglobulins which prevent infections (Mannis and Smolin 1996). In addition, the dense glycocalyx made up of mucins MUC1, MUC4 and MUC16 on the corneal epithelium acts as a physical barrier and prevents the adhesion of the bacteria (Govindarajan and Gipson 2010). But damage of the epithelial cells is accompanied by the production of antimicrobial peptides *β*-defensins and cathelicidin (LL-37) (McDermott 2009). *β*-Defensins 2 and 3 protect against *P. aeruginosa* infection (Augustin et al. 2011; Wu et al. 2009). Tam et al. (2012) identified cytokeratin-derived antimicrobial peptides from corneal epithelial cells. In addition, the corneal epithelial cells secrete cytokines (IL-1a) to activate immune defences and thus protect against microbial invasion (Niederkorn et al. 1989). Stromal keratocytes also secrete both IL-1 and IL-6 (Cubitt et al. 1995) and defensins (Gottsch et al. 1998) which exhibit antimicrobial activity (against several fungi, bacteria and viruses) and facilitate epithelial healing (Ganz et al. 1985). Corneal nerves also participated in the innate defence of the cornea (Müller et al. 1996) by releasing neuropeptides (like the calcitonin gene-related peptide and substance P) (Tran et al. 2000a, b). In addition, complement which generates enzymes, opsonins, anaphylatoxins and chemotaxins could be involved in corneal inflammation, and so also interferons (IFN) made by host cells could protect specifically against viral infection. Several other cells of innate immunity like neutrophils (Burg and Pillinger 2001), eosinophils, macrophages and NK cells which are phagocytic, possess antigen-presenting capabilities and also secrete inflammatory cytokines could also participate in the innate immune response against microbial infections (Underhill and Ozinsky 2002; Moretta et al. 2002).

10.2.1 Toll-Like Receptors in Innate and Adaptive Immune Responses at the Ocular Surface

Different types of TLRs like TLR-1, TLR-2, TLR-3, TLR-4, TLR-7, TLR-9 and TLR-10 are known to be associated with the normal as well as in the diseased and allergic states of the eye (Micera et al. 2005) (Fig. 10.1). In addition, the costimulatory CD14 and MD2 molecules in the human retinal pigment epithelial (RPE) cells are also known to take part in the immune response of the retina (Kumar et al. 2004). The induction of the different types of TLRs varies depending on the agonist involved. For instance, expression of TLR-2, TLR-3 and TLR-4 by RPE

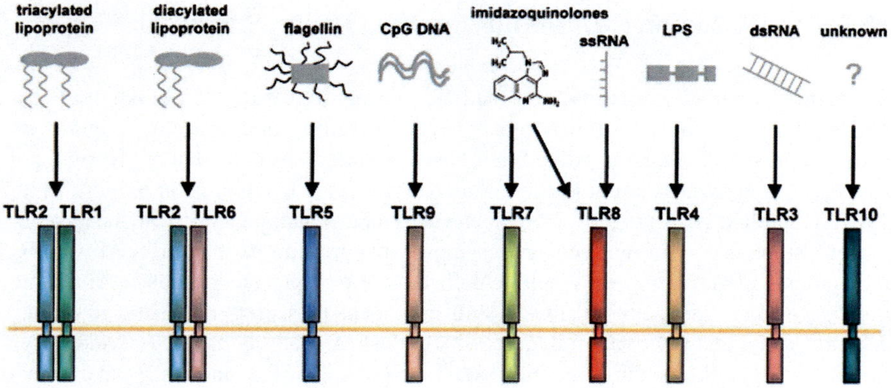

Fig. 10.1 Toll-like receptors. (File is licensed under the Creative Commons Attribution-ShareAlike 4.0 (license, English; URL: https://commons.wikimedia.org/wiki/user:BQUB1819-MBallve))

cells is upregulated by TLR-3 agonists or IFN-g (Kumar et al. 2004), whereas in corneal epithelial cells TLR-4 agonists induce the expression of TLR-4, CD14 and MD2, and as a consequence induction of pro-inflammatory cytokines and chemokine secretion occur (Song et al. 2001; Chang et al. 2004). TLR agonist like *Pseudomonas aeruginosa* could also cause bacterial keratitis in contact lens wearers (Zhang et al. 2003) as a result of specific neutrophil recruitment and local inflammatory response (Dabbagh and Lewis 2003). Thus, activation of corneal TLR-4 could be both beneficial and detrimental, and the fragile balance between these two opposing actions determines corneal scarring and blindness (Song et al. 2001). The role of TLRs in LPS-induced *Pseudomonas aeruginosa* keratitis (Kumagai et al. 2005) is discussed separately under the section bacterial keratitis. In mice it was demonstrated that specific expression of TLR-4, TLR-2 and TLR-9 in keratitis animals was entirely MyD88-dependent (Johnson et al. 2005), and high levels of TLR-9 expression were observed in the healthy conjunctival epithelium compared with TLR-2 and TLR-4, thus implying that TLR-9 acts as a first line of defence on the conjunctiva (Bonini et al. 2005). In contrast, in patients with vernal keratoconjunctivitis (VKC), the expression of TLRs was different compared to the mice, and the conjunctiva showed upregulation of TLR-4, downregulation of TLR-9 and a slight reduction in TLR-2 compared with healthy conjunctiva (Bonini et al. 2005). High levels of TLR-4 expression in VKC tissues also correlated with allergic phenotype (Bonini et al. 2005). It is suggested that research on TLRs could eventually lead to a better understanding of ocular infection and the role of immunity (Micera et al. 2005).

10.3 Cell-Mediated Immunity on the Ocular Surface

In addition to innate immunity, cell-mediated immunity exerts its role when microbial replication is out of control and when the invading microorganism and/or its antigens persist. The cells involved include the Langerhans cells that are recognized as a part of the corneal immune surveillance system and T helper cells (Gillette et al. 1982; Chandler et al. 1985). The Langerhans cells mediate their effects through various steps involving antigen processing, transportation to the surface by MHC molecules and activation of T cells which have a two-pronged attack to eliminate pathogens. The T cells either directly kill pathogens or secrete cytokines to induce effector cells, mainly macrophages, which destroy pathogens or activate other inflammatory cells for the same purpose. T helper cells (Th1 and Th2) with differential cytokine production profiles (Mosmann et al. 1986; Cherwinski et al. 1987) have been involved in this process. Th1 cells are cytolytic and participate in inflammatory immune defence at the ocular surface, whereas Th2 cells are involved in persistent antibody production (Akpek and Gottsch 2003).

10.4 Cellular and Molecular Basis of Bacterial Keratitis

Approximately 90% of all microbial keratitis cases are caused by bacterial keratitis (Musa et al. 2010), and worldwide *P. aeruginosa* is the most common causative agent (Dutta et al. 2012). Keratitis if left untreated could become severe and could cause endophthalmitis, perforation and loss of vision (Wong et al. 2012; Callegan et al. 1994). Microbial invasion of the cornea is discussed with respect to specific bacteria that cause keratitis like *Pseudomonas aeruginosa*, *Serratia* spp., *Staphylococcus aureus* and *Streptococcus pneumonia* (Wong et al. 2012).

10.4.1 Pseudomonas aeruginosa *Keratitis*

P. aeruginosa is a common cause of keratitis in soft contact lens users and is often refractive to treatment, resulting in visual outcomes that are more severe than other bacterial ulcers (Sy et al. 2012; Green et al. 2008). Contact lens wear is a risk factor for *Pseudomonas* keratitis, and it is likely that contact lens wear causes migration of Langerhans cells into the central cornea (Hazlett et al. 1999) subsequently triggering cytokine release and influx of inflammatory cells (Hazlett 2002). In infectious keratitis caused by Gram-negative bacteria, lipopolysaccharide (LPS) is an important virulence factor. LPS mediates the attachment of the bacteria to the contact lenses and the cornea, entry of the bacteria and their survival in the corneal epithelial cells (Lakhundi et al. 2017). LPS also stimulates neutrophil migration and infiltration into the cornea which results in corneal scarring and opacification and confers bacterial resistance to the complement-mediated killing by the host cells. Exopolysaccharide from both Gram-negative and Gram-positive bacteria also influences immunosuppression as well as interferes with phagocytosis. LPS when released produces stromal rings of polymorphonuclear leucocytes in the stroma of the cornea Lakhundi et al. 2017). *P. aeruginosa* LPS induces corneal ulcers when injected intra-stromally into rabbit cornea, causing an intense neutrophil infiltrate (Schultz et al. 1997) in the absence of any live bacteria. In a mice model, it was observed that when a cornea following abrasion was infected with *P. aeruginosa*, macrophages invaded the cornea as the most predominant cell type and expressed TLR-2, TLR-4 and TLR-5. The activation of TLR-4/TLR-5 in the resident corneal macrophages following *P. aeruginosa* infection leads to a sequence of events including NF-kB translocation to the nucleus, transcription of CXCL1 chemokines, neutrophil recruitment to the corneal stroma and finally bacterial killing and tissue damage (Sun et al. 2010). The exact mechanism by which neutrophils kill bacteria in bacterial keratitis has yet to be determined and has been attributed to the ability of neutrophils to produce ROS and MMPs including collagenases that prevent microbial dissemination by disrupting the extracellular matrix (Leal et al. 2012; Nathan 2006). Disrupting the collagen matrix is known to affect corneal clarity and impair vision. In *P. aeruginosa* keratitis, TLR-4 and other TLRs on the surface of human corneal epithelial cells (Roy et al. 2011) were implicated in the expression of antimicrobial peptides (β-defensins and cathelicidin) (Kumar et al. 2007; Redfern et al. 2011) which facilitated bacterial killing in addition to neutrophil infiltration which also effectively killed bacteria (Huang et al. 2006; McClellan et al. 2003). TLR-4 also regulates endotoxin-induced keratitis by increased expression of macrophage inflammatory factor (MIP)-2 and platelet adhesion molecule (PECAM)-1 (Song et al. 2001). MIP-2 recruitment of neutrophils was also observed in *Pseudomonas*-infected corneas in mice (Kernacki et al. 2000). In addition, IL-1b and IL-1R on other resident cells in the cornea (Fig. 10.2) (Sun et al. 2010; Pearlman et al. 2013) also have an important role in the host response to *P. aeruginosa* corneal infection. Mice experiments confirmed that these cytokines or its receptor is important to clear infection since in mice depleted of or deficient in this cytokine or its receptor are unable to clear the bacteria (Rudner et al. 2000; Karmakar et al. 2012). In *P. aeruginosa* the type III secretion system has also been demonstrated as required for corneal disease elicited in corneal infections (Lee et al. 2003; Stover et al. 2000).

10.4.2 Streptococcus pneumoniae *Keratitis*

In the developing world, *S. pneumoniae* is the leading cause of bacterial keratitis and exhibits antimicrobial resistance (AMR) against various antibiotics (Bharathi et al. 2010; Marquart and O'Callaghan 2013; Mulet et al. 2010). Ocular trauma and prior surgery are the predisposing factors of pneumococcal keratitis (Wagoner et al. 2007; Marquart and O'Callaghan 2013; Rehany et al. 2004; Mulet et al. 2010). Attachment of streptococci to biotic surface is facilitated by a diverse group of bacterial surface

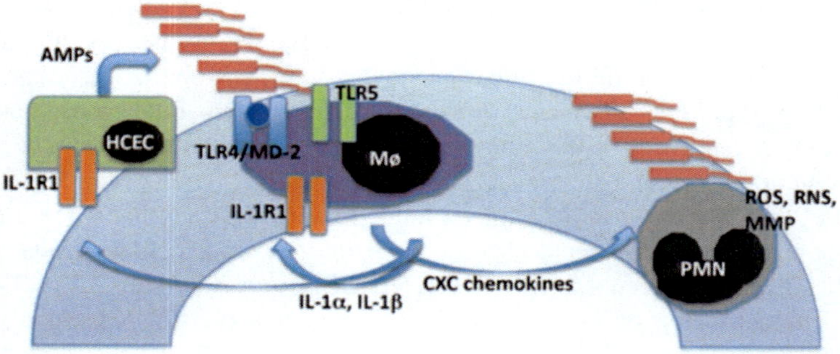

Fig. 10.2 Cellular and molecular basis of *Pseudomonas aeruginosa* keratitis. When *P. aeruginosa* infects an abraded cornea, specific TLR-4/MD-2 and TLR-5 are activated leading to production of CXC chemokines and pro-inflammatory cytokines by resident macrophages (Mø) including IL-1α and IL-1β. These cytokines mediate neutrophil (PMN) recruitment from the cornea, which secrete reactive oxygen species (ROS) and reactive nitrogen species (RNS) and matrix metalloproteinases (MMP). ROS, RNS and MMPs disrupt the corneal stromal architecture, and simultaneously corneal clarity is compromised. IL-1α and IL-1β have autocrine and paracrine effects on human corneal epithelial cells (HCEC) expressing IL-1R1, resulting in the production of chemokines and antimicrobial peptides (AMP), which exacerbate the inflammatory response. (This diagram is based primarily from data in Sun et al. 2010 and modified from Pearlman et al. 2013)

proteins MSCRAMMS (microbial surface components recognizing adhesive matrix molecules) (Hammerschmidt 2006) and includes plasmin and fibronectin binding protein A (PfbA), pneumolysin (ply), pneumococcal surface protein A (PspA), pneumococcal adherence and virulence factor A (PavA), pneumococcal surface adhesin A (PsaA), putative protease maturation protein A (PpmA), IgAI protease (IgAIp), choline-binding protein A (CbpA/PcpA) and the streptococcal lipoprotein rotamase A (SIsA). These proteins are associated with pneumococcal adherence and virulence (Rajam et al. 2008). Pili and fibrils on the streptococcal cell surface may also mediate attachment to cell surface. Among these proteins, pneumolysin is a bacterial cholesterol-dependent cytolysin which forms pores in the host cell membranes and thus damages the cells (Tilley et al. 2005). Pneumolysin also initiates immune-mediated damage by activating complement system and inducing inflammation (Paton et al. 1984). PspA and PspC and neuraminidase A (NanA) are also involved in non-ocular infections. Recently, ZmpC, a metalloprotease, has been shown to enhance internalization of the bacterium probably by inducing ectodomain shedding of the membrane-associated mucin (MAM) from conjunctival and corneal epithelial cells leading to the loss of glycocalyx barrier function (Govindarajan et al. 2012).

10.4.3 Serratia marcescens *Keratitis*

Serratia marcescens are routinely isolated from lenses, lens cases (Johnson et al. 2005) and corneal surface of keratitis patients (Parment 1997) and have the potential to induce keratitis. These bacteria produce pro-inflammatory cytokines and chemotactic cytokines with chemotactic properties and play a role in neutrophil recruitment to the corneal stroma leading to loss of corneal clarity (Dong et al. 2011). The potential of *S. marcescens* to induce keratitis was not consistent; though it was observed in a rabbit model, it did not infect BALB/c mice (Hume et al. 1999; Willcox and Hume 1999). In contradiction, clinical isolates of *S. marcescens* could infect C57BL/6 mice, and the severity of disease was less than that reported for T3SS expressing *P. aeruginosa* (Zhou et al. 2012; Sun et al. 2012). *Serratia marcescens* possess type 1 fimbriae, a mannose-specific adhesin on the flagella which bind to mannose containing glycoproteins on the corneal epithelial cells (Labbate et al. 2007). Adhesion to human corneal epithelial (HCE) cells is also involved in the expression of two *N*-acylhomoserine lactone (AHLs)-regulated genes, *bsmA* and *bsmB*. In addition to type 1 fimbriae genes, the *oxyR*gene was also involved in cell-cell and cell-biotic surface interactions (Shanks et al. 2007). Another factor that influenced attachment to epithelial cells included mannose-resistant and mannose-sensitive pili (Hejazi and Falkiner 1997). Following attachment, *Serratia* metalloprotease induces extensive corneal tissue damage by causing descemetocele formation, liquefactive necrosis, solubilization of the stromal proteoglycan ground substance and dispersal of normal collagen fibrils (Lyerly and Kreger 1979; Lyerly et al. 1981). A protease of 56-kDa is one of the major factors contributing to the pathogenesis of keratitis caused by *S. marcescens* (Kreger et al. 1986; Kamata et al. 1985), and it was confirmed that mutants of *S. marcescens* deficient in the 56 kDa metalloprotease are less cytotoxic to mammalian cells (Marty et al. 2002). As in *P. aeruginosa* keratitis, it was observed that *S. marcescens* corneal infection was also dependent on TLR-4/MD-2 and IL-1R1 activation of MyD88 (Zhou et al. 2012).

10.4.4 *Staphylococcus aureus Keratitis*

S. aureus is also a common aetiological agent of keratitis (Jeng et al. 2010; Marquart and O'Callaghan 2013). The predisposing factors include viral infection, contact lens use, ocular surgery or trauma or other eye diseases (Coster and Badenoch 1987; Cheung and Slomovic 1995). In *S. aureus* also the surface adhesins, collectively known as MSCRAMMS, mediate adhesion of bacteria to host extracellular matrix components, fibronectin, fibrinogen, collagen, elastin and laminin (Jett and Gilmore 2002). One of these adhesins, the collagen-binding adhesin (Can), in *S. aureus* is a virulence factor involved in the early events of pathogenesis (Rhem et al. 2000), whereas FnBPs bind to HCE cells and influence the host-parasite interactions (Jett and Gilmore 2002). *S. aureus* also produces an alpha- and gamma-toxin involved in keratitis. The former alpha-toxin interacts with cavoline-1 and forms pores in the membrane (Pany et al. 2004; Menestrina et al. 2001) and destroys the infected eye's corneal tissues (Callegan et al. 1994). Gamma-toxin consists of an F and S components that are non-toxic on their own, but the F-S pairs cause cell lysis for which purpose they form a ringlike structure that could penetrate the membrane (Szmigielski et al. 1999; Gravet et al. 1998). The strain deficient in either alpha- or gamma-toxins was considerably less virulent. *S. aureus* also secretes several other two component toxins (Gravet et al. 1998). Another protein produced by *setnm-1* gene, which codes for a protease, has also been shown to cause extensive corneal damage and is considered to confer virulence in *S. aureus* and cause keratitis (Caballero et al. 2008). Mutant deficient in *setnm-1* gene is considerably less virulent compared to the parent or the rescued strain (Caballero et al. 2010).

10.5 Cellular and Molecular Basis of Mycotic Keratitis

Mycotic keratitis is a fungal infection, affecting greater than 50% of infectious keratitis in the tropics and subtropical regions of the world (Thomas and Kaliamurthy 2013; Nath et al. 2011). *Candida* spp. and *Candida albicans* are the dominant causative agents of mycotic keratitis in temperate climates, whereas the filamentous fungi (*Phaeohyphomycetes*, *Fusarium*, *Curvularia*, *Aspergillus*, *Scedosporium apiospermum* and *Paecilomyces*) (Thomas and Kaliamurthy 2013) are more common in tropical latitudes (Leck et al. 2002). The most important predisposing factor is trauma of the eye (Gopinathan et al. 2009; Thomas and Kaliamurthy 2013; Nath et al. 2011), and the other factors include ocular surgery, use of corticosteroids, contact lens use and ocular surface diseases. Clinical manifestations of filamentous fungal keratitis include photophobia, pain, ocular discharge, opacity of the cornea and reduced vision (Ansari et al. 2013) and may involve any part of the cornea (Thomas 1994; Thomas and Kaliamurthy 2013). Keratitis caused by yeasts like *Candida* resembles bacterial keratitis with infiltrate, epithelial defect and slow progression (Sun et al. 2007).

The mechanism of pathogenicity proceeds through adhesion, invasiveness and morphogenesis. Fungal adhesins (mannoproteins) (Imwidthaya 1995; Thomas 2003) on the surface of the fungi adhere to various mannose glycoproteins which are upregulated on the surface of damaged corneal epithelial cells (Clarke and Niederkorn 2006). Laminin, fibronectin and collagen are the proteins on the corneal epithelium which can bind to fungi and thus are involved in keratitis (Thomas 2003). In *Aspergillus fumigatus*, the fungal conidia bind to wells coated with laminin, collagen, fibrinogen and fibronectin substrates (Coulot et al. 1994). In addition, in *Aspergillus* and *Penicillium* conidia possess sialic acid-dependent lectins which recognize and bind to laminin and fibrinogen (Thau et al. 1994; Parta et al. 1994; Scholtmeijer et al. 2001). Following adhesion, invasiveness of the cornea occurs. Fungal invasion of the cornea results in the fungus entering the anterior chamber of the eye, increasing the intraocular pressure and affecting the normal drainage of aqueous humour causing fungal malignant glaucoma as observed in the case of *Fusarium* and *Aspergillus* keratitis (Kuriakose and Thomas 1991; Jain et al. 2007). Studies have also indicated that fungal invasiveness is related to the fungal load and inversely proportional to the intensity of inflammatory response (Vemuganti et al. 2002). Phenotypic switching of morphology as observed in *F. solani* has also been implicated in keratitis and is normally seen when treated with antifungal drugs (Jackson et al. 2007; Kiryu et al. 1991; Thomas et al. 1991). Jackson et al. (2007) demonstrated that in *C. albicans*, morphogenic transformation into invasive filamentous forms was associated with keratitis. This morphogenic transformation in *C. albicans* is dependent on the expression of *SAP6* gene which encodes a unique secreted aspartyl proteinase. A clinical isolate of *A. flavus* from a keratitis patient secretes a variety of proteases including cysteine proteases, serine proteases and metalloproteases, and these fungal collagenases are the mediators of severe corneal damage observed in keratitis patients (Zhu et al. 1990; Gopinathan et al. 2001). Matrix metalloproteinases are upregulated in rabbits as well as in the tear film of horses with ulcerative keratitis and are thought to be involved in the pathogenesis of fungal keratitis (Gopinathan et al. 2001; Ollivier et al. 2004).

10.6 Cellular and Molecular Basis of Herpes Simplex Keratitis

In herpes simplex keratitis (HSK), the virus forms a dendrite-like structure in the corneal epithelium, whereas the virus infiltrates and causes vascularization, ulceration or oedema in the stroma and endothelium. Innate immunity leading to clearing of viral particles by macrophages appears to be a strategy (Cheng et al. 2000). Neutrophils and interferon α (IFNα) play a central role in viral suppression (Tumpey et al. 1996; Hendricks et al. 1991; Noisakran and Carr 2000). NK cells indirectly influence the severity of HSK (Bouley et al. 1996). Cytokine MIP-1a induces influx of T cells and neutrophils in HSV keratitis (Tumpey et al. 1998a). In accordance, in knockout mice for MIP-1a, it was observed that inflammatory infiltrates are decreased with HSV infections. HSK patients have an increased expression of Th1 cytokines IFN gamma and IL-2 (Niemialtowski and Rouse 1992) and expression of IL10, which suppresses the expression of the cytokines and ameliorates the development of HSK (Tumpey et al. 1998b; Yan et al. 2001).

10.7 Cellular and Molecular Basis of Acanthamoeba Keratitis

Acanthamoeba is an opportunistic protist pathogen which could cause keratitis which could become sight threatening (Dart et al. 2009) if the treatment is delayed. In *Acanthamoeba* keratitis the cornea becomes ulcerated leading to stromal abscess, perforation, loss of visual acuity, ring infiltration, blindness and eventually enucleation. Acanthamoeba was first detected in the eye in the early 1970s in the USA (Clarke et al. 2012) and the UK (Nagington et al. 1974). Contact lens wear, non-sterile contact lens rinsing, corneal trauma and biofilm formation on contact lenses (Trabelsi et al. 2012) were identified as the risk factors. Patients with *Acanthamoeba* keratitis present with excruciating pain, redness of the eye, lacrimation, epiphora, conjunctival hyperaemia, eyelid ptosis, photophobia and foreign body sensation (Alizadeh et al. 1994a, b; Niederkorn et al. 1999; Lakhundi et al. 2017). Acanthamoebae are also known to cluster around the corneal nerves producing radial keratoneuritis (Alizadeh et al. 2005), and in vitro the trophozoites can kill nerve cells (Pettit et al. 1996).

Pathogenesis of *Acanthamoeba* keratitis occurs through different stages (Fig. 10.3). In the adhesion stage, acanthamoeba trophozoite membrane protein, the mannose-binding protein (MBP), binds to mannose glycoproteins on the corneal surface (Lakhundi et al. 2017; Clarke and Niederkorn 2006; Panjwani 2010). Normally, adhesion is preceded by mild trauma or corneal abrasion (Clarke and Niederkorn 2006) or prior contact lens use (Alizadeh et al. 2005). Mannose induces acanthamoeba trophozoites to release MIP133, which is extremely cytolytic to corneal epithelial cells in vitro, and probably destroy epithelial cells by either phagocytosis, cytolysis or apoptosis (Alizadeh et al. 2001). Cytolysis of epithelial cells is dependent on cytoskeletal elements and calcium channels (Taylor et al. 1995), whereas phagocytosis is mediated by amoebastomes or food cups on the surface of acanthamoeba (Khan 2001). Acanthamoeba also induces apoptosis of different types of cells including keratocytes, corneal epithelial cells, retinal pigment epithelial cells, iris ciliary body cells, neuroblastoma cells etc. (Niederkorn et al. 1999; Clarke et al. 2005). Several types of enzymes including proteases (serine proteases, cysteine proteases, metalloproteinases, elastases and collagenolytic enzymes), a plasminogen activator (aPA) and phospholipase A secreted by acanthamoeba trophozoites (Cao et al. 1998; Cho et al. 2000) are involved in the final step of penetration and dissolution of the underlying collagenous stroma. aPA activates host plasminogen to plasmin, which in turn activates matrix metalloproteinases (MMPs) to degrade the components of extracellular matrix (Berman 1993) including type I and II collagens, laminin and fibronectin (Parks 1999). This degradation results in ulceration of the normal quiescent cornea (Parks 1999). Ecto-ATPases secreted by pathogenic isolates of acanthamoeba also protect the cells from effector cells of the immune system, in repolarization of the membrane after depolarization, and thus influence adhesion to the host cell (Siddiqui and Khan 2012). Trophozoites often cluster around corneal nerves, causing extreme pain and

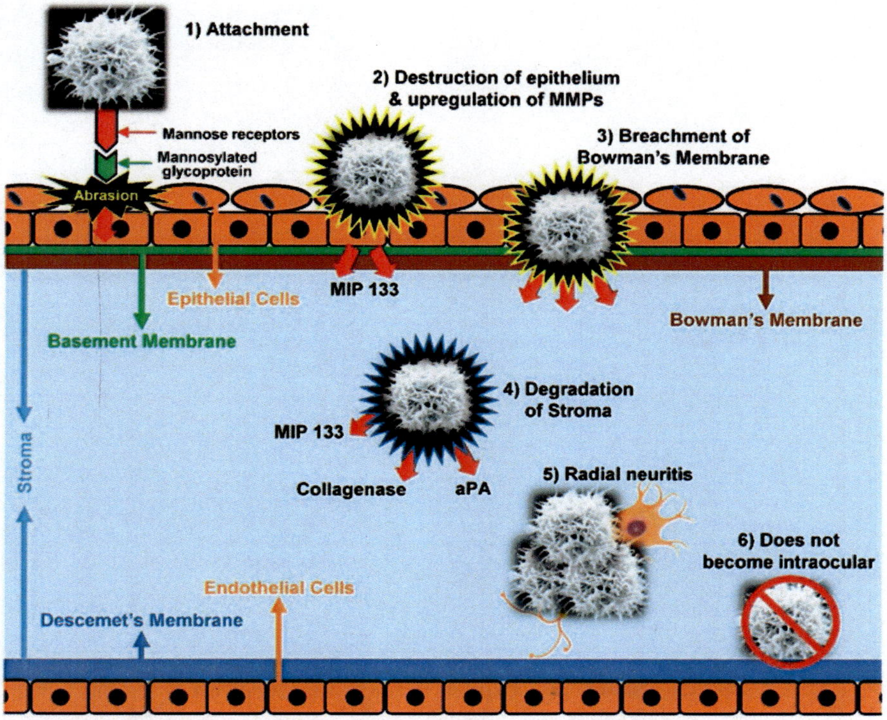

Fig. 10.3 Diagram depicting the six stages in the pathogenesis of *Acanthamoeba* keratitis. (*1*) Acanthamoeba trophozoites adhere to the corneal surface by binding to the mannose-binding protein (MBP). (*2*) This binding triggers the trophozoites to release MIP133, which induces cytolysis in corneal epithelial cells. (*3*) Following cytolysis the trophozoites breach the Bowman's membrane, enter the stroma and dissolve the collagenous stroma by producing several proteases (MIP133, a collagenase, an MMP and a novel acanthamoeba plasminogen activator) (*4*). (*5*) Trophozoites then cluster around corneal nerves, producing radial keratoneuritis causing extreme pain but rarely progresses beyond the corneal endothelium to produce intraocular infections (*6*). (Modified from Clarke and Niederkorn 2006)

producing radial keratoneuritis. *Acanthamoeba* keratitis rarely progresses beyond the corneal endothelium to produce intraocular infections (Fig. 10.3).

10.8 Cellular and Molecular Basis of Helminthic Keratitis

Parasitic helminthic microfilariae also cause keratitis. These parasitic microfilariae from the conjunctiva migrate into the cornea, become immotile and die. These dead microfilariae cause an intense host immune response and repeated infestation with the dead microfilariae causing sclerosing keratitis and loss of vision. Several proteins

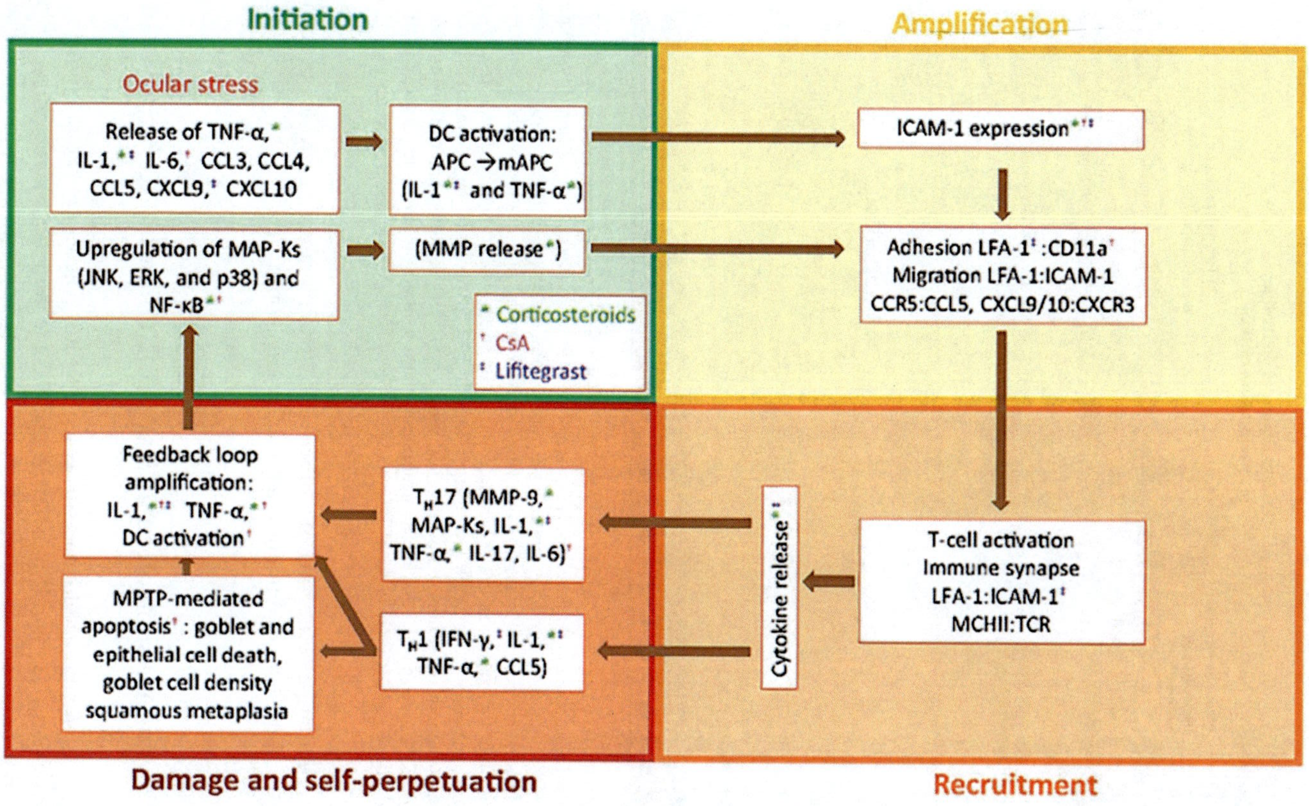

Fig. 10.4 The four phases of pathogenesis of dry eye disease (DED). DED treatment with corticosteroids, CsA and lifitegrast is colour-coded in green, red and purple as indicated in the initiation phase, and the DED effectors that are influenced are indicated with the corresponding specific colour. The abbreviations are as follows: APC (antigen-presenting cell); chemokines (CXCL and CCL); chemokine receptors (CXCR and CCR); cluster of designation molecule 11a (CD11a); cyclosporin A (CsA); dendritic cell (DC); extracellular signal-related kinase (ERK); intercellular adhesion molecule 1 (ICAM-1); interferon gamma

(IFN-γ); interleukin (IL); c-Jun N-terminal kinase (JNK); lymphocyte function-associated antigen 1 (LFA-1); mitogen-activated protein kinase (MAP-K); mature antigen-presenting cell (mAPC); major histocompatibility complex II (MHCII); matrix metalloproteinase (MMP); 1-methyl-4-phenyl-1,2,3,6-tetrahydropyridine (MPTP); nuclear factor kappa B (NF-κB); T cell receptor (TCR); tumour necrosis factor alpha (TNF-α). (Data and figure from Periman et al. 2020) including calgranulin C, a neutrophil-protein, and paramyosin, a muscle protein (Akpek et al. 2002), form a highly antigenic complex of paramyosin-calgranulin C which can induce stromal keratitis very similar to the human onchocercal corneal disease in helminthic keratitis patients (Gallin et al. 1988). When inflammation peaks the most prominent inflammatory cells observed are eosinophils and neutrophils (Pearlman et al. 1998; Kaifi et al. 2001). In keratitis following injection with *O. volvulus*, sensitized CD4+ T cells expressing IL-4 and IL-5 have been observed implying their role in keratitis (Pearlman et al. 1995). For instance, in an IL-4 gene knockout model, infiltration of eosinophils was noted indicating a reduced inflammatory response. In a murine model, when soluble extracts of filarial nematodes were injected into the corneal stroma, the predominant inflammatory response in the cornea was increased abundance of endosymbiotic *Wolbachia* bacteria (Saint Andre et al. 2002); the inflammatory response by *Wolbachia* was dependent on the expression of TLR-4 on host cells, providing evidence that the innate immune system plays a crucial role in the initial inflammatory response.

10.9 Cellular and Molecular Basis of Sterile Corneal Inflammation

Corneal inflammation has also been detected in the absence of cultivable bacteria. Such inflammations of the cornea without any infective agent are considered sterile corneal inflammation. These are less severe than active keratitis and occur in contact lens wearers who are estimated to be 134 million worldwide. Flagellin stimulation of TLR-2 and TLR-5 and nucleic acid stimulation of TLR-3 and TLR-9 can induce corneal inflammation (Johnson et al. 2005, 2008).

10.10 Cellular and Molecular Basis of Uveitis

Uveitis may be caused by autoimmune conditions, and 50% of the cases are associated with rheumatologic conditions, HL A-B27-associated disease such as seronegative arthropathies, juvenile rheumatoid arthritis and Reiter's syndrome or could be idiopathic. Infectious uveitis (20%) is comparatively less, and infection spreads to the highly vascular uvea from another part of the body (Durand 2015). The common infectious aetiologies of uveitis include herpetic infections, cytomegalovirus (Karavellas et al. 2000) retinitis, acute retinal necrosis and toxoplasmosis, and the less common are syphilis, TB, Lyme disease, sarcoidosis, multiple sclerosis and *Candida*. Breach in the blood-eye barrier is related to the pathophysiology of uveitis, and as a consequence of this breakdown in the barrier, WBCs enter the eye. Neutrophils predominate in chronic uveitis, whereas mononuclear cells predominate in acute uveitis cases (Durand 2015). This is like the findings of Horai and Caspi

(2019) who demonstrated that dysbiosis in the gut microbiome of uveitic mice triggered the activation of retina-specific autoreactive T cells in the gut which migrate to the eye, cross the blood retinal barrier and cause inflammation in the retina.

10.11 Cellular and Molecular Basis of Scleritis

It is assumed that the immune system is crucial to the pathogenesis of non-infectious scleritis which is recognized as an inflammation of the sclera which is sight threatening, with patients complaining of redness of the eye and severe pain. In the tears and blood of scleritis patients, several inflammatory factors are increased in abundance (Sainz-de-la-Maza et al. 2016; Vergouwen et al. 2020; Seo et al. 2006). Several cellular infiltrates like polymorpho-nuclear granulocytes, macrophages, $CD3^{+}$ T cells, CD138+ plasma cells, $CD20^{+}$ B cells and B-lymphoid follicles (Usui et al. 2008) are seen in the scleritis patients. Immune complexes of IgG and IgM types (Wakefield et al. 2013a) and apoptotic markers FasL (CD95L) and Fas (CD95) were in abundance in biopsies of necrotizing scleritis (Díaz-Valle et al. 1998; Wakefield et al. 2013a). It is suggested that the observed extensive necrosis in some scleritis cases is either due to ischemia or due to the breakdown of proteoglycan matrix and collagen fibrils facilitated by proteolytic enzymes including matrix metalloproteinases (MMPs) (Wakefield et al. 2013b). Thus, several studies indicated an autoimmune origin of scleritis (Wakefield et al. 2013a, b). For instance, antibodies directed against two sclera-specific antigens were present in idiopathic scleritis patients (Wakefield et al. 2013b). The process that initiates the autoimmune cascade in scleritis is so far not known though immunological factors, genetics and environmental factors are likely to be involved, including viral or bacterial infections including trauma (Watson and Young 2004; Wakefield et al. 2013b). Immunosuppressive therapy is recommended for scleritis treatment and involves three steps. First, the patients are treated with non-steroidal anti-inflammatory drugs. In the second step, the patients are treated with temporary use of anti-rheumatic drugs and/or prednisone which enhances T cell apoptosis and cytokine production and reduces B and T cell proliferation, leucocyte transmigration and antibody production (Daniel Diaz et al. 2016; Beardsley et al. 2013). In the third step, infliximab and adalimumab which inhibit the cytokine TNF-α (de Fidelix et al. 2015) and anakinra, tocilizumab and gevokizumab which inhibit several interleukins (Vergouwen et al. 2020; Bottin et al. 2018) and CD20 therapy (rituximab) were effective in scleritis patients (You et al. 2018; Cao et al. 2016). These treatment regimens involving B cells, T cells or cytokines suggested that the pathogenesis of scleritis involves both innate and adaptive immunity. Future investigations to unravel the pathogenesis of scleritis should include studies that lead to the identification of autoantibodies and/or autoreactive T cells and clinically relevant biomarkers and understanding the role of MMPs and gene loci conferring genetic susceptibility to scleritis.

10.12 Cellular and Molecular Basis of Conjunctivitis

Conjunctival tissue of conjunctivitis patients exhibited the presence of eosinophils, mast cells and goblet cells which may together constitute the cellular basis of the observed conjunctival inflammation (Munitz and Levi-Schaffer 2004; Miyazaki et al. 2008). Quantitative studies indicated that in the normal population, 6% of the individuals presented with low-grade conjunctival eosinophilia (Kari and Haahtela 1992), whereas it was significantly increased from 13 to 43% depending on whether it was perennial allergic conjunctivitis or acute non-allergic conjunctivitis (Kari 1988; Kari and Haahtela 1992). Eosinophils find their way to the conjunctival tissue from the bone marrow via peripheral blood circulation (Morgan et al. 1991; Kari and Haahtela 1992). Studies have also confirmed that the major effector cells are the mast cells in allergic conjunctivitis patients, and several factors like histamine, prostaglandins, tryptase, cytokines and leukotrienes play important roles in the immunopathological mechanisms in allergic conjunctivitis. The clinical features of allergic conjunctivitis include conjunctival hyperaemia, tearing, chemosis, itching, mucus discharge, papillae and eyelid oedema. The early phase of the allergic immune response is mediated by histamines, whereas cytokines and lipid mediators are involved in the late phase of the immunopathology of allergic conjunctivitis (De Gaulle and Minhas 2018).

10.13 Molecular Basis of Blepharitis

The pathogenesis of blepharitis was investigated by using proteomics to identify proteins in the tears of patients. Nine proteins were downregulated in blepharitis patients and included serum albumin precursor, lacritin precursor, alpha-1 antitrypsin, cystatin-SA III, lysozyme, prolactin inducible protein (PIP/GCDFP-15), Ig-kappa chain VIII, pyruvate kinase and an unnamed protein. Thus, these proteins that are different in the tears in blepharitis patients compared to healthy subjects may provide insights into the pathogenesis of blepharitis (Koo et al. 2005).

10.14 Cellular and Molecular Basis of Endophthalmitis

In endophthalmitis patients the posterior segment of the eye is infected, and such infections frequently cause loss of vision. A characteristic feature of endophthalmitis is that the patients exhibit a dense polymorphonuclear infiltrate in the vitreous cavity and other regions of the posterior segment of the eye like the trabecular meshwork, ciliary body, anterior and posterior lens capsule, choroids and retina (Engstrom et al. 1991). The onset of infiltration could be slow and delayed, as in the case of *S. epidermidis* endophthalmitis (Pleyer et al. 1992), or more fulminant, as in

B. cereus, *E. faecalis* and *S. aureus* endophthalmitis (Callegan et al. 1999). In addition to infiltration of cells, cell envelope, peptidoglycan, lipopolysaccharide and teichoic acid are also shed into these spaces during intraocular growth (Callegan et al. 2002). These infectious organisms and their other components stimulate resident inflammatory immune cells to produce pro-inflammatory cytokines and immune mediators in addition to toxins and reactive oxygen species involved in phagocytosis (Callegan et al. 2002). Callegan et al. (2007) suggested that endophthalmitis could cause mild inflammation which recovers on effective treatment or the inflammation could be severe and ineffective to treatment and may cause vision loss due to toxin production. The ultimate result is that the toxins cause RPE dysfunction, unimpeded infiltration of inflammatory cells and mediators into the posterior segment and breakdown of the blood retinal barrier and as a consequence upregulate inflammatory cytokines, resulting in severe inflammation (Callegan et al. 2002, 2007).

10.15 Cellular and Molecular Basis of Retinitis

Retinitis pigmentosa (RP) is a degenerative blinding condition caused by mutations in rhodopsin, the rod visual pigment of the eye (Athanasiou et al. 2018). Over 150 different mutations have been characterized in rhodopsin and are the most common cause of autosomal dominant RP (adRP), recessive RP (arRP) and congenital stationary night blindness (adCSNB). In-depth analyses of the rhodopsin mutations revealed seven discrete classes of mutations that affect primarily the protein structure and function; the observe defects range from misfolding, ER retention and instability, disruption of endocytosis, inappropriate localization, proteostasis, disruption in intracellular traffic, altered post-translational modifications, instability, dimerization deficiency and other altered functions (Table 10.1) (Athanasiou et al. 2018). Clinically two major classes of mutations are recognized based on the onset of rod dysfunction. Class A mutants had early-onset severe rod dysfunction, and class B had a slow-onset, less severe phenotype with a slower progression (Table 10.2). Thus, rhodopsin mutations are the basis for retinal degeneration, and it was suggested that it is important to develop mutation mechanism-dependent therapeutic strategies (Athanasiou et al. 2018). Ferrari et al. (2011) suggested that both genetics and clinical data together might provide better understanding of disease development and choices of therapy (Ferrari et al. 2011).

10.16 Cellular and Molecular Basis of Dry Eye Disease

Dry eye disease (DED) is a multifactorial disease affecting the tear film in which there is a loss of homeostasis of the tear film, thus altering the tear film stability and hyperosmolarity. In addition to these effects on the tear film, the ocular surface is

Table 10.1 Seven categories of rhodopsin mutations associated with autosomal dominant RP (adRP) and congenital stationary night blindness (adCSNB)[a]

Rhodopsin mutations	Key features	Mutation
1	Post-Golgi trafficking and OS targeting	*L328P,* Q344R/P/ter, *T342M*, A346P, V345L/M, ter349/Q/EP347A/R/Q/L/S/T
2	Misfolding, ER retention and instability	N15S[b], V20G[b], T17M[c], Q28H[b], P23A/H/L[b], P53R,G51R/V, V87D, T58R/*M*,L125R, G89D, G106*R*/W, E113K, C110F/R/S/Y, W161R,C167R/W, A164E/V, Y178N/D/CP171Q/L/S, E181K, C185R,G182S/V, C187G/Y, G188R/E, H211R/P, D190N/G/Y, C222R,S270R, P267R/L, 296N/E/M
3	Disrupted vesicular traffic and endocytosis	R135G/L/P/W
4	Altered post-translational modifications and reduced stability	T4K[c], N55K, T17M[c], M39R, G90V
5	Altered transducing activation	M44T, V137M
6	Constitutive activation	A292E, G90D,A295V, T94I
7	Dimerization deficiency	*F45L, F220C, V209M*
Unclassified	No observed biochemical or cellular defect or not studied in detail	F52Y,F56Y,P12R, R21C, Q28H, L40R, L46R, L47R, L57R, *R69H*,Y60ter, N78I, Q64ter, V87L,L79P, L88P, V104F, T92I, T97I, G114D/V, G109R, E122G, W126L/ter, S127F, L131P, Y136ter, C140S, T160T, M163T, A169P, P170H/R, S176F, P180A/S, Q184P, S186P/W, Y191C, *T193M*, M207R/K, *V210F*, I214N, P215L/T, M216R/L/K, *R252P*, T289P, S297R, A298D, K311E, N315ter, E341K, S343C

[a]From Athanasiou et al. (2018). For details refer to http://www.hgmd.cf.ac.uk/
[b]Might behave as class 4 after 11-cis-retinal rescue
[c]Might behave as class 2 on overexpression but class 4 in vivo

Table 10.2 Clinical classification of rhodopsin mutations

Phenotype	Mutation	Reference
Class A	L88P[1], L131P[1], R135L/G/W[2], E181K[2], Q344P[1], V345L[2], P347L[2]	Audo et al. (2010), Cideciyan et al. (1998), Iannaccone et al. (2006)
Class B	N15S[1], T17M[2], P23H[2], T58R[2], Q64ter[2], V87D[2], G89D[2], G106R[2], P180A[3], G188R[3], D190G/Y[2], T193M[2], M207K[2], Q312ter[2], T342M[2], Q344ter[2]	Audo et al. (2010), Cideciyan et al. (1998), Iannaccone et al. (2006)

From Athanasiou et al. (2018)
Class A: severe early onset of loss of rod function
Class B: retained rod function at least in certain retinal regions

inflamed and damaged, and neurosensory abnormalities are seen (Craig et al. 2017). Several proteins (such as S100A8, S100A9, secretory phospholipase A2, lipocalin-1, some chemokines, cytokines and metalloproteinase-9) and mucin-1 mRNA served as good DED biomarkers. Conjunctival cell proliferation capacity and viability have also been proposed as DED biomarkers. Neuromediator levels in tears were also involved in both DED and ocular allergy (Enríquez-de-Salamanca et al. 2012). In graft-versus-host disease-associated DED, a total of 33 genes associated with inflammatory pathways were significantly upregulated in the conjunctival cells compared to healthy controls. Of the 33 genes identified, NAMPT, EGFR, IL-6 and IL-9 were identified as the best markers of GVH-DED (Cocho et al. (2015). Lam et al. based on lipidome analyses of the meibum of DED Asian patients (Lam et al. 2011) observed that sphingomyelin and phosphatidylcholine were also significantly increased in DED patients. In moderate-DED patients, a significant lower level of triacylglyceride was detected compared with mild-DED patients, and the fatty acid profile was significantly different among the mild-, moderate- and severe-DED patients. But this data should be looked at with caution since the control and DED patients were not age-matched and sex-matched (Enríquez-de-Salamanca et al. 2012). Periman et al. (2020) reviewed the immunopathology of DED and suggested that the immunopathogenesis of DED is a four-part process (Fig. 10.4). In the initiation phase, acute ocular inflammation is observed, and it involved upregulation of MAP-K JNK, ERK and p38 stimulation of NF-κB. In addition, a combination of cytokines (IL-1, IL-6 and TNF-α), T cell-attracting chemokines (CCL5, CXCL9, CXCL10) and inflammatory chemokines (CCL3 and CCL4), which are pro-inflammatory mediators, is significantly increased in the conjunctival and corneal epithelium (Luo et al. 2004). The increase in IL-1 and TNF-α triggers the activation and maturation of APCs (antigen-presenting dendritic cells) (Hamrah et al. 2003). Upregulated ICAM-1 acts on lymphatic endothelial cells and facilitates mAPC adhesion (Johnson et al. 2006). The mAPCs migrate to regional lymph nodes and induce differentiation of TH0 cells into T_H1, T_H2, T_H17 and Tregs. Effector T cells then migrate to the conjunctival stroma (Johnson et al. 2006; Pflugfelder et al. 2017; Rouzaut et al. 2010) and lacrimal gland and release cytokines that cause ocular surface epithelial damage and tear dysfunction related to DED (El Annan et al. 2009). T_H1 cells secrete the hallmark cytokine IFN-γ which promotes apoptosis and conjunctival goblet cell loss of the ocular surface (De Paiva et al. 2007) and is associated with conjunctival squamous epithelial hyperplasia (De Paiva et al. 2007). T_H17 cells have a dual function of secreting cytokine IL-17 and also promote the production of the MMPs which are known to disrupt corneal epithelial barrier function, thus causing further ocular insults (Zhang et al. 2012). IL-17 and IFN-γ promote the production of MMPs, chemokines, pro-inflammatory cytokines, cell adhesion molecules and prolymphangiogenic molecules and thus exert pathogenic effects (Barabino et al. 2012). These modulators further damage the ocular surface and thus amplify the cycle of inflammation. Integrin LFA-1 interacts specifically with ICAM-1, and this interaction is important to the immunoinflammatory pathway of DED (Pflugfelder et al. 2017). Interaction between LFA-1 and ICAM-1 induces migration of dendritic cells to regional lymph nodes (Perez et al. 2016), aids in the migration of activated T cells to the site of inflammation from the blood (Muller 2013), facilitates immune synapse between T_H0 cells and mAPCs (Hogg et al. 2003), activates effector T cells at the ocular surface (Schaumburg et al. 2011), releases cytokines and chemokines (Schaumburg et al. 2011; Pflugfelder et al. 2017) and recruits T cells at the conjunctival epithelium and ocular surface (Pflugfelder 2008). Thus, considering that LFA-1/ICAM-1 interaction influences the ocular immune response and the ocular surface inflammation, targeting their association has emerged as therapeutic targets (Pflugfelder et al. 2017).

10.17 Cellular and Molecular Basis of Meibomian Gland Dysfunction

Recently Dietrich et al. (2021) reviewed the pathophysiology of meibomian gland dysfunction (MGD) and suggested that evaporative dry eye (EDE) is common in patients with MGD, leading to dry eye disease (DED) (Craig et al. 2017) characterized by lipid layer instability and increased breakup of the tear film (Chhadva et al. 2017). At these breakup points, excessive evaporation of the tears occurs leading to hyperosmolarity, which induces apoptosis in conjunctival and corneal epithelial cells and triggers ocular surface inflammation (Bron et al. 2017). Despite the high prevalence of DED in MGD patients (5–50%), the underlying molecular mechanisms of MGD are not clear (Posa et al. 2014). In MGD patients, androgen deficiency or anti-androgen treatment is associated with meibomian gland dysfunction and EDE (Sullivan et al. 2002). Gene expression analysis of human meibomian gland epithelial cells (hMGEC) treated with androgen dihydrotestosterone revealed that genes involved in lipid biosynthesis, lipid transport and binding were upregulated, while genes associated with cell cycle and RNA transport and keratinization together with pro-inflammatory genes were downregulated (Khandelwal et al. 2012). Overall, androgens appeared to have a positive influence on the function, differentiation and inflammatory status of the meibomian glands. Conversely, oestrogens tend to have a negative influence on meibomian gland function. A detailed gene expression analysis of lids obtained from patients with MGD revealed alterations in about 400 genes (Liu et al. 2011). Remarkably, genes associated with keratinization, such as small proline-rich proteins, S100 proteins, cornified envelope components, psoriasin and hornerin, were significantly upregulated, suggesting their responsibility for the hyper-keratinization of ductal epithelial cells (Liu et al. 2011; Garreis et al. 2011; Paulsen and Garreis 2014). The hMGEC showed that serum-induced differentiation leads to an upregulation of progesterone and oestrogen receptors. CK6 expression, calprotectin and neutrophil gelatinase-associated lipocalin (NGAL, Lcn2) were significantly upregulated in patients with meibomian gland dysfunction (Liu et al. 2011; Dale et al. 1983). Lee et al. (2018) demonstrated that markers related to keratinization, inflammation and lipid metabolism including small proline-rich protein 2A, keratin-10, S100-A8 and

S100-A9 were upregulated in the MGD group (Liu et al. 2011), and compared to healthy controls, MGD patients also had significantly elevated MMP-9, IL-1b, interleukin-8 (IL-8) and TGF-b1 gene expression (Zhang et al. 2015). Under physiological conditions, the transcription factor PPARγ plays an important role within the pathological processes that cause acinar atrophy, gland dropout and hyposecretion, leading to age-related meibomian gland dysfunction (Dietrich et al. 2021). An increase in leukocyte infiltration was found within the meibomian gland acini of elderly persons, which correlated with meibum quality (Nien et al. 2011) and gland dropout (Obata 2002). Recently, neutrophil extracellular traps (NETs) have been shown to reduce the differentiation and proliferation of hMGEC, which may contribute to inflammation-induced meibomian gland dysfunction (An et al. 2019). In addition, aggregated NETs have recently been shown to exacerbate the obstruction of meibomian gland orifices, acinar atrophy and ductal dilatation in a mouse model of allergic eye disease (Mahajan et al. 2021). Lipid composition changes of meibum or tear film were detected in patients suffering from meibomian gland-related blepharitis, meibomian gland dysfunction or EDE, which is likely to be responsible for burning eyes (Green-church et al. 2011). Various changes including a decrease in triglycerides and cholesterol (Shine and McCulley 1991, 1996) and a decrease in oleic acid, palmitic acid and stearic acid were also observed (Shine and McCulley 2000). Furthermore, meibum of patients with blepharitis and dry eye symptoms has generally low levels of phosphatidyl ethanolamine and sphingomyelin (Shine and McCulley 1998). Overall, research on the underlying mechanisms of meibomian gland pathology needs to be intensified.

10.18 Conclusions

The involvement of various immune cells and the up- and downregulation of several specific proteins, lipids and other molecules are highlighted in the diseased state of the eye compared to the normal healthy eye indicating the role of adaptive and cellular immunity in the pathogenesis of ocular diseases. But the sequence of cellular and molecular events that cause pathogenesis is yet to be established for all the diseases. Such studies would eventually help in targeted therapy of the molecules/pathways involved in the pathogenesis of the eye.

References

Akpek EK, Gottsch JD (2003) Immune defense at the ocular surface. Eye (Lond) 17:949–956. https://doi.org/10.1038/sj.eye.6700617

Akpek EK, Liu SH, Thompson R et al (2002) Identification of paramyosin as a binding protein for Calgranulin C in experimental helminthic keratitis. Invest Ophthalmol Vis Sci 43:2677–2684

Alberts B, Johnson A, Lewis J et al (2002) The adaptive immune system in molecular biology of the cell, 4th edn. Garland Science, New York

Alizadeh H, Pidherney MS, McCulley JP et al (1994a) Apoptosis as a mechanism of cytolysis of tumor cells by a pathogenic free-living amoeba. Infect Immun 62:1298–1303. https://doi.org/10.1128/iai.62.4.1298-1303.1994

Alizadeh H, Pidherney MS, McCulley JP et al (1994b) Acanthamoeba keratitis. In: Pepose JS, Holland GN, Wilhelmus KR (eds) Ocular infection and immunity. Mosby, St. Louis, pp 1062–1071

Alizadeh H, Apte S, El-Agha MS et al (2001) Tear IgA and serum IgG antibodies against Acanthamoeba in patients with Acanthamoeba keratitis. Cornea 20:622–627. https://doi.org/10.1097/00003226-200108000-00013

Alizadeh H, Neelam S, Hurt M et al (2005) Role of contact lens wear, bacterial flora, and mannose-induced pathogenic protease in the pathogenesis of amoebic keratitis. Infect Immun 73:1061–1068. https://doi.org/10.1128/IAI.73.2.1061-1068.2005

An S, Raju I, Surenkhuu B et al (2019) Neutrophil extracellular traps (NETs) contribute to pathological changes of ocular graft-vs.-host disease (oGVHD) dry eye: implications for novel biomarkers and therapeutic strategies. Ocul Surf 17:589–614. https://doi.org/10.1016/j.jtos.2019.03.010

Ansari Z, Miller D, Galor A (2013) Current thoughts in fungal keratitis: diagnosis and treatment. Curr Fungal Infect Rep 7:209–218. https://doi.org/10.1007/s12281-013-0150-110.1007/s12281-013-0150-1

Aoki T, Kitazawa K, Deguchi H et al (2021) Current evidence for corynebacterium on the ocular surface. Microorganisms 9(2):254. https://doi.org/10.3390/microorganisms9020254

Athanasiou D, Aguila M, Bellingham J et al (2018) The molecular and cellular basis of rhodopsin retinitis pigmentosa reveals potential strategies for therapy. Prog Retin Eye Res 62:1–23. https://doi.org/10.1016/j.preteyeres.2017.10.002

Audo I, Manes G, Mohand-Said S et al (2010) Spectrum of rhodopsin mutations in French autosomal dominant rod-cone dystrophy patients. Invest Ophthalmol Vis Sci 51:3687–3700. https://doi.org/10.1167/iovs.09-4766

Augustin DK, Heimer SR, Tam C et al (2011) Role of defensins in corneal epithelial barrier function against Pseudomonas aeruginosa traversal. Infect Immun 79:595–605. https://doi.org/10.1128/IAI.00854-10

Barabino S, Chen Y, Chauhan S et al (2012) Ocular surface immunity: homeostatic mechanisms and their disruption in dry eye disease. Prog Retin Eye Res 31:271–285. https://doi.org/10.1016/j.preteyeres.2012.02.003

Beardsley RM, Suhler EB, Rosenbaum JT et al (2013) Pharmacotherapy of scleritis: current paradigms and future directions. Expert Opin Pharmacother 14:411–424. https://doi.org/10.1517/14656566.2013.772982

Berman MB (1993) Regulation of corneal fibroblast MMP-1 collagenase secretion by plasmin. Cornea 12:420–432. https://doi.org/10.1097/00003226-199309000-00009

Bharathi MJ, Ramakrishnan R, Shivakumar C et al (2010) Etiology and antibacterial susceptibility pattern of community-acquired bacterial ocular infections in a tertiary eye care hospital in south India. Indian J Ophthalmol 58:497–507. https://doi.org/10.4103/0301-4738.71678

Bonini S, Micera A, Iovieno A et al (2005) Expression of Toll-like receptors in healthy and allergic conjunctiva. Ophthalmology 112:1528. ; discussion 1548-1529. https://doi.org/10.1016/j.ophtha.2005.04.009

Bottin C, Fel A, Butel N et al (2018) Anakinra in the treatment of patients with refractory scleritis: a pilot study. Ocul Immunol Inflamm 26:915–920. https://doi.org/10.1080/09273948.2017.1299869

Bouley DM, Kanangat S, Rouse BT (1996) The role of the innate immune system in the reconstituted SCID mouse model of herpetic stromal keratitis. Clin Immunol Immunopathol 80:23–30. https://doi.org/10.1006/clin.1996.0090

Bron AJ, de Paiva CS, Chauhan SK et al (2017) TFOS DEWS II pathophysiology report. Ocul Surf 15:438–510. https://doi.org/10.1016/j.jtos.2017.05.011

Burg ND, Pillinger MH (2001) The neutrophil: function and regulation in innate and humoral immunity. Clin Immunol 99:7–17. https://doi.org/10.1006/clim.2001.5007

Caballero AR, McCormick CC, Tang A et al (2008) Isolation and characterization of a new Staphylococcus aureus protease. Invest Ophthalmol Vis Sci 49:5514–5514

Caballero AR, Foster TJ, Monk IR et al (2010) Ocular pathology of a staphylococcus aureus mutant lacking a recently discovered virulence factor. Invest Ophthalmol Vis Sci 51:3891–3891

Callegan MC, Engel LS, Hill JM et al (1994) Corneal virulence of Staphylococcus aureus: roles of alpha-toxin and protein A in pathogenesis. Infect Immun 62:2478–2482. https://doi.org/10.1128/iai.62.6.2478-2482.1994

Callegan MC, Booth MC, Jett BD et al (1999) Pathogenesis of gram-positive bacterial endophthalmitis. Infect Immun 67:3348–3356. https://doi.org/10.1128/IAI.67.7.3348-3356.1999

Callegan MC, Engelbert M, Parke DW 2nd et al (2002) Bacterial endophthalmitis: epidemiology, therapeutics, and bacterium-host interactions. Clin Microbiol Rev 15:111–124. https://doi.org/10.1128/CMR.15.1.111-124.2002

Callegan MC, Gilmore MS, Gregory M et al (2007) Bacterial endophthalmitis: therapeutic challenges and host-pathogen interactions. Prog Retin Eye Res 26:189–203. https://doi.org/10.1016/j.preteyeres.2006.12.001

Cao Z, Jefferson DM, Panjwani N (1998) Role of carbohydrate-mediated adherence in cytopathogenic mechanisms of Acanthamoeba. J Biol Chem 273:15838–15845. https://doi.org/10.1074/jbc.273.25.15838

Cao JH, Oray M, Cocho L et al (2016) Rituximab in the Treatment of Refractory Noninfectious Scleritis. Am J Ophthalmol 164:22–28. https://doi.org/10.1016/j.ajo.2015.12.032

Chandler JW, Cummings M, Gillette TE (1985) Presence of Langerhans cells in the central corneas of normal human infants. Invest Ophthalmol Vis Sci 26:113–116

Chang JH, McCluskey P, Wakefield D (2004) Expression of toll-like receptor 4 and its associated lipopolysaccharide receptor complex by resident antigen-presenting cells in the human uvea. Invest Ophthalmol Vis Sci 45:1871–1878. https://doi.org/10.1167/iovs.03-1113

Cheng H, Tumpey TM, Staats HF et al (2000) Role of macrophages in restricting herpes simplex virus type 1 growth after ocular infection. Invest Ophthalmol Vis Sci 41:1402–1409

Cherwinski HM, Schumacher JH, Brown KD et al (1987) Two types of mouse helper T cell clone. III. Further differences in lymphokine synthesis between Th1 and Th2 clones revealed by RNA hybridization, functionally monospecific bioassays, and monoclonal antibodies. J Exp Med 166: 1229–1244. https://doi.org/10.1084/jem.166.5.1229

Cheung J, Slomovic AR (1995) Microbial etiology and predisposing factors among patients hospitalized for corneal ulceration. Can J Ophthalmol 30:251–255

Chhadva P, Goldhardt R, Galor A (2017) Meibomian gland disease: the role of gland dysfunction in dry eye disease. Ophthalmology 124:S20–S26. https://doi.org/10.1016/j.ophtha.2017.05.031

Cho JH, Na BK, Kim TS et al (2000) Purification and characterization of an extracellular serine proteinase from Acanthamoeba castellanii. IUBMB Life 50:209–214. https://doi.org/10.1080/152165400300001534

Cideciyan AV, Hood DC, Huang Y et al (1998) Disease sequence from mutant rhodopsin allele to rod and cone photoreceptor degeneration in man. Proc Natl Acad Sci U S A 95:7103–7108. https://doi.org/10.1073/pnas.95.12.7103

Clarke DW, Niederkorn JY (2006) The pathophysiology of Acanthamoeba keratitis. Trends Parasitol 22:175–180. https://doi.org/10.1016/j.pt.2006.02.004

Clarke DW, Alizadeh H, Niederkorn JY (2005) Failure of Acanthamoeba castellanii to produce intraocular infections. Invest Ophthalmol Vis Sci 46:2472–2478. https://doi.org/10.1167/iovs.05-0140

Clarke B, Sinha A, Parmar DN et al (2012) Advances in the diagnosis and treatment of acanthamoeba keratitis. J Ophthalmol 2012:484892. https://doi.org/10.1155/2012/484892

Cocho L, Fernandez I, Calonge M et al (2015) Gene expression-based predictive models of graft versus host disease-associated dry eye. Invest Ophthalmol Vis Sci 56:4570–4581. https://doi.org/10.1167/iovs.15-16736

Coster DJ, Badenoch PR (1987) Host, microbial, and pharmacological factors affecting the outcome of suppurative keratitis. Br J Ophthalmol 71:96–101. https://doi.org/10.1136/bjo.71.2.96

Coulot P, Bouchara JP, Renier G et al (1994) Specific interaction of Aspergillus fumigatus with fibrinogen and its role in cell adhesion. Infect Immun 62:2169–2177. https://doi.org/10.1128/iai.62.6.2169-2177.1994

Craig JP, Nichols KK, Akpek EK et al (2017) TFOS DEWS II Definition and classification report. Ocul Surf 15:276–283. https://doi.org/10.1016/j.jtos.2017.05.008

Cubitt CL, Lausch RN, Oakes JE (1995) Differences in interleukin-6 gene expression between cultured human corneal epithelial cells and keratocytes. Invest Ophthalmol Vis Sci 36:330–336

Dabbagh K, Lewis DB (2003) Toll-like receptors and T-helper-1/T-helper-2 responses. Curr Opin Infect Dis 16:199–204. https://doi.org/10.1097/00001432-200306000-00003

Dale I, Fagerhol MK, Naesgaard I (1983) Purification and partial characterization of a highly immunogenic human leukocyte protein, the L1 antigen. Eur J Biochem 134:1–6. https://doi.org/10.1111/j.1432-1033.1983.tb07522.x

Daniel Diaz J, Sobol EK, Gritz DC (2016) Treatment and management of scleral disorders. Surv Ophthalmol 61:702–717. https://doi.org/10.1016/j.survophthal.2016.06.002

Dart JK, Saw VP, Kilvington S (2009) Acanthamoeba keratitis: diagnosis and treatment update 2009. Am J Ophthalmol 148(487-499):e482. https://doi.org/10.1016/j.ajo.2009.06.009

Dartt DA (2011) Tear lipocalin: structure and function. Ocul Surf 9:126–138. https://doi.org/10.1016/s1542-0124(11)70022-2

De Gaulle C, Minhas B (2018) Immunopathology of allergic conjunctivitis. Eur Med J 3(1):76–83

De Paiva CS, Villarreal AL, Corrales RM et al (2007) Dry eye-induced conjunctival epithelial squamous metaplasia is modulated by interferon-gamma. Invest Ophthalmol Vis Sci 48:2553–2560. https://doi.org/10.1167/iovs.07-0069

Díaz-Valle D, del Castillo JMB, Castillo A et al (1998) Immunologic and clinical evaluation of postsurgical necrotizing sclerocorneal ulceration. Cornea 17

Dietrich J, Garreis F, Paulsen F (2021) Pathophysiology of meibomian glands—an overview. Ocul Immunol Inflamm 29:803–810. https://doi.org/10.1080/09273948.2021.1905856

Dong Q, Brulc JM, Iovieno A et al (2011) Diversity of bacteria at healthy human conjunctiva. Invest Ophthalmol Vis Sci 52:5408–5413. https://doi.org/10.1167/iovs.10-6939

Durand ML (2015) 117 - Infectious causes of uveitis. In: Bennett JE, Dolin R, Blaser MJ (eds) Mandell, Douglas, and Bennett's principles and practice of infectious diseases, 8th edn. W.B. Saunders, pp 1423–1431.e2. https://doi.org/10.1016/B978-1-4557-4801-3.00117-X. ISBN 9781455748013

Dutta D, Cole N, Willcox M (2012) Factors influencing bacterial adhesion to contact lenses. Mol Vis 18:14–21

El Annan J, Chauhan SK, Ecoiffier T et al (2009) Characterization of effector T cells in dry eye disease. Invest Ophthalmol Vis Sci 50:3802–3807. https://doi.org/10.1167/iovs.08-2417

Engstrom RE Jr, Mondino BJ, Glasgow BJ et al (1991) Immune response to Staphylococcus aureus endophthalmitis in a rabbit model. Invest Ophthalmol Vis Sci 32:1523–1533

Enríquez-de-Salamanca A, Bonini S, Calonge M (2012) Molecular and cellular biomarkers in dry eye disease and ocular allergy. Curr Opin Allergy Clin Immunol 12

Ferrari S, Di Iorio E, Barbaro V et al (2011) Retinitis pigmentosa: genes and disease mechanisms. Curr Genomics 12:238–249. https://doi.org/10.2174/138920211795860107

de Fidelix TS, Vieira LA, de Freitas D et al (2015) Biologic therapy for refractory scleritis: a new treatment perspective. Int Ophthalmol 35:903–912. https://doi.org/10.1007/s10792-015-0124-0

Flanagan JL, Willcox MD (2009) Role of lactoferrin in the tear film. Biochimie 91:35–43. https://doi.org/10.1016/j.biochi.2008.07.007

Gallin MY, Murray D, Lass JH et al (1988) Experimental interstitial keratitis induced by Onchocerca volvulus antigens. Arch Ophthalmol 106:1447–1452. https://doi.org/10.1001/archopht.1988.01060140611033

Ganz T, Selsted ME, Szklarek D et al (1985) Defensins. Natural peptide antibiotics of human neutrophils. J Clin Invest 76:1427–1435. https://doi.org/10.1172/JCI112120

Garreis F, Gottschalt M, Paulsen FP (2010) Antimicrobial peptides as a major part of the innate immune defense at the ocular surface. Dev Ophthalmol 45:16–22. https://doi.org/10.1159/000315016

Garreis F, Gottschalt M, Schlorf T et al (2011) Expression and regulation of antimicrobial peptide psoriasin (S100A7) at the ocular surface and in the lacrimal apparatus. Invest Ophthalmol Vis Sci 52:4914–4922. https://doi.org/10.1167/iovs.10-6598

Gillette TE, Chandler JW, Greiner JV (1982) Langerhans cells of the ocular surface. Ophthalmology 89:700–711. https://doi.org/10.1016/s0161-6420(82)34737-5

Gopinathan U, Ramakrishna T, Willcox M et al (2001) Enzymatic, clinical and histologic evaluation of corneal tissues in experimental fungal keratitis in rabbits. Exp Eye Res 72:433–442. https://doi.org/10.1006/exer.2000.0971

Gopinathan U, Sharma S, Garg P et al (2009) Review of epidemiological features, microbiological diagnosis and treatment outcome of microbial keratitis: experience of over a decade. Indian J Ophthalmol 57:273–279. https://doi.org/10.4103/0301-4738.53051

Gottsch JD, Li Q, Ashraf MF et al (1998) Defensin gene expression in the cornea. Curr Eye Res 17: 1082–1086. https://doi.org/10.1076/ceyr.17.11.1082.5235

Govindarajan B, Gipson IK (2010) Membrane-tethered mucins have multiple functions on the ocular surface. Exp Eye Res 90:655–663. https://doi.org/10.1016/j.exer.2010.02.014

Govindarajan B, Menon BB, Spurr-Michaud S et al (2012) A metalloproteinase secreted by Streptococcus pneumoniae removes membrane mucin MUC16 from the epithelial glycocalyx barrier. PLoS One 7:e32418. https://doi.org/10.1371/journal.pone.0032418

Gravet A, Colin DA, Keller D et al (1998) Corrigendum to: characterization of a novel structural member, LukE-LukD, of the bi-component staphylococcal leucotoxins family (FEBS 20890). FEBS Lett 441:342–342. https://doi.org/10.1016/s0014-5793(98)01472-0

Green M, Apel A, Stapleton F (2008) Risk factors and causative organisms in microbial keratitis. Cornea 27:22–27. https://doi.org/10.1097/ICO.0b013e318156caf2

Green-Church KB, Butovich I, Willcox M et al (2011) The international workshop on meibomian gland dysfunction: report of the subcommittee on tear film lipids and lipid-protein interactions in health and disease. Invest Ophthalmol Vis Sci 52:1979–1993. https://doi.org/10.1167/iovs.10-6997d

Hammerschmidt S (2006) Adherence molecules of pathogenic pneumococci. Curr Opin Microbiol 9:12–20. https://doi.org/10.1016/j.mib.2005.11.001

Hamrah P, Liu Y, Zhang Q et al (2003) Alterations in corneal stromal dendritic cell phenotype and distribution in inflammation. Arch Ophthalmol 121:1132–1140. https://doi.org/10.1001/archopht.121.8.1132

Hazlett LD (2002) Pathogenic mechanisms of P. aeruginosa keratitis: a review of the role of T cells, Langerhans cells, PMN, and cytokines. DNA Cell Biol 21:383–390. https://doi.org/10.1089/10445490260099665

Hazlett LD, McClellan SM, Hume EB et al (1999) Extended wear contact lens usage induces Langerhans cell migration into cornea. Exp Eye Res 69:575–577. https://doi.org/10.1006/exer.1999.0728

Hejazi A, Falkiner FR (1997) Serratia marcescens. J Med Microbiol 46:903–912. https://doi.org/10.1099/00222615-46-11-903

Hendricks RL, Weber PC, Taylor JL et al (1991) Endogenously produced interferon alpha protects mice from herpes simplex virus type 1 corneal disease. J Gen Virol 72(Pt 7):1601–1610. https://doi.org/10.1099/0022-1317-72-7-1601

Hogg N, Laschinger M, Giles K et al (2003) T-cell integrins: more than just sticking points. J Cell Sci 116:4695–4705. https://doi.org/10.1242/jcs.00876

Horai R, Caspi RR (2019) Microbiome and autoimmune uveitis. Front Immunol 10:232. https://doi.org/10.3389/fimmu.2019.00232. Published 2019 Feb 19

Huang X, Du W, McClellan SA et al (2006) TLR4 is required for host resistance in Pseudomonas aeruginosa keratitis. Invest Ophthalmol Vis Sci 47:4910–4916. https://doi.org/10.1167/iovs.06-0537

Hume EB, Conerly LL, Moreau JM et al (1999) Serratia marcescens keratitis: strain-specific corneal pathogenesis in rabbits. Curr Eye Res 19:525–532. https://doi.org/10.1076/ceyr.19.6.525.5283

Iannaccone A, Man D, Waseem N et al (2006) Retinitis pigmentosa associated with rhodopsin mutations: correlation between phenotypic variability and molecular effects. Vision Res 46: 4556–4567. https://doi.org/10.1016/j.visres.2006.08.018

Imwidthaya P (1995) Mycotic keratitis in Thailand. J Med Vet Mycol 33:81–82. https://doi.org/10.1080/02681219580000171

Jackson BE, Wilhelmus KR, Hube B (2007) The role of secreted aspartyl proteinases in Candida albicans keratitis. Invest Ophthalmol Vis Sci 48:3559–3565. https://doi.org/10.1167/iovs.07-0114

Jain V, Maiti A, Shome D et al (2007) Aspergillus-induced malignant glaucoma. Cornea 26:762–763. https://doi.org/10.1097/ICO.0b013e31805fc37b

Jeng BH, Gritz DC, Kumar AB et al (2010) Epidemiology of ulcerative keratitis in Northern California. Arch Ophthalmol 128:1022–1028. https://doi.org/10.1001/archophthalmol.2010.144

Jett BD, Gilmore MS (2002) Host-parasite interactions in Staphylococcus aureus keratitis. DNA Cell Biol 21:397–404. https://doi.org/10.1089/10445490260099683

Johnson AC, Heinzel FP, Diaconu E et al (2005) Activation of toll-like receptor (TLR)2, TLR4, and TLR9 in the mammalian cornea induces MyD88-dependent corneal inflammation. Invest Ophthalmol Vis Sci 46:589–595. https://doi.org/10.1167/iovs.04-1077

Johnson LA, Clasper S, Holt AP et al (2006) An inflammation-induced mechanism for leukocyte transmigration across lymphatic vessel endothelium. J Exp Med 203:2763–2777. https://doi.org/10.1084/jem.20051759

Johnson AC, Li X, Pearlman E (2008) MyD88 functions as a negative regulator of TLR3/TRIF-induced corneal inflammation by inhibiting activation of c-Jun N-terminal kinase. J Biol Chem 283:3988–3996. https://doi.org/10.1074/jbc.M707264200

Kaifi JT, Diaconu E, Pearlman E (2001) Distinct roles for PECAM-1, ICAM-1, and VCAM-1 in recruitment of neutrophils and eosinophils to the cornea in ocular onchocerciasis (river blindness). J Immunol 166:6795–6801. https://doi.org/10.4049/jimmunol.166.11.6795

Kamata R, Matsumoto K, Okamura R et al (1985) The Serratial 56K Protease as a Major Pathogenic Factor in Serratial Keratitis. Ophthalmology 92:1452–1459. https://doi.org/10.1016/s0161-6420(85)33855-1

Karavellas MP, Song M, Macdonald JC, Freeman WR (2000) Longterm posterior and anterior segment complications of immune recovery uveitis associated with cytomegalovirus retinitis. Am J Ophthalmol 130:57–64

Kari O (1988) Atopic conjunctivitis. A cytologic examination. Acta Ophthalmol (Copenh) 66:381–386. https://doi.org/10.1111/j.1755-3768.1988.tb04027.x

Kari O, Haahtela T (1992) Conjunctival eosinophilia in atopic and non-atopic external eye symptoms. Acta Ophthalmol 70:335–340. https://doi.org/10.1111/j.1755-3768.1992.tb08574.x

Karmakar M, Sun Y, Hise AG et al (2012) Cutting edge: IL-1beta processing during Pseudomonas aeruginosa infection is mediated by neutrophil serine proteases and is independent of NLRC4 and caspase-1. J Immunol 189:4231–4235. https://doi.org/10.4049/jimmunol.1201447

Kernacki KA, Barrett RP, Hobden JA et al (2000) Macrophage inflammatory protein-2 is a mediator of polymorphonuclear neutrophil influx in ocular bacterial infection. J Immunol 164:1037–1045. https://doi.org/10.4049/jimmunol.164.2.1037

Khan NA (2001) Pathogenicity, morphology, and differentiation of Acanthamoeba. Curr Microbiol 43:391–395. https://doi.org/10.1007/s002840010325

Khandelwal P, Liu S, Sullivan DA (2012) Androgen regulation of gene expression in human meibomian gland and conjunctival epithelial cells. Mol Vis 18:1055–1067

Kiryu H, Yoshida S, Suenaga Y et al (1991) Invasion and survival of Fusarium solani in the dexamethasone-treated cornea of rabbits. J Med Vet Mycol 29:395–406. https://doi.org/10.1080/02681219180000631

Koo BS, Lee DY, Ha HS et al (2005) Comparative analysis of the tear protein expression in blepharitis patients using two-dimensional electrophoresis. J Proteome Res 4:719–724. https://doi.org/10.1021/pr0498133

Kreger AS, Lyerly DM, Hazlett LD et al (1986) Immunization against experimental Pseudomonas aeruginosa and Serratia marcescens keratitis. Vaccination with lipopolysaccharide endotoxins and proteases. Invest Ophthalmol Vis Sci 27:932–939

Kumagai N, Fukuda K, Fujitsu Y et al (2005) Lipopolysaccharide-induced expression of intercellular adhesion molecule-1 and chemokines in cultured human corneal fibroblasts. Invest Ophthalmol Vis Sci 46:114–120. https://doi.org/10.1167/iovs.04-0922

Kumar MV, Nagineni CN, Chin MS et al (2004) Innate immunity in the retina: toll-like receptor (TLR) signaling in human retinal pigment epithelial cells. J Neuroimmunol 153:7–15. https://doi.org/10.1016/j.jneuroim.2004.04.018

Kumar A, Yin J, Zhang J et al (2007) Modulation of corneal epithelial innate immune response to pseudomonas infection by flagellin pretreatment. Invest Ophthalmol Vis Sci 48:4664–4670. https://doi.org/10.1167/iovs.07-0473

Kuriakose T, Thomas PA (1991) Keratomycotic malignant glaucoma. Indian J Ophthalmol 39:118–121

Labbate M, Zhu H, Thung L et al (2007) Quorum-sensing regulation of adhesion in Serratia marcescens MG1 is surface dependent. J Bacteriol 189:2702–2711. https://doi.org/10.1128/JB.01582-06

Lakhundi S, Siddiqui R, Khan NA (2017) Pathogenesis of microbial keratitis. Microb Pathog 104: 97–109. https://doi.org/10.1016/j.micpath.2016.12.013

Lam SM, Tong L, Yong SS et al (2011) Meibum lipid composition in Asians with dry eye disease. PLoS One 6:e24339. https://doi.org/10.1371/journal.pone.0024339

Leal SM Jr, Vareechon C, Cowden S et al (2012) Fungal antioxidant pathways promote survival against neutrophils during infection. J Clin Invest 122:2482–2498. https://doi.org/10.1172/JCI63239

Leck AK, Thomas PA, Hagan M et al (2002) Aetiology of suppurative corneal ulcers in Ghana and south India, and epidemiology of fungal keratitis. Br J Ophthalmol 86:1211–1215. https://doi.org/10.1136/bjo.86.11.1211

Lee EJ, Cowell BA, Evans DJ et al (2003) Contribution of ExsA-regulated factors to corneal infection by cytotoxic and invasive Pseudomonas aeruginosa in a murine scarification model. Invest Ophthalmol Vis Sci 44:3892–3898. https://doi.org/10.1167/iovs.02-1302

Lee L, Garrett Q, Flanagan J et al (2018) Genetic factors and molecular mechanisms in dry eye disease. Ocul Surf 16:206–217. https://doi.org/10.1016/j.jtos.2018.03.003

Liu S, Richards SM, Lo K et al (2011) Changes in gene expression in human meibomian gland dysfunction. Invest Ophthalmol Vis Sci 52:2727–2740. https://doi.org/10.1167/iovs.10-6482

Luo L, Li DQ, Doshi A et al (2004) Experimental dry eye stimulates production of inflammatory cytokines and MMP-9 and activates MAPK signaling pathways on the ocular surface. Invest Ophthalmol Vis Sci 45:4293–4301. https://doi.org/10.1167/iovs.03-1145

Lyerly D, Kreger A (1979) Purification and characterization of a Serratia marcescens metalloprotease. Infect Immun 24:411–421. https://doi.org/10.1128/iai.24.2.411-421.1979

Lyerly D, Gray L, Kreger A (1981) Characterization of rabbit corneal damage produced by Serratia keratitis and by a serratia protease. Infect Immun 33:927–932. https://doi.org/10.1128/iai.33.3.927-932.1981

Mahajan A, Hasikova L, Hampel U et al (2021) Aggregated neutrophil extracellular traps occlude Meibomian glands during ocular surface inflammation. Ocul Surf 20:1–12. https://doi.org/10.1016/j.jtos.2020.12.005

Mannis MJ, Smolin G (1996) Natural defense mechanisms of the ocular surface. In: Pepose JS, Holland GN, Wilhelmeus KR (eds). Ocular infection and immunity, Mosby: St Louis, MO, pp 185-190

Marquart ME, O'Callaghan RJ (2013) Infectious keratitis: secreted bacterial proteins that mediate corneal damage. J Ophthalmol 2013:369094. https://doi.org/10.1155/2013/369094

Marty KB, Williams CL, Guynn LJ et al (2002) Characterization of a cytotoxic factor in culture filtrates of Serratia marcescens. Infect Immun 70:1121–1128. https://doi.org/10.1128/IAI.70.3.1121-1128.2002

McClellan SA, Huang X, Barrett RP et al (2003) Macrophages restrict Pseudomonas aeruginosa growth, regulate polymorphonuclear neutrophil influx, and balance pro- and anti-inflammatory cytokines in BALB/c mice. J Immunol 170:5219–5227. https://doi.org/10.4049/jimmunol.170.10.5219

McDermott AM (2009) The role of antimicrobial peptides at the ocular surface. Ophthalmic Res 41: 60–75. https://doi.org/10.1159/000187622

Menestrina G, Serra MD, Prevost G (2001) Mode of action of beta-barrel pore-forming toxins of the staphylococcal alpha-hemolysin family. Toxicon 39:1661–1672. https://doi.org/10.1016/s0041-0101(01)00153-2

Micera A, Stampachiacchiere B, Aronni S et al (2005) Toll-like receptors and the eye. Curr Opin Allergy Clin Immunol 5:451–458. https://doi.org/10.1097/01.all.0000182537.55650.99

Miller D, Iovieno A (2009) The role of microbial flora on the ocular surface. Curr Opin Allergy Clin Immunol 9:466–470. https://doi.org/10.1097/ACI.0b013e3283303e1b

Miyazaki D, Tominaga T, Yakura K et al (2008) Conjunctival mast cell as a mediator of eosinophilic response in ocular allergy. Mol Vis 14:1525–1532

Moretta L, Bottino C, Pende D et al (2002) Human natural killer cells: their origin, receptors and function. Eur J Immunol 32:1205–1211. https://doi.org/10.1002/1521-4141(200205)32:5<1205::Aid-immu1205>3.0.Co;2-y

Morgan SJ, Williams JH, Walls AF et al (1991) Mast cell hyperplasia in atopic keratoconjunctivitis. An immunohistochemical study. Eye (Lond) 5(Pt 6):729–735. https://doi.org/10.1038/eye.1991.134

Mosmann TR, Cherwinski H, Bond MW et al (1986) Two types of murine helper T cell clone. I Definition according to profiles of lymphokine activities and secreted proteins. J Immunol 136: 2348–2357

Mulet ME, Perez-Santonja JJ, Ferrer C et al (2010) Microbial keratitis after intrastromal corneal ring segment implantation. J Refract Surg 26:364–369. https://doi.org/10.3928/1081597X-20090617-06

Muller WA (2013) Getting leukocytes to the site of inflammation. Vet Pathol 50:7–22. https://doi.org/10.1177/0300985812469883

Müller LJ, Pels L, Vrensen GF (1996) Ultrastructural organization of human corneal nerves. Invest Ophthalmol Vis Sci 37:476–488

Munitz A, Levi-Schaffer F (2004) Eosinophils: 'new' roles for 'old' cells. Allergy 59:268–275. https://doi.org/10.1111/j.1398-9995.2003.00442.x

Musa F, Tailor R, Gao A et al (2010) Contact lens-related microbial keratitis in deployed British military personnel. Br J Ophthalmol 94:988–993. https://doi.org/10.1136/bjo.2009.161430

Nagington J, Watson PG, Playfair TJ et al (1974) Amoebic infection of the eye. Lancet 304:1537–1540. https://doi.org/10.1016/s0140-6736(74)90285-2

Nath R, Baruah S, Saikia L et al (2011) Mycotic corneal ulcers in upper Assam. Indian J Ophthalmol 59:367–371. https://doi.org/10.4103/0301-4738.83613

Nathan C (2006) Neutrophils and immunity: challenges and opportunities. Nat Rev Immunol 6: 173–182. https://doi.org/10.1038/nri1785

Niederkorn JY, Peeler JS, Mellon J (1989) Phagocytosis of particulate antigens by corneal epithelial cells stimulates interleukin-1 secretion and migration of Langerhans cells into the central cornea. Reg Immunol 2:83–90

Niederkorn JY, Alizadeh H, Leher H et al (1999) The pathogenesis of Acanthamoeba keratitis. Microbes Infect 1:437–443. https://doi.org/10.1016/s1286-4579(99)80047-1

Niemialtowski MG, Rouse BT (1992) Predominance of Th1 cells in ocular tissues during herpetic stromal keratitis. J Immunol 149:3035–3039

Nien CJ, Massei S, Lin G et al (2011) Effects of age and dysfunction on human meibomian glands. Arch Ophthalmol 129:462–469. https://doi.org/10.1001/archophthalmol.2011.69

Noisakran S, Carr DJ (2000) Plasmid DNA encoding IFN-alpha 1 antagonizes herpes simplex virus type 1 ocular infection through CD4+ and CD8+ T lymphocytes. J Immunol 164:6435–6443. https://doi.org/10.4049/jimmunol.164.12.6435

Obata H (2002) Anatomy and histopathology of human meibomian gland. Cornea 21:S70–S74. https://doi.org/10.1097/01.ico.0000263122.45898.09

Ollivier FJ, Brooks DE, Van Setten GB et al (2004) Profiles of matrix metalloproteinase activity in equine tear fluid during corneal healing in 10 horses with ulcerative keratitis. Vet Ophthalmol 7: 397–405. https://doi.org/10.1111/j.1463-5224.2004.04052.x

Panjwani N (2010) Pathogenesis of acanthamoeba keratitis. Ocul Surf 8:70–79. https://doi.org/10.1016/s1542-0124(12)70071-x

Pany S, Vijayvargia R, Krishnasastry MV (2004) Caveolin-1 binding motif of alpha-hemolysin: its role in stability and pore formation. Biochem Biophys Res Commun 322:29–36. https://doi.org/10.1016/j.bbrc.2004.07.073

Parks WC (1999) Matrix metalloproteinases in repair. Wound Repair Regen 7:423–432. https://doi.org/10.1046/j.1524-475x.1999.00423.x

Parment PA (1997) The role of Serratia marcescens in soft contact lens associated ocular infections. A review. Acta Ophthalmol Scand 75:67–71. https://doi.org/10.1111/j.1600-0420.1997.tb00253.x

Parta M, Chang Y, Rulong S et al (1994) HYP1, a hydrophobin gene from Aspergillus fumigatus, complements the rodletless phenotype in Aspergillus nidulans. Infect Immun 62:4389–4395. https://doi.org/10.1128/iai.62.10.4389-4395.1994

Paton JC, Rowan-Kelly B, Ferrante A (1984) Activation of human complement by the pneumococcal toxin pneumolysin. Infect Immun 43:1085–1087. https://doi.org/10.1128/iai.43.3.1085-1087.1984

Paulsen F, Garreis F (2014) What drives Meibomian gland disease? Arch Soc Esp Oftalmol 89: 175–176. https://doi.org/10.1016/j.oftal.2014.05.001

Pearlman E, Lass JH, Bardenstein DS et al (1995) Interleukin 4 and T helper type 2 cells are required for development of experimental onchocercal keratitis (river blindness). J Exp Med 182:931–940. https://doi.org/10.1084/jem.182.4.931

Pearlman E, Hall LR, Higgins AW et al (1998) The role of eosinophils and neutrophils in helminth-induced keratitis. Invest Ophthalmol Vis Sci 39:1176–1182

Pearlman E, Sun Y, Roy S et al (2013) Host defense at the ocular surface. Int Rev Immunol 32:4–18. https://doi.org/10.3109/08830185.2012.749400

Perez VL, Pflugfelder SC, Zhang S et al (2016) Lifitegrast, a novel integrin antagonist for treatment of dry eye disease. Ocul Surf 14:207–215. https://doi.org/10.1016/j.jtos.2016.01.001

Periman LM, Perez VL, Saban DR et al (2020) The immunological basis of dry eye disease and current topical treatment options. J Ocul Pharmacol Ther 36:137–146. https://doi.org/10.1089/jop.2019.0060

Petrillo F, Pignataro D, Lavano MA et al (2020) Current evidence on the ocular surface microbiota and related diseases. Microorganisms. 8(7):1033. https://doi.org/10.3390/microorganisms8071033

Pettit DA, Williamson J, Cabral GA et al (1996) In vitro destruction of nerve cell cultures by Acanthamoeba spp.: a transmission and scanning electron microscopy study. J Parasitol 82: 769–777

Pflugfelder SC (2008) Prevalence, burden, and pharmacoeconomics of dry eye disease. Am J Manag Care 14:S102–S106

Pflugfelder SC, Stern M, Zhang S et al (2017) LFA-1/ICAM-1 Interaction as a therapeutic target in dry eye disease. J Ocul Pharmacol Ther 33:5–12. https://doi.org/10.1089/jop.2016.0105

Pleyer U, Mondino BJ, Adamu SA et al (1992) Immune response to Staphylococcus epidermidis-induced endophthalmitis in a rabbit model. Invest Ophthalmol Vis Sci 33:2650–2663

Posa A, Sel S, Dietz R et al (2014) Updated incidence of dry eye syndrome in Germany. Klin Monbl Augenheilkd 231:42–46. https://doi.org/10.1055/s-0032-1328749

Rajam G, Anderton JM, Carlone GM et al (2008) Pneumococcal surface adhesin A (PsaA): a review. Crit Rev Microbiol 34:131–142. https://doi.org/10.1080/10408410802275352

Redfern RL, Reins RY, McDermott AM (2011) Toll-like receptor activation modulates antimicrobial peptide expression by ocular surface cells. Exp Eye Res 92:209–220. https://doi.org/10.1016/j.exer.2010.12.005

Rehany U, Balut G, Lefler E et al (2004) The prevalence and risk factors for donor corneal button contamination and its association with ocular infection after transplantation. Cornea 23:649–654. https://doi.org/10.1097/01.ico.0000139633.50035.cf

Rhem MN, Lech EM, Patti JM et al (2000) The collagen-binding adhesin is a virulence factor in Staphylococcus aureus keratitis. Infect Immun 68:3776–3779. https://doi.org/10.1128/IAI.68.6.3776-3779.2000

Rouzaut A, Garasa S, Teijeira A et al (2010) Dendritic cells adhere to and transmigrate across lymphatic endothelium in response to IFN-alpha. Eur J Immunol 40:3054–3063. https://doi.org/10.1002/eji.201040523

Roy S, Sun Y, Pearlman E (2011) Interferon-gamma-induced MD-2 protein expression and lipopolysaccharide (LPS) responsiveness in corneal epithelial cells is mediated by Janus tyrosine kinase-2 activation and direct binding of STAT1 protein to the MD-2 promoter. J Biol Chem 286:23753–23762. https://doi.org/10.1074/jbc.M111.219345

Rudner XL, Kernacki KA, Barrett RP et al (2000) Prolonged elevation of IL-1 in Pseudomonas aeruginosa ocular infection regulates macrophage-inflammatory protein-2 production, polymorphonuclear neutrophil persistence, and corneal perforation. J Immunol 164:6576–6582. https://doi.org/10.4049/jimmunol.164.12.6576

Saint Andre A, Blackwell NM, Hall LR et al (2002) The role of endosymbiotic Wolbachia bacteria in the pathogenesis of river blindness. Science 295:1892–1895. https://doi.org/10.1126/science.1068732

Sainz-de-la-Maza M, Molins B, Mesquida M et al (2016) Interleukin-22 serum levels are elevated in active scleritis. Acta Ophthalmol 94:e395–e399. https://doi.org/10.1111/aos.13005

Schaumburg CS, Siemasko KF, De Paiva CS et al (2011) Ocular surface APCs are necessary for autoreactive T cell-mediated experimental autoimmune lacrimal keratoconjunctivitis. J Immunol 187:3653–3662. https://doi.org/10.4049/jimmunol.1101442

Scholtmeijer K, Wessels JG, Wosten HA (2001) Fungal hydrophobins in medical and technical applications. Appl Microbiol Biotechnol 56:1–8. https://doi.org/10.1007/s002530100632

Schultz CL, Morck DW, McKay SG et al (1997) Lipopolysaccharide induced acute red eye and corneal ulcers. Exp Eye Res 64:3–9. https://doi.org/10.1006/exer.1996.0190

Seo KY, Lee HK, Kim EK et al (2006) Expression of tumor necrosis factor alpha and matrix metalloproteinase-9 in surgically induced necrotizing scleritis. Ophthalmic Res 38:66–70. https://doi.org/10.1159/000090010

Shanks RM, Stella NA, Kalivoda EJ et al (2007) A Serratia marcescens OxyR homolog mediates surface attachment and biofilm formation. J Bacteriol 189:7262–7272. https://doi.org/10.1128/JB.00859-07

Shine WE, McCulley JP (1991) The role of cholesterol in chronic blepharitis. Invest Ophthalmol Vis Sci 32:2272–2280

Shine WE, McCulley JP (1996) Meibomian gland triglyceride fatty acid differences in chronic blepharitis patients. Cornea 15:340–346. https://doi.org/10.1097/00003226-199607000-00002

Shine WE, McCulley JP (1998) Keratoconjunctivitis sicca associated with meibomian secretion polar lipid abnormality. Arch Ophthalmol 116:849–852. https://doi.org/10.1001/archopht.116.7.849

Shine WE, McCulley JP (2000) Association of meibum oleic acid with meibomian seborrhea. Cornea 19:72–74. https://doi.org/10.1097/00003226-200001000-00014

Siddiqui R, Khan NA (2012) Biology and pathogenesis of Acanthamoeba. Parasit Vectors 5:6. https://doi.org/10.1186/1756-3305-5-6

Song PI, Abraham TA, Park Y et al (2001) The expression of functional LPS receptor proteins CD14 and toll-like receptor 4 in human corneal cells. Invest Ophthalmol Vis Sci 42:2867–2877

Stover CK, Pham XQ, Erwin AL et al (2000) Complete genome sequence of Pseudomonas aeruginosa PAO1, an opportunistic pathogen. Nature 406:959–964. https://doi.org/10.1038/35023079

Sullivan DA, Sullivan BD, Evans JE et al (2002) Androgen deficiency, Meibomian gland dysfunction, and evaporative dry eye. Ann N Y Acad Sci 966:211–222. https://doi.org/10.1111/j.1749-6632.2002.tb04217.x

Sun RL, Jones DB, Wilhelmus KR (2007) Clinical characteristics and outcome of Candida keratitis. Am J Ophthalmol 143:1043–1045. https://doi.org/10.1016/j.ajo.2007.02.016

Sun Y, Karmakar M, Roy S et al (2010) TLR4 and TLR5 on corneal macrophages regulate Pseudomonas aeruginosa keratitis by signaling through MyD88-dependent and -independent pathways. J Immunol 185:4272–4283. https://doi.org/10.4049/jimmunol.1000874

Sun Y, Karmakar M, Taylor PR et al (2012) ExoS and ExoT ADP ribosyltransferase activities mediate Pseudomonas aeruginosa keratitis by promoting neutrophil apoptosis and bacterial survival. J Immunol 188:1884–1895. https://doi.org/10.4049/jimmunol.1102148

Sy A, Srinivasan M, Mascarenhas J et al (2012) Pseudomonas aeruginosa keratitis: outcomes and response to corticosteroid treatment. Invest Ophthalmol Vis Sci 53:267–272. https://doi.org/10.1167/iovs.11-7840

Szmigielski S, Prevost G, Monteil H et al (1999) Leukocidal toxins of staphylococci. Zentralbl Bakteriol 289:185–201. https://doi.org/10.1016/s0934-8840(99)80105-4

Tam C, Mun JJ, Evans DJ et al (2012) Cytokeratins mediate epithelial innate defense through their antimicrobial properties. J Clin Invest 122:3665–3677. https://doi.org/10.1172/JCI64416

Taylor WM, Pidherney MS, Alizadeh H et al (1995) In vitro characterization of Acanthamoeba castellanii cytopathic effect. J Parasitol 81:603–609

Thau N, Monod M, Crestani B et al (1994) rodletless mutants of Aspergillus fumigatus. Infect Immun 62:4380–4388. https://doi.org/10.1128/iai.62.10.4380-4388.1994

Thomas PA (1994) Mycotic keratitis—an underestimated mycosis. J Med Vet Mycol 32:235–256. https://doi.org/10.1080/02681219480000321

Thomas PA (2003) Current perspectives on ophthalmic mycoses. Clin Microbiol Rev 16:730–797. https://doi.org/10.1128/CMR.16.4.730-797.2003

Thomas PA, Kaliamurthy J (2013) Mycotic keratitis: epidemiology, diagnosis and management. Clin Microbiol Infect 19:210–220. https://doi.org/10.1111/1469-0691.12126

Thomas PA, Garrison RG, Jansen T (1991) Intrahyphal hyphae in corneal tissue from a case of keratitis due to Lasiodiplodia theobromae. J Med Vet Mycol 29:263–267. https://doi.org/10.1080/02681219180000381

Tilley SJ, Orlova EV, Gilbert RJ et al (2005) Structural basis of pore formation by the bacterial toxin pneumolysin. Cell 121:247–256. https://doi.org/10.1016/j.cell.2005.02.033

Trabelsi H, Dendana F, Sellami A et al (2012) Pathogenic free-living amoebae: epidemiology and clinical review. Pathol Biol (Paris) 60:399–405. https://doi.org/10.1016/j.patbio.2012.03.002

Tran MT, Lausch RN, Oakes JE (2000a) Substance P differentially stimulates IL-8 synthesis in human corneal epithelial cells. Invest Ophthalmol Vis Sci 41:3871–3877

Tran MT, Ritchie MH, Lausch RN et al (2000b) Calcitonin gene-related peptide induces IL-8 synthesis in human corneal epithelial cells. J Immunol 164:4307–4312. https://doi.org/10.4049/jimmunol.164.8.4307

Tumpey TM, Chen SH, Oakes JE et al (1996) Neutrophil-mediated suppression of virus replication after herpes simplex virus type 1 infection of the murine cornea. J Virol 70:898–904. https://doi.org/10.1128/JVI.70.2.898-904.1996

Tumpey TM, Cheng H, Cook DN et al (1998a) Absence of macrophage inflammatory protein-1alpha prevents the development of blinding herpes stromal keratitis. J Virol 72:3705–3710. https://doi.org/10.1128/JVI.72.5.3705-3710.1998

Tumpey TM, Cheng H, Yan XT et al (1998b) Chemokine synthesis in the HSV-1-infected cornea and its suppression by interleukin-10. J Leukoc Biol 63:486–492. https://doi.org/10.1002/jlb.63.4.486

Ueta M (2008) Innate immunity of the ocular surface and ocular surface inflammatory disorders. Cornea 27(Suppl 1):S31–S40. https://doi.org/10.1097/ICO.0b013e31817f2a7f

Underhill DM, Ozinsky A (2002) Phagocytosis of microbes: complexity in action. Annu Rev Immunol 20:825–852. https://doi.org/10.1146/annurev.immunol.20.103001.114744

Usui Y, Parikh J, Goto H et al (2008) Immunopathology of necrotising scleritis. Br J Ophthalmol 92:417–419. https://doi.org/10.1136/bjo.2007.126425

Vemuganti GK, Garg P, Gopinathan U et al (2002) Evaluation of agent and host factors in progression of mycotic keratitis: a histologic and microbiologic study of 167 corneal buttons. Ophthalmology 109:1538–1546. https://doi.org/10.1016/s0161-6420(02)01088-6

Vergouwen DPC, Rothova A, Berge JCT et al (2020) Current insights in the pathogenesis of scleritis. Exp Eye Res 197:108078. https://doi.org/10.1016/j.exer.2020.108078

Wagoner MD, Al-Swailem SA, Sutphin JE et al (2007) Bacterial keratitis after penetrating keratoplasty: incidence, microbiological profile, graft survival, and visual outcome. Ophthalmology 114:1073–1079. https://doi.org/10.1016/j.ophtha.2006.10.015

Wakefield D, Di Girolamo N, Thurau S et al (2013a) Scleritis: challenges in immunopathogenesis and treatment. Discov Med 16:153–157

Wakefield D, Di Girolamo N, Thurau S et al (2013b) Scleritis: immunopathogenesis and molecular basis for therapy. Prog Retin Eye Res 35:44–62. https://doi.org/10.1016/j.preteyeres.2013.02.004

Watson PG, Young RD (2004) Scleral structure, organisation and disease. A review. Exp Eye Res 78:609–623. https://doi.org/10.1016/s0014-4835(03)00212-4

Willcox MD, Hume EB (1999) Differences in the pathogenesis of bacteria isolated from contact-lens-induced infiltrative conditions. Aust N Z J Ophthalmol 27:231–233. https://doi.org/10.1046/j.1440-1606.1999.00189.x

Wong RL, Gangwani RA, Yu LW et al (2012) New treatments for bacterial keratitis. J Ophthalmol 2012:831502. https://doi.org/10.1155/2012/831502

Wu T, Mitchell B, Carothers T et al (2003) Molecular analysis of the pediatric ocular surface for fungi. Curr Eye Res 26:33–36. https://doi.org/10.1076/ceyr.26.1.33.14253

Wu M, McClellan SA, Barrett RP et al (2009) Beta-defensins 2 and 3 together promote resistance to Pseudomonas aeruginosa keratitis. J Immunol 183:8054–8060. https://doi.org/10.4049/jimmunol.0902140

Yan XT, Zhuang M, Oakes JE et al (2001) Autocrine action of IL-10 suppresses proinflammatory mediators and inflammation in the HSV-1-infected cornea. J Leukoc Biol 69:149–157. https://doi.org/10.1189/jlb.69.1.149

You C, Ma L, Lasave AF et al (2018) Rituximab induction and maintenance treatment in patients with scleritis and granulomatosis with polyangiitis (Wegener's). Ocul Immunol Inflamm 26:1166–1173. https://doi.org/10.1080/09273948.2017.1327602

Zhang J, Xu K, Ambati B et al (2003) Toll-like receptor 5-mediated corneal epithelial inflammatory responses to Pseudomonas aeruginosa flagellin. Invest Ophthalmol Vis Sci 44:4247–4254. https://doi.org/10.1167/iovs.03-0219

Zhang X, Volpe EA, Gandhi NB et al (2012) NK cells promote Th-17 mediated corneal barrier disruption in dry eye. PLoS One 7:e36822. https://doi.org/10.1371/journal.pone.0036822

Zhang L, Su Z, Zhang Z et al (2015) Effects of azithromycin on gene expression profiles of proinflammatory and anti-inflammatory mediators in the eyelid margin and conjunctiva of patients with meibomian gland disease. JAMA Ophthalmol 133:1117–1123. https://doi.org/10.1001/jamaophthalmol.2015.2326

Zhou R, Zhang R, Sun Y et al (2012) Innate immune regulation of Serratia marcescens-induced corneal inflammation and infection. Invest Ophthalmol Vis Sci 53:7382–7388. https://doi.org/10.1167/iovs.12-10238

Zhu WS, Wojdyla K, Donlon K et al (1990) Extracellular proteases of Aspergillus flavus. Fungal keratitis, proteases, and pathogenesis. Diagn Microbiol Infect Dis 13:491–497. https://doi.org/10.1016/0732-8893(90)90081-6

Chapter 11
What We Need to Know About the Human Ocular Microbiome

Abstract This concluding chapter lists gaps in our understanding of the identity and possible functions of the ocular microbiome, mycobiome and virome. The focus has been on identifying the gaps that would help to enhance our understanding of the identity, dynamics and function of the microbiota with reference to the normal and the diseased eye.

Keywords Microbiome · Mycobiome · Virome · Ocular surface · Gaps

11.1 Introduction

The origins of this book titled *The Human Ocular Microbiome: Bacteria, Fungi and Viruses in the Human Eye* could be traced to the Human Microbiome Project and the multitude of questions that emerged such as the following: does the ocular surface possess a microbiome? Does the ocular microbiome play a role in disease physiology? Is the ocular surface microbiome niche-specific? Does the ocular surface microbiome influence immune tolerance? Can ocular surface microbes be used to overcome eye diseases? Over the last decade, reasonable number of studies have addressed some of the above queries (Dong et al. 2011; Miller and Iovieno 2009), and the relevant data is captured in this comprehensive book. Fungi and viruses which constitute the mycobiome and virome respectively of the ocular surface also need to be included. Further, at the molecular level, the ocular biome has been implicated in ocular disorders, and this needs to be understood with respect to the cellular and molecular basis of ocular diseases.

This book documents the historical discovery of microbes (bacteria, fungi and viruses) on the ocular surface (Keilty 1930; Köllner 1906; Von Graefe and Mittheilungen 1857; Rowe et al. 1953). Further, conventional cultivable method confirmed the presence of microbes on the ocular surface and indicated that the numbers were few and barely exceeded one or two microbes. The situation improved with the advent of next-generation sequencing (NGS) technologies which demonstrated a ten-to-hundred-fold increase in the paucimicrobial community of the ocular surface. More importantly it was observed that the dynamics, abundance and diversity of the microbiomes changed in the eye impacted with diseases compared

S. Shivaji, *Human Ocular Microbiome*,
https://doi.org/10.1007/978-981-19-1754-7_11

to the normal healthy eye. Dysbiotic microbiomes, mycobiomes and viromes are discussed with respect to their specific relevance to an ocular disease, and wherever possible the cellular and molecular basis of pathogenesis with respect to a specific ocular disorder is discussed. Such studies on the ocular microbiome, mycobiome and virome may eventually help to understand the basis of the little understood ocular idiopathic diseases and develop targeted therapies for overcoming specific ocular diseases.

11.2 Gaps in Ocular Surface Microbiome

Despite the overwhelming information on the role of the microbiomes, mycobiomes and viromes in sustaining a healthy ocular surface or in influencing the pathogenesis of the ocular disease, several gaps exist which need to be addressed. Few of these gaps include the following:

1. Gut microbiome studies have unequivocally proved an association between the gut microbiome and a few ocular diseases (Shivaji 2017, 2019, 2021) and have opened up several avenues for therapy. But, the importance of the gut microbiome in ocular health would be more appreciated if a connection is established between the gut and ocular microbiomes. A likely approach would be to understand the mechanism by which inflammatory cells of the gut and gut metabolites, inflammatory molecules and cytokines influence ocular surface disorders.
2. Ocular bacteria and fungi need to be identified up to the species level. Physiological attributes of these ocular microbes with respect to resistance to drugs, the reasons for the resistance and strategies to overcome resistance and to resolve the infection need to be addressed separately.
3. Apart from culture-dependent identification of ocular microbes from clinical samples which is considered as a gold standard for the identification of the microbes and for subsequent diagnostics and therapy, the NGS approach to analyse the microbiome should be adopted as a routine approach especially when the ocular sample yields just one or two growing cultures as in the case of fungi.
4. Biofilm formation and AMR need to be addressed in specific ocular isolates rather than in type strains.
5. The concept of polymicrobial biofilms and their response to antibiotics is an area that needs to be investigated.
6. Competition between microbes in a biofilm and the impact it has on AMR needs to be understood.
7. Screening of synthetic and naturally occurring compounds for dispersing biofilms to overcome AMR needs to be continued.
8. The identification of role of fungi in several ocular diseases is scanty and needs to be done.

9. Mycobiome analysis of ocular surface using NGS needs to be extended to all ocular diseases including conjunctivitis, blepharitis, retinitis, scleritis, dry eye disease, orbital cellulitis, uveitis, Meibomian gland dysfunction, etc.
10. Laboratory testing for viral infections should shift from traditional culture methods to PCR- and NGS-based technologies.
11. Research into discovering antiviral compounds for the treatment of viral diseases of the eye needs to be strengthened.
12. Ocular surface viromes need to be characterized in the diseased eye, and the confounding factors that influence the abundance and diversity of the eye by host factors such as ethnicity, age, gender, diet and systemic health condition need to be identified.
13. Viruses are known to cause several ocular disorders but as yet the virome has been analysed only from the vitreous fluid of PFR and retinitis patients (Arunasri et al. 2021). Similar studies on the virome of the diseased eye would shed light on the involvement of viruses in other ocular diseases.
14. The sequence of cellular and molecular events that cause pathogenesis is yet to be established for all the ocular diseases. Such studies would eventually help in targeted therapy of the molecules/pathways involved in the pathogenesis of the eye.

11.3 Conclusions

The identified gaps in the ocular surface microbiome, mycobiome and virome would eventually help to instigate more research into the ocular surface with the primary purpose of appreciating the role of the microorganisms in ocular health. Additionally, such studies would help to improve our strategies to treat ocular diseases.

References

Dong Q, Brulc JM, Iovieno A et al (2011) Diversity of bacteria at healthy human conjunctiva. Invest Ophthalmol Vis Sci 52:5408–5413. https://doi.org/10.1167/iovs.10-6939

Keilty RA (1930) The bacterial flora of the normal conjunctiva with comparative nasal culture study. Am J Ophthalmol 13:876–879. https://doi.org/10.1016/s0002-9394(30)92437-3

Köllner H (1906) III. Schimmelpilzerkrankung der Sklera. Ophthalmologica 16:441–447. https://doi.org/10.1159/000291135

Kotakonda A, Gumpili SP, Mudit T, Pappuru Rajeev R, Sisinthy S (2021) Intraocular viral communities associated with post-fever retinitis. Front Med 8:1–13. https://doi.org/10.3389/fmed.2021.724195. ISSN 2296-858X. URL https://www.frontiersin.org/article/10.3389/fmed.2021.724195

Miller D, Iovieno A (2009) The role of microbial flora on the ocular surface. Curr Opin Allergy Clin Immunol 9:466–470. https://doi.org/10.1097/ACI.0b013e3283303e1b

Rowe WP, Huebner RJ, Gilmore LK et al (1953) Isolation of a cytopathogenic agent from human adenoids undergoing spontaneous degeneration in tissue culture. Proc Soc Exp Biol Med 84: 570–573. https://doi.org/10.3181/00379727-84-20714

Shivaji S (2017) We are not alone: a case for the human microbiome in extra intestinal diseases. Gut Pathog 9:13. https://doi.org/10.1186/s13099-017-0163-3

Shivaji S (2019) Connect between gut microbiome and diseases of the human eye. J Biosci 44:110. https://doi.org/10.1007/s12038-019-9931-1

Shivaji S (2021) A systematic review of gut microbiome and ocular inflammatory diseases: are they associated? Indian J Ophthalmol 69:535–542. https://doi.org/10.4103/ijo.IJO_1362_20

Von Graefe A, Mittheilungen K (1857) Verschwa¨rung der Sclera nach einer Schieloperation. Arch Ophthalmol 3:409–411